T0213441

Ethical Engineering

Ethical Engineering: A Practical Guide with Case Studies provides detailed and practical guidance in making decisions about the many ethical issues practicing engineers may face in their professional lives. It outlines a decision-making procedure and helps engineers construct an ethics toolkit consisting of professional models, a comprehensive set of ethical considerations and factors that help in weighing those considerations, and analyses of particular issues, such as reverse engineering a patented process. Illustrating case studies, both brief and detailed, are provided.

Features:

- Introduces the nature of ethical decision-making as applied to engineering values and issues.
- Helps readers develop a detailed ethics toolkit that identifies options and solutions and allows them to monitor and adjust as necessary.
- Features topics such as safety, sustainability, bioethics, diversity and equality, information technology and AI, as well as critical areas often overlooked in engineering texts, such as mentoring, advertising (for consulting firms), engineering sales, and much more.
- Includes 85 case studies to illustrate a variety of scenarios.
- Offers an international perspective with links to codes of ethics from around the world, including Saudi Arabia, India, New Zealand, Chile, and Japan.

Emphasizing the importance of the moral life and of engineering as an occupation with high ideals, this book helps readers navigate a variety of real-world ethical issues they are likely to face in this increasingly interdisciplinary, global, and diverse profession.

What Every Engineer Should Know
Series Editor

Phillip A. Laplante
Pennsylvania State University

What Every Engineer Should Know about Career Management
Mike Ficco

What Every Engineer Should Know about Starting a High-Tech Business Venture
Eric Koester

What Every Engineer Should Know about MATLAB® and Simulink®
Adrian B. Biran

Green Entrepreneur Handbook: The Guide to Building and Growing a Green and Clean Business
Eric Koester

What Every Engineer Should Know about Cyber Security and Digital Forensics
Joanna F. DeFranco

What Every Engineer Should Know about Modeling and Simulation
Raymond J. Madachy and Daniel Houston

What Every Engineer Should Know about Excel, Second Edition
J.P. Holman and Blake K. Holman

Technical Writing: A Practical Guide for Engineers, Scientists, and Nontechnical Professionals, Second Edition
Phillip A. Laplante

What Every Engineer Should Know About the Internet of Things
Joanna F. DeFranco and Mohamad Kassab

What Every Engineer Should Know about Software Engineering
Phillip A. Laplante and Mohamad Kassab

What Every Engineer Should Know About Cyber Security and Digital Forensics
Joanna F. DeFranco and Bob Maley

Ethical Engineering: A Practical Guide with Case Studies
Eugene Schlossberger

What Every Engineer Should Know About Data-Driven Analytics
Phillip A. Laplante and Satish Mahadevan Srinivasan

For more information about this series, please visit: www.routledge.com/What-Every-Engineer-Should-Know/book-series/CRCWEESK

Ethical Engineering
A Practical Guide with Case Studies

Eugene Schlossberger

CRC Press
Taylor & Francis Group
Boca Raton London New York

CRC Press is an imprint of the
Taylor & Francis Group, an **informa** business

Designed cover image: © Shutterstock

First edition published 2023
by CRC Press
6000 Broken Sound Parkway NW, Suite 300, Boca Raton, FL 33487-2742

and by CRC Press
4 Park Square, Milton Park, Abingdon, Oxon, OX14 4RN

CRC Press is an imprint of Taylor & Francis Group, LLC

Library of Congress Cataloging-in-Publication Data

Names: Schlossberger, Eugene, 1951- author.
Title: Ethical engineering : a practical guide with case studies / Eugene Schlossberger.
Description: First edition. | Boca Raton : CRC Press, 2023. | Series: What every engineer should know | Includes bibliographical references and index. | Summary: "Ethical Engineering: A Practical Guide with Case Studies provides detailed and practical guidance in making decisions about the many ethical issues practicing engineers may face in their professional lives. It outlines a decision-making procedure and helps engineers construct an ethics toolkit consisting of professional models, a comprehensive set of ethical considerations and factors that help in weighing those considerations, and analyses of particular issues, such as reverse engineering a patented process. Illustrating case studies, both brief and detailed, are provided. Features: Introduces the nature of ethical decision making as applied to engineering values and issues. Helps readers develop a detailed ethics toolkit that identifies options and solutions and allows them to monitor and adjust as necessary. Features topics such as safety, sustainability, bioethics, diversity and equality, information technology and AI, as well as critical areas often overlooked in engineering texts, such as mentoring, advertising (for consulting firms), engineering sales, and much more. Includes more than 50 case studies to illustrate a variety of scenarios. Offers an international perspective with codes of ethics from around the world, including Saudi Arabia, India, New Zealand, Chile, and Japan. Adds further cases and samples for discussion and a summary of key ideas. Emphasizing the importance of the moral life and of engineering as an occupation with high ideals, this book helps readers navigate a variety of real-world ethical issues they are likely to face in this increasingly interdisciplinary, global, and diverse profession"-- Provided by publisher.
Identifiers: LCCN 2022045262 (print) | LCCN 2022045263 (ebook) | ISBN 9781032151137 (hardback) | ISBN 9781032151120 (paperback) | ISBN 9781003242574 (ebook)
Subjects: LCSH: Engineering ethics.
Classification: LCC TA157 .S38323 2023 (print) | LCC TA157 (ebook) | DDC 620.001--dc23/eng/20221202
LC record available at https://lccn.loc.gov/2022045262
LC ebook record available at https://lccn.loc.gov/2022045263

ISBN: 9781032151137 (hbk)
ISBN: 9781032151120 (pbk)
ISBN: 9781003242574 (ebk)

DOI: 10.1201/9781003242574

Typeset in Minion
by KnowledgeWorks Global Ltd.

For my family: Kurt, Else, Norman, Maricar,

Erin, Noah, Rachel, and Moss

Contents

What Every Engineer Should Know: Series Statement xv
Preface xvii
Acknowledgments xix
About the Author... xxi

PART I   Introduction and Ethical Decision-Making

Chapter 1   The Nature of Engineering Ethics 3

What This Book Is About 3
Why Ethics for Engineers? 4
A Revealing Case 9
Cut-Throat v. Community Workplaces............................... 10
The Consumer Life v. The Life of Values 12
What Should I Do? 15

Chapter 2   Ethical Decision-Making... 21

The Nature of Ethical Decision-Making 22
Making Ethical Decisions 24
A Template for Moral Decision-Making 26
Preparation 27
Creating a Toolkit Rationally: Looking at the
Arguments 30
Three Rational Responses to an Argument................. 30
Four Rational Responses to a Counterexample............ 31
Picture Building and the Moral Edifice........................ 33
Relativism... 35
Making Particular Decisions: The Five-Step Process....... 37

PART II Sources of Ethical Decision-Making

Chapter 3 The Engineering Way 49

What is Engineering..49
Design Ethics and Cultural Appropriation...........................52
Thinking like an Engineer...54
    Characteristics of the Compleat Engineer 55
    The Engineering Process..59
Models of the Profession...61
    Models of the Engineering Profession 63
Engineering and Business 65

Chapter 4 Values of the Engineering Profession 75

Technology as Practical Wisdom 75
Safety 79
    Extent of a Risk 79
    Balancing Risks against Benefits 80
    Nature of Risks..82
    Publicizing Risks...82
Human Progress 90
Clean, Clear Decision-Making 91
Community 92
Partnership with Nature... 96

Chapter 5 Additional Ethical Sources (Part 1) 109

When to Fight a Battle...109
Treating Others Fairly and Well...112
    Illustrating Cases 113
The Duty to Leave the World No Worse 116
World Religious Traditions.. 122
Promoting Good Consequences.. 128
    Illustrating Example...130
The Golden Rule 130
Universality 133
Proclamative Principle (Duty to Set a Moral
Precedent) 137
Case Studies..138

Respect for Persons...140
Rights...141
Autonomy 145

Chapter 6 Additional Ethical Sources (Part 2) 153

Personal Values, Moral Beauty, and
The Good Life..153
Virtues...156
The Precautionary Principle...156
Responsibility (Dual-Use) 158
  The Nature of Moral Responsibility...................................159
  Promoting Good Consequences (The Utilitarian
  Perspective)...160
  Safety and Future Generations 161
  Leaving the World No Worse
  Than One Found it...162
  The Weak Samaritan Principle 163
  Respect for Autonomy...164
  The Doctrine of Intervening Wills
  (*Novus Actus Interveniens*) 165
  Four Concluding Questions 166
Two Principles of Institutional Responsibility.....................167
Institutional Duties 168
When to Break the Rules..170
Codes of Ethics..177
  Contents of Codes of Ethics 178
  The Role of Codes of Ethics...179

PART III Problems and Issues in Engineering

Chapter 7 Honesty and Professionalism.......................................185

Whistleblowing..185
Competence..197
Keeping Accurate Records and Obeying the Law.................198
Sales, Honesty, and Disclosure of
Product's Liabilities...202
  Consulting v. Adversarial Sales...203

Chapter 8   Good Faith...209

    Conflict of Interest..209
    Confidentiality and Trade Secrets..................212
    Patents and Copyrights......................................215
    Bidding...216

Chapter 9   Employee-Employer Relations.........................227

    Types of Work Relationships............................227
    Leadership and Healthy Work Environments...................229
    Protégés...232
    Dealing with Subordinates...............................233
    Equality/Equity, Diversity, and Inclusion (EDI)................234
    Sexual Harassment, Favoritism, and Professional
    Relations...238
    Hiring Practices..244
    Interdepartmental Dealings and Hiring Away from
    Another Firm...246

Chapter 10 The Environment, Climate Change, and
              Sustainability......................................255

    Climate Change..255
    Pollution...261
    Sustainability...262
    Conservation and Preservation........................264
    Solutions...264
        Balancing Environmental Risk and Benefit.................264
        Recycle, Reuse, and Reduce..............................267
        Nine General Responses to Environmental Problems...271
    Conclusion...277

Chapter 11 Appropriate Technology and Less Developed
              Regions (LDRs)....................................281

    Substituting Cash for Subsistence Crops.............282
    Products Banned at Home but Legal Abroad..........284
    Depleting Natural Resources (Such as Cutting
    Down Rainforests)...286

Less Stringent Environmental Regulation and
Unsustainable Practices 286
Products Safe at Home but Unsafe Under Local
Conditions 287
Cultural Disruption 288
Doing as the Romans Do 289
Implementing Regionally Appropriate Technology 290
Ethical Factors Useful in Dealing with these Concerns 292

Chapter 12 Bioengineering and Medical Engineering 297

Overview 297
Broad Issues Arising in Multiple Contexts 299
Safety and the Environment 299
Social Implications, Playing God, Respecting
Humanity, and the Slippery Slope 299
The Enhancement/Treatment Distinction 301
Access and Social Justice 305
Areas of Ethical Concern 306
Designer Babies and Human Gene Engineering 306
GMOs in and outside of the Food Chain 309
New Reproductive Technologies 313
Biotechnology to Mitigate Disability or Enhance
Human Beings, Brain Research, and Psychological
Control 314
Genetic Testing 315
Further Issues 317
Summary 318

Chapter 13 Information Technology, Artificial Intelligence,
and Software Engineering 325

Overview 325
Characteristics of Information Technology
Producing Special Ethical Concerns 326
List of Issues in Information Technology 327
Three General Reminders (Value-Sensitive Design,
Broader Context, Anti-Patterns) 327
Issues about Software and Technology Development 329

Rushing to Market and Cutting Costs.................................329

Results of Failure and Unintended Consequences..........330

Dual Use..333

Intellectual Property/Copyright/Patents/
Trade Secrets..333

Issues About Content (Data)...334

Bias...334

Privacy, Security, and FRT 335

Hosts' Responsibility for User/Client
Code and Content.. 343

Information Technology, AI, and Society............................. 346

Autonomous Technologies.. 346

Job Displacement and the Digital Divide....................... 348

Additional Resources and Illustrating Cases...................... 350

Chapter 14 Consulting Engineering...................................... 355

Advertising.. 355

Dishonest or Misleading Advertising............................. 356

Unseemly or Demeaning Advertising 357

Advertising with Morally Questionable Messages.......... 358

New Forms of Advertising.. 359

Competing with Other Firms... 360

Competitive Bidding 360

Contingency Fees.. 360

Bribes and Kickbacks 360

Derogatory Remarks about Other Engineers 360

Reviewing the Work of Others 361

Reviewing, Checking, and Stamping the Work of
Unlicensed Individuals... 361

Reviewing the Work of Other Engineers 362

Safety and Liability.. 362

Appendix I: Links to Codes of Ethics from across the World......... 365

Appendix II: Summary of Major Western Ethical Theories 369

Appendix III: Additional Cases... 377

Appendix IV: Two Sample Suggestions..................................... 389

Formation of an Environmental and Community
Issues Advisory Board (ECIAB). 389
Ethical Ombudsperson 390

List of Cases 391

Index 395

What Every Engineer Should Know: Series Statement

What every engineer should know amounts to a bewildering array of knowledge. Regardless of the areas of expertise, engineering intersects with all the fields that constitute modern enterprises. The engineer discovers soon after graduation that the range of subjects covered in the engineering curriculum omits many of the most important problems encountered in the line of daily practice – problems concerning new technology, business, law, and related technical fields.

With this series of concise, easy-to-understand volumes, every engineer now has within reach a compact set of primers on important subjects such as patents, contracts, software, business communication, management science, and risk analysis, as well as more specific topics such as embedded systems design. These are books that require only a lay knowledge to understand properly, and no engineer can afford to remain uniformed of the fields involved.

Preface

The materials in this volume discuss the ethical not the legal aspects of engineering and technologically oriented practice. *These materials must not be used as a substitute for legal advice.* Laws change and vary from place to place. *Engineers must stay up to date about all relevant laws and regulations.* Because the engineer and their organization both have a general duty to obey the law, *every recommendation made in these pages should be understood as containing the phrase "provided there is no legal duty to do otherwise."*

Eugene Schlossberger

Acknowledgments

This volume revises, supersedes, and vastly expands *The Ethical Engineer* (Temple University Press 1993). I am grateful to Temple University Press for giving permission to incorporate, with revisions, much of the material from the earlier volume into the present work. I would also like to thank my students over the years, as their feedback has proven beneficial, as well as those, such as Michael Davis, who commented on or discussed with me relevant conference papers and articles of mine since 1993.

About the Author

Eugene Schlossberger, Professor Emeritus of Philosophy at Purdue University Northwest, is the author of four previous books (*Moral Responsibility and Persons*; *The Ethical Engineer*; *A Holistic Approach to Rights: Affirmative Action, Reproductive Rights, Censorship, and Future Generations*; and *Moral Responsibility Beyond Our Fingertips: Collective Responsibility, Leaders, and Attributionism*) as well as numerous articles on business and engineering ethics, environmental ethics, and political philosophy in journals such as *Mind*, *Philosophical Studies*, *Analysis*, and *Science and Engineering Ethics*.

Part I

Introduction and Ethical Decision-Making

1

The Nature of Engineering Ethics

WHAT THIS BOOK IS ABOUT

This book is a practical guide to ethical decision-making for practicing engineers, students, and others in technically oriented business and industry. It will help you to make ethical decisions yourself, understand the reasons behind organizational or company[1] policies, legal rules, and professional codes, and give you a broader picture of some of the issues and controversies in engineering. It is also meant to change the way you think and feel about engineering, so that you can be a better and happier engineer. And it may just cause you to take a new look at the ethical dimensions of life generally.

Unlike many ethics books, this one is geared to finding answers. The emphasis is on clearly and economically presenting factors to be used in making decisions. Kapuska (2020) says that "Ethics for engineers is not about theory and philosophy: it is about the application of moral and ethical principles to solve practical engineering issues." While there is some theory and philosophy in this volume, it is primarily about solving those practical issues. Of course, there is no such thing as an ethics rulebook. Rather, what you need is a sort of ethics construction kit, giving you tools and materials to construct your own solutions to problems you may face. This book gives you some specific recommendations, many examples of ethical decisions, several techniques for ethical decision-making, and a variety of rules, principles, and values you can draw upon. It explains the reasons behind many of the rules and principles and gives you guidance in weighing them. It also gives you context to help understand both specific issues and broader issues within which engineering decisions occur. Finally, it provides a wealth of cases to think about.

Thus, Part I talks about how to make ethical decisions. Part II gives you a large number of ethical considerations to put in your ethics toolkit, from

DOI: 10.1201/9781003242574-2

ideas from world religious traditions (such as Ahimsa, Ubuntu, Qist, and Mino Pimatisiwin) to advice on when to break the rules or fight a battle. Part III talks about the issues in engineering about which decisions must be made. So, for example, you'll find a template for ethical decision-making in Part I, the limitations of the golden rule and advice about when to fight a battle in Part II, and a discussion of genetically modified organisms (GMOs) in Part III. The Appendices give you, among other things, additional cases, links to codes of ethics from around the world, and a brief overview of Western ethical theory. If this book is being used as a textbook, some instructors will place Appendix II on their syllabus between Chapters 1 and 2 or between Chapters 2 and 3, as some (not all) applied ethics courses begin with a short unit on ethical theory. Non-Western traditions are discussed in Chapter 5.

I've tried to capture and summarize much of what is best in the engineering ethics literature and to add much that is new. Some of the recommendations made are quite specific, others fairly general. Some of the material will raise new ideas, while some of it is a clear and thorough summary of "common sense." Some of the recommendations are controversial; others are widely agreed upon. Ultimately, you must make your own decisions on these issues. *You should not accept my recommendations blindly: you have a responsibility, as an ethical person, to ask yourself about every sentence in this book "is this true? Do I agree?"*

Together, these materials present a comprehensive picture of and guide to engineering ethics, from safety and whistleblowing to cultural appropriation and self-driving cars, from views about humanity's place in the natural world to mentoring and reviewing the work of other engineers, from bidding and advertising to the nature of privacy, from software piracy and the digital divide to moral beauty and the engineering way (though of course no single volume can cover everything).

WHY ETHICS FOR ENGINEERS?

Practicing engineers sometimes think that ethical problems are not really their concern. Many ethical decisions are not really up to you. After all, much of what an engineer may or may not do is determined by legal and company rules, and many of the ethical questions that do arise are decided by committees or executives far removed from the average engineer. Of what use, then, is a book or program or course on engineering ethics?

There are several reasons why it is important that individual engineers have a clear grasp of engineering ethics. The "power to change the world comes with ethical obligations" (Peterson 2020, 3). One reason individual engineers should be trained in ethics is that *many of the ethical decisions that individual engineers must make are not settled by rules*. After all, rules do not cover every situation: often the rules only set limits within which decisions must be made, and some situations are not covered at all. As Hanson (1983) notes, "Not every ethical conflict and concern that will arise can be anticipated …. Therefore it [is important to] establish firmly in all employee's minds a few key values that can be applied to any situation that will arise" (p. 186). In addition, rules require interpretation in a way that is not always obvious. No set of rules or policies can anticipate every ethical problem that may arise, and even the sincerest engineers may need help in understanding the ethical aspects of some situations. "Good ethical conduct implies more than following a set of rules …. Good ethical conduct is not the result of checklists, rules, codes" (Baggini 2019). Rather, "because of the complexities involved in ethical dilemmas, engineers must develop their ability to apply moral intelligence (knowledge of what is right) when we are under pressure in real-life situations" (Veach 2006, 97).

So, only ethically aware engineers can correctly apply ethical rules to complex situations, keeping to the spirit as well as the letter of ethical rules. Moreover, as Budinger and Budinger (2006, xiii) note, "ethical dilemmas may have more than one 'right solution.'"

A second reason is that *individual engineers do sometimes confront ethical issues in their own work*. "In a 1997 survey of 100 practicing engineers, 70% of the individuals indicated that yes, they had been faced with an ethical issue in the course of their engineering practice" (Bielefeldt and Canney 2016). Even when an engineer does not make the decision, the engineer can influence the decision in various ways. "For example, the designer of a search engine for an online retailer might choose to display the most expensive items first. This choice might favor the interest of the retailer, to maximize profit, over the interest of the customer, to minimize cost" (Loui and Miller 2008). When the ultimate decision is made higher up, the engineer can be a voice for ethics within the company. In the worst case, if the problem is severe enough, she[2] can resign or even blow the whistle. So, engineers cannot just say "not my problem." Moreover, ethics may play an increasing role as engineers advance. Haigh

(2013) illustrated the increasing role of ethics in his job as his career progressed, from project engineer to group coach, with a series of pie charts. As a group coach, Haigh indicated, ethics accounted for approximately 60% of his role.

A third reason is that *organizations function best when the values implicit in rules and executive decisions are widely understood and discussed within the organization.* The best employment situation, for both employer and employees, is a community atmosphere in which everyone works together for a common goal in which everyone believes. Organizations flourish when their people have common values. This is not possible if the rules are just bureaucratic mandates handed down from above, rather than ways of working toward commonly understood goals. Furthermore, without communication between all levels of an organization, ethical problems may slip between the cracks. Such "synergistic" effects arise when employees are not sensitive to complex ethical dimensions of company operations. It is less easy for ethical problems to fall between the cracks when each engineer is aware of and sensitive to ethical concerns and potential problems. In general, large decisions are often the resultant of many small decisions at different levels. Thus, the large decisions reflect the values and ethical aims of the company only when those making the small decisions understand those values and aims and understand how their decisions fit in with the ethical outlook of the company. In addition, engineers who understand the moral basis of the rules have a greater *motivation* to obey them. This is both because they see the rules as morally sound directives rather than as annoying and senseless restrictions, and because there is less incentive to make one's own "numbers" look better if the company functions as a community rather than as an arena for individual "gladiators" to better their own prospects. Finally, the single most powerful influence on the operation of a firm or company is its *corporate climate.* "An ethical climate involves the total organization and must pervade and infuse all who work for the company" (Fleming, 1983, 217). Although the corporate climate is heavily influenced by those at the top, people who advance in the organization tend to take with them the attitudes they have learned before being promoted. And so "the best time to give ethics training to senior managers is before they become senior executives" (Jones 1982, 8).

A fourth reason for engineers to be sensitive to ethical questions is that *engineers who understand the ethical dimension of engineering are better and happier engineers.* Engineers will be happier in a company in which every engineer understands the value of community, since a community atmosphere requires everyone's participation, and it is more pleasant and rewarding to work in a place with a community atmosphere. Engineers who understand the "engineering way," and the values of engineering as a profession, who see engineering as serving a high social goal, will put more into (and take more satisfaction in) their work than those who see engineering merely as a way of drawing a paycheck.

From the employer's standpoint, employing ethically sensitive engineers has several benefits. First, *ethics is good business*. Ethical sensitivity often avoids costly situations. *While acting ethically often costs more in the short run, acting unethically usually costs more in the long run.* Engineers and businesspersons are sometimes tempted to act unethically because the benefits of unethical conduct are immediate and highly visible, while the benefits of ethical conduct are often long-term and hard to calculate. For example, it is easy to show exactly how much money is being saved when you deny a worker a $2,000 bonus to which they are entitled. What you lose by treating employees unfairly, while real, is hard to calculate. It is hard to calculate the precise financial effects of a community atmosphere, of employee loyalty, of good engineers coming to work for the firm because the firm has a good reputation, and so forth. While employee motivation certainly affects productivity, it is impossible to point to the particular extra widgets produced by greater motivation. It is hard to calculate precisely how many people didn't buy Fords because of negative publicity surrounding the Pinto.[3] Acting ethically and responsibly may prevent a costly government regulation but there is no line in the budget for preventing regulations. There is no line in the budget that shows the money saved by forestalling litigation as a result of good and fruitful communication with environmental groups. So, a standard "dollars and cents" analysis will often fail to take account of the hidden but real costs of unethical decisions. An Institute of Business Ethics (IBE) study found that companies "who are explicit about business ethics, out-perform in financial and other indicators those companies who say they do not have a code." For example, "companies with a code of ethics generated significantly more EVA [Economic Value Added] and MVA [Market Value Added]" (Webley and More 2003). Thus, a good understanding of the ethical dimensions of engineering decisions is a business necessity. "Neglecting to take steps to insure an ethical corporate climate has proven to be an ill-considered risk for many organizations" (Ottoson 1982, 155). (See also Schlossberger 2016.)

A second reason for firms and companies to hire ethically trained engineers is increased marketability. Due to an increasing concern about ethical issues on the part of the public and regulatory agencies, a company whose engineers are ethically trained has a considerable public relations advantage. Companies are increasingly more likely to hire consulting firms that show a strong awareness of ethical issues. Government is more likely to deal with a company that has a strong ethical profile. Regulatory agencies are not blind to a firm or company's efforts to hire ethical engineers or train the engineers it does employ. Moreover, the public perceives every employee as a representative of the firm or company, and the company or firm is judged by what its engineers say and do, both on the job and off. *Ethically aware engineers are ambassadors of goodwill, both for their firms and for engineering as a profession.*

Another advantage of ethical training is greater understanding of the concerns of environmental groups. A partnership between engineers and environmentalists, based on mutual understanding, is more productive than an adversary relationship.

A fourth reason for companies to desire ethical engineers is that *ethics makes for better and more productive engineers.* Ethically aware engineers are more likely to create a community atmosphere and show trust and fairness in the workplace. In addition, engineering is not just a job but a profession with an ethical dimension. Engineers, who see their work as part of a moral project take greater pride and satisfaction in their work, are more productive, show greater loyalty to the firm or company, and are more willing to go the extra mile. When employees regard their work as meaningful, turnover, accidents, and employee theft decrease, while quality of work output improves.

Fifth, since firms function best as a *community* with a shared set of goals and values, it is important that the values implicit in firm decisions are widely understood and discussed within the firm. *Firms make better decisions when there is free and informed input from all engineers.*

Finally, *ethical engineers should work toward having a greater role in corporate and civil decision-making.* "One concern that people have about rail safety, for example, is that it seems to be the case that fewer engineers are being employed at the higher levels of management resulting in important decisions being made without appropriate input from engineers. If this concern is right, this is a trend that engineers should be active in trying to reverse" (Royal Academy of Engineering 2011, 65).

Hiring and promotion practices illustrate some of these points. The ethical company has a model of what sort of engineer it wants to hire and promote. An organization that wishes to make ethical decisions, for example, should give serious thought to the integrity and ethical commitment of an employee. A company that wishes to function as a community will, while scrupulously avoiding discrimination, seek to hire and promote team-spirited engineers, who consider the needs and welfare of those whom they supervise, and will be less eager to promote abrasive, self-centered engineers. If these goals and criteria are not clearly understood by all those involved in hiring and promotion decisions, the company is not likely to achieve the personnel profile it seeks. After all, the personnel profile of a company is the result of many smaller decisions at different levels. Moreover, traits such as team spirit are not easily described or quantized, and so decision-makers who do not clearly understand the moral outlook of the company cannot apply these criteria correctly. Finally, traits valuable to the company as a whole are not always as clearly and immediately useful to a supervisor. So, the motivation for hiring or promoting the right people is increased when those doing the hiring understand the reasons for looking for people with these traits.

In sum, the ethical profile of a company depends not only on the rules set by upper-level executives, but on the ethical outlook and understanding of each and every employee.

A REVEALING CASE

Here is an example of an ethical problem that no simple rule will solve. It is a problem that any engineer might have to face. And the problem must be solved by the individual engineer, not by a committee or superior. The case also serves to illustrate why ethics should matter to you and how to go about thinking about an ethical problem.

Case 1: The Overheard Remark

While eating lunch in the cafeteria, you overhear Smith, the supervisor of another department, speaking to another engineer. Smith says that Jones appears to be on a fast track and may soon prove a rival for promotion. So, Smith intends to lie to Garcia, Jones' supervisor. Smith is going to tell Garcia that he continually hears Jones saying that Garcia is incompetent and that Jones tells everyone that she would do the job much better than Garcia. You realize that this false "confidential" information will not appear in Jones' evaluations or personnel file, and that Jones will have no chance to answer this false accusation. What, if anything, should you do about this?

Discussion: Your first response might be, "This is not my problem. Why should I care what happens to Jones?" In a broader sense, you are asking the question, "why should I care about ethics in the workplace?" This is a question whose answer is relevant to just about everything talked about in this book. I suggest that if you do not care about Jones and if you do not care about injustice, then you are not only doing wrong: you are also hurting yourself. It is extremely important to understand how you hurt yourself by not caring about ethics, for the two points I'm going to make are, in a sense, at the center of engineering ethics. In the short run, ignoring ethics is often profitable. It sometimes seems more practical not to do the right thing. But, as I suggested above, in the long run, if you're not ethical, you will pay for it. From the corporate standpoint, as we saw, ethics is good business. From the individual's standpoint, you're better off doing the right thing. So, before we talk about what do to in Case 1, let's pause for a minute to consider two major reasons why you're losing out if you don't care deeply about ethics.

Cut-Throat v. Community Workplaces

There are, basically, two sorts of workplaces, the cut-throat workplace and the community workplace. In a cut-throat workplace, people are concerned only about themselves, about advancing their careers or getting by as easily as possible. In a cut-throat workplace, people will lie, stab each other in the back, sabotage projects, and exploit each other. In a community workplace, people care about each other because they are working toward common goals based on common values.

Clearly, there are major disadvantages to working in a cut-throat workplace. First of all, it is not much fun, nor is it conducive to good work, to deal with people who distrust each other. If you are always wondering who is about to do what behind your back, and if your fellow workers are only out for themselves, then you won't enjoy coming to work and you won't do your best. If nothing else, you will spend too much time protecting your back. This is especially true for engineers, since good work requires teamwork, and teamwork requires trust. Workbenching, leaving your ideas out for everyone to make informal comments and offer suggestions, often on a design wall, conversation channel, or other shared space (physical or electronic), makes you a better engineer. In a cut-throat environment where others will steal or sabotage your ideas to get ahead, workbenching is shooting yourself in the foot. Moreover, in a cut-throat workplace, where work is merely a means of drawing a paycheck, the forty or so hours a week one spends at work are unrewarding drudgery. This is because, in a cut-throat workplace, engineering is just a series of tasks, rather than a worthwhile project. After all, few would put changing the oil in her car on her list of favorite ways to spend a Saturday afternoon. Yet, for many people, being part of a winning racing team is exciting, even if one's job in the team is to change the oil. Regarded merely as a *task*, changing the oil is a chore. Regarded as part of a cherished *project*, it can be fun and rewarding. The same is true of engineering. The engineer who takes pride in being part of the team developing a socially useful product will enjoy her work. The engineer who views her work as just pushing keys on a computer will not. Since work constitutes such a large part of one's life, the difference between a community workplace and a cut-throat workplace can mean the difference between a happy life and an unhappy one. "Engineers spend at least a third of their waking lives at work and derive a good part of their self-esteem and sense of self from their jobs" (Perlman and Varma 2002, 41). After all, for most people, a typical workday is largely consumed by work, preparing for work, commuting, and doing household, financial, and other chores. So, if you hate your job, you live for the weekend. Those two days have to be quite incredible to make up for five days of drudgery. (See also "Community" in Chapter 4.)

Thus, any thoughtful engineer cares very much about the work environment. It should be important to you whether the organization for which you work is a place where people stab each other in the back with lies, or a place characterized by honesty and fair dealing. It makes a big difference whether you work in a cut-throat or a community workplace. But establishing a good community atmosphere requires everyone's participation. So, by caring about other people and acting to promote fairness and decency in the workplace, you are helping yourself. By contrast, if you don't care about Jones, and don't care about injustice in the workplace, you are fostering a cut-throat, not a community workplace.

The Consumer Life v. The Life of Values

There is another important reason that Jones is your problem. There are two kinds of life one might lead. One way is a dead end and the other way requires caring about what happens to Jones.

Consider two different ways of leading one's life. We can live what I'll call "consumer lives," or we can lead lives dedicated to intrinsic value. The consumer life is dedicated to getting personal enjoyments or "goodies" of one sort or another: the consumer life is just one "kick" after another. The consumer need not be "materialistic": the commodities they seek may be love or personal excellence. The point is that they regard them as commodities. They do not, for example, regard human love as a good thing for its own sake, as intrinsically valuable whenever it occurs. Rather, they want to be loved as one might desire to own a television or a trophy, so as to feel prosperous in the goods of life. They want love as one might desire a sofa, a massage, or a piece of chocolate cake, namely for the good feeling one gets from consuming or owning them.

Of course, everyone enjoys an occasional "goodie." Enjoying your jacuzzi, flan, or Lakme perfume doesn't mean you're living a "consumer life." The question is what really drives you, what really matters to you, what your life is fundamentally about. The "consumer" is driven by the restless need to satisfy personal desires for a Mercedes, fashionable clothes, power, excellence at tennis, or whatever. They see fame, love, and personal accomplishment as goods to be obtained to satisfy their personal desires, and the thrill of obtaining them is what their life is all about.

Put another way, the consumer has desires but no values. What we *desire* is what we want for ourselves, what it makes us feel good to have. For example, if I desire courage, then I want to be courageous, and it makes me feel good when I'm courageous. What we *value* is what we think good for its own sake, quite apart from ourselves. If I value courage, then I want others to be courageous, and I am pleased by courageous acts that have nothing at all to do with me. In short, because I believe in courage, I want the world to be a courageous place. Of course, we usually *also* desire what we value. If courage is a valuable thing, then of course I want to be courageous. The point is that some people want to be loved *because* they value love. The consumer does not value love for its own sake, he *merely* wants it.

Now the consumer life is the life we are constantly being urged to pursue. If we listen to TV ads, we might think that life is about getting sexy nails, or fancy jewelry, or about getting a six pack of beer and a six pack of women (or men), as an ad once suggested. But the consumer life is a recipe for failure. The logic of the consumer life entails that anyone who lives a consumer life is necessarily isolated, and anyone who lives the consumer life is necessarily unhappy, either dissatisfied or bored.

The consumer is necessarily isolated because no one else can feel what you feel, and so enjoying "goodies" is basically a private project. If you are a consumer, if your goal in life is getting feel-goods, then there is only one role other people can play in your life: they are a tool for or impediment to getting feel-goods. If your goal in life is to enjoy "goodies," what kind of marriage can you have? Since all you really care about is obtaining good-ies, all your spouse can be for you is a tool to help you obtain consumables, to help you feel more personally excellent, or provide sex, or financial sup-port, or affection, or amusing company, etc. In short, you and your spouse are allies only insofar as you are useful to each other in getting your own private goodies. And so, the moment you become a net liability in your spouse's search for feel-goods, you become a tool that doesn't work. And a tool that doesn't work is garbage. Why keep a tool that no longer works? Thus, you are basically alone in life, no matter how many friends you seem to have.

The consumer is necessarily unhappy, since the consumer life is by nature a losing game. As Thomas Hobbes said, we are restless desire machines. Once a goodie is obtained, it loses its allure, and so the con-sumer is always restlessly moving on to the next goodie. There is no such thing as success, since feel-goods don't last: you can't put half your feel-good in the bank or invest it to grow more feel-goods. A feel-good, once felt, is gone forever. Perhaps you worked hard through college, hoping to earn enough to buy an expensive sports car. Finally, you earn $80,000 a year. You buy the sports car, and every morning you walk out on the driveway and say, with a thrill in your voice, "this baby is mine." You feel great – for a month. But before long, you take the sports car for granted: it's just your car. You need a new kick. If only you earned $120,000 a year, you could buy a new house. So you work hard, and finally you buy the house. You wake up in the morning, look around you, and say "this house is mine." You feel great – for a week. But before long, you need a new kick. If only you earned $150,000, you could get an Olympic-sized swimming pool. The cycle is endless because a commodity, once enjoyed, is gone, and you must start over from scratch. And the more you have, the more it takes to produce the same thrill. Your first car is more of a thrill than your tenth. Your first fantastic dress is a greater feel-good than your twentieth. There is a reason why the metaphor for the rush of excitement is a kid in a candy store. Adults have far more resources than kids do, but it is hard to match that level of feel-good. The life of the consumer, though more respectable, is in many ways like the life of the drug addict – you always need a new high, and it takes more to give you the same high. So, you will be forever dissatisfied, no matter what you have. It is never enough because the thrill wears off and you need more to match the same level of excitement. Unless, of course, you have everything, in which case you are bored. That is one reason so many celebrities turn to drugs. The sixth time you are on the cover of a magazine is not as exciting as the first. To sustain that buzz after so many high-voltage feel-goods, they turn to drugs, or adultery, or other self-destructive behavior. Either you never have enough, or you have so much that you're desperately looking for something else to want. It is a losing game. There is no structure to the consumer life and so no progress. For example, if my goal were to further knowledge for its own sake, then each small step I take would have meaning in a larger enterprise, and I could find satisfaction in knowing that I had played some part in expanding the horizons of knowledge. The consumer life does not allow for this: buying two sports cars brings one no closer to a goal, for the only goal of the consumer life is having more goodies, and the number of goodies is endless. The consumer life, thus, is a barren life. The consumer is either dissatisfied, always hungry for more no matter how much they have, or jaded and bored when there is nothing left to hunger for. There is simply no way to win at this game.

The alternative is to be driven by a recognition of value, to be committed to things that are good for their own sake, independent of me having them. Thrills are fleeting, while values endure. The thrill of seeing my first article in print is long gone. It is nice when something I write comes out in print but it cannot match the thrill of seeing my first publication. Nonetheless, the value of what I contributed, and what other people have contributed to human understanding, remains. If I value understanding as a good thing, a thing that is worthwhile for its own sake, and dedicate my life to understanding, I become immediately invested in other people's lives. After all, if I value understanding for its own sake, then what I want is not just for me to understand many things, but for this world to be a world full of understanding. And so my goals and projects mean that I care about other people's understanding, about the frontiers of human knowledge. I have a stake in other people's education. I am happy and satisfied to the extent that I learn something, because I view learning Spanish or quantum physics as a step in a larger process. And I am happy and satisfied to the extent that others learn, since I value understanding itself, quite apart from anything that happens inside my own head. It means something to me if scientists I will never meet learn something about the structure of the atom that I will never understand. Similarly, if I value excellence at tennis for its own sake, I appreciate tennis excellence in other players, and have a stake in their becoming good at the sport. Put another way, I see myself not as an isolated player playing my own game but as part of a communal project, a joint human enterprise pushing toward the unknown, or pushing toward new heights of tennis excellence. My fellow citizens and I are mutually invested in each other because we are dedicated to a common goal, and because we all see the knowledge/tennis skills each of us gains as good things in their own right.

This is why it is so important for the engineer to see engineering as a calling, to embrace the values of the engineering profession. The engineer who does so is not alone, for they are part of the engineering community. Every engineering advance, no matter who makes it, is a personal victory. Because they value human excellence for its own sake, they take pleasure in every form of human excellence, and see their own life not as a collection of meaningless events, but as progress toward a worthwhile goal, the search for excellence. And every other engineer who values excellence wants them to succeed, because they are all part of the same team, striving for a common goal. Their life is meaningful and happy. Looking back near the end of one's life, it is not the size of your house that makes you content with life you have led. (For a third reason why you should care, see "Personal Values, Moral Beauty, and the Good Life" in Chapter 6.)

I suggest, then, that the only life not doomed to misery and failure is the life of commitment to value, a life that gives us a stake in each other because we are committed to a common enterprise.

Now, if I lead a life committed to value, I care about things like injustice. So, I will care about what Smith proposes to do to Jones. (See also the discussion of moral beauty and the good life in Chapter 6.)

What Should I Do?

To avoid being lonely and frustrated, and to avoid working in a cut-throat workplace, you have to care about justice and fairness, and you have to care about the work environment. These are things that affect you and that are morally important. To the extent that you are committed to these things, Smith's plan is your enemy. So, in a very real sense, this *is* your problem.

That doesn't necessarily mean you ought to *do* anything. Part of being an ethical engineer is knowing which battles to fight and which to sit out. (See "When to Fight a Battle" in Chapter 5.) In some cases, doing nothing may be the ethically appropriate choice.

The point is that you have to make an ethical choice here – you must give serious thought to what is right. A careful moral decision is called for. And a wide range of choices confront you, including doing nothing, sending an anonymous note to Jones, having an informal chat with Jones, speaking with or sending an anonymous note to Garcia telling them that Smith's claim is false, threatening to expose Smith if he carries out his plan, speaking to your own supervisor, and lodging a formal complaint against Smith. What should you do?

To simplify matters, we will consider four options: do nothing, speak first to Garcia, speak first to Smith, and speak first to Jones. (Anonymous notes have ethical drawbacks[4] and we can assume, for the sake of discussion, that your supervisor declines to become involved in another department's issues.) There are, of course, many ways you could do each of these. Even if you decide, say, to speak to Jones, you must decide how you're going to put it to her. In important ways, approaching an ethical problem is like approaching a design problem. *Part of ethical thinking is engineering a solution*, finding the best balance of moral preferences and requirements within the constraints of the situation. Ethical thinking requires creativity in part because of the enormous variety of circumstances and possibilities that engineers confront. In fact, in virtually all of the cases we will look at, there are more options and factors than I can discuss. All I can do is show you some of the relevant factors and possibilities, and how they play out.

If the threat to Jones is not serious, and if you are not in a position to do much about it, you may decide to sit this one out since you are not directly involved (it is not your department, you are not being asked to carry out an immoral order, and you are not responsible for overseeing personnel matters). For example, if Garcia is very fair-minded, knows what Smith is like, and so is likely to ignore what Smith says, the best course may be to stay out of it; "don't fix what ain't broke," as the saying goes. (See "When To Fight A Battle" in Chapter 5 for further guidance.)

Suppose, however, that the threat to Jones is a real one, you are in a position to do something about it, and you decide that you cannot, in good conscience, do nothing. Which approach is morally best?

There is no simple rule for deciding this but you can evaluate ethically the options open to you. One thing you could do is speak directly to Garcia. There are at least four reasons for not doing this. First, Garcia may not know what to make of your claim, since it will be your word against Smith's. Second, fairness to Smith requires that you hear his side of things before reporting him. (See "Treating Others Fairly and Well" in Chapter 5.) After all, you may not know the whole story: it is always possible that something Smith has to say could change your whole view of things. For example, we all vent by saying things we don't really mean. Smith may be mouthing off but have no intention of actually doing anything. Thus, by going directly to Garcia, you are being unfair to Smith. Third, Jones ought to have some voice in what happens since she is most directly involved; by going to Garcia, you deprive Jones of the opportunity to take responsibility for her own fate. Thus, going directly to Garcia would violate Jones' autonomy. (See "Autonomy" in Chapter 5.) Fourth, by going directly to Garcia, you disrupt the normal operation of that department and its channels. While it is sometimes necessary to do this, it is always preferable, when feasible, to respect proper channels.

These reasons suggest speaking to Smith or Jones before going to Garcia. Should you speak first to Smith or to Jones? If you speak first to Jones, you are being unfair to Smith, since you're not giving him a chance to present his side. However, if you speak first to Smith, you're violating Jones' autonomy. Since there is a good reason not to speak first to Smith and a good reason not to speak first to Jones, we have to weigh these reasons and see which, in this case, is more important. Now, because Jones is not Smith's peer or superior, speaking to Jones first is not likely to cause severe harm to Smith: usually, fairness to Smith is not severely compromised by speaking to Jones before speaking to Smith. However, since speaking to Smith might have important consequences for Jones' career, it is very important that it is Jones, not you, who decides whether and how Smith is confronted. Usually, Jones' autonomy would be severely undermined if you spoke to Smith before you spoke to Jones. However, you might weigh these factors differently if, for example, you have a reason to believe that telling Jones might seriously hurt Smith (e.g., Jones and Garcia are close personal friends). The higher up in the organization Jones and Smith are, in most cases, the less disparity there is in their ability to harm each other. A vice president has more access to the board, normally, than a secretary has to a district manager. (Again, this is only a generalization. It may not be true in a particular case.)

Moreover, in some circumstances, it would be better to speak to Smith privately, and, if you are not satisfied, then speak to Jones. For example, if you are confident that a few informal words reminding Smith of his ethical obligations would deter him, that would be the best solution for all involved. So, ordinarily, the best course would be to speak to Jones first, telling Jones that you are willing to speak to Smith and/or Garcia if Jones wants you to. In doing so, it is important to be as responsive to the other factors as possible. For example, keeping fairness to Smith in mind, you should stress that Jones should not jump to conclusions about Smith and should, if feasible, get Smith's side of things before taking action. You might ask Jones if she would like you to speak discretely to Smith and get his side. Again, although it is generally better to handle this situation informally, there are circumstances in which using formal channels is preferable. You must ask yourself to what extent there are significant legal ramifications, and to what extent using formal channels would protect you, Smith, and Jones.

In general, your decision must be influenced by factors such as: how responsive to ethical concerns are the individuals involved? Are there likely to be vindictive repercussions for you, for Jones, for Smith? Are you likely to do more harm than good by taking action? Do you know any of the individuals well? For example, if you know Garcia well and know that they will not jump to conclusions, it might prove best to voice your concern to Garcia, knowing that Garcia will not take any action until they investigate thoroughly, in a way that is fair to both Smith and Jones. If you know Garcia is very by-the-book and will simply ignore what Smith says, doing nothing might be the best course.

In short, although there are general considerations that provide guidance, your decision must take into account the character of the organizational setting and the individuals involved. Thus, each case is somewhat different and requires judgment in weighing a variety of factors. No rulebook concerning personnel matters can tell you what to do in every case.

How, then, do we go about doing this – how do we engineer a solution for a particular case? That is the subject of Chapter 2.

NOTES

1. In this volume, for convenience, I will sometimes speak simply of "company," when I am also speaking about government or nonprofit agencies, multiple or single engineer firms, or other engineering organizations.

2. In the interest of inclusion, singular personal pronouns used will switch between "he," "she," and "they."
3. Sales of the Pinto declined from 479,688 in 1973 to 187,708 in 1979 (Sherefkin 2008). It is impossible to gauge precisely the effect on other Ford products.
4. Sending an anonymous note is problematic in several ways. To name just three, it does not give the subject of the note an opportunity to respond, it does not give the recipient of the note a chance to ask questions, and anonymous notes create paranoia and foster distrust in the workplace.

REFERENCES

Baggini, Julian (2019) "Corporate Ethics Is for the Best," *Philonomist* https://www.philonomist.com/en/article/corporate-ethics-best

Bielefeldt, Angela R. and Canney, Nathan E. (2016) "Perspectives of Engineers on Moral Dilemmas in the Workplace," ASEE Conference New Orleans, LA June 26–29, 2016 available from https://peer.asee.org/perspectives-of-engineers-on-ethical-dilemmas-in-the-workplace

Budinger, Thomas F. and Budinger, Miriam D. (2006) *Ethics of Emerging Technologies: Scientific Facts and Moral Challenges* (Hoboken, NJ: John Wiley & Sons).

Fleming, John E. (1983) "Managing the Corporate Ethical Climate," in W. Michael Hoffman et al., eds., *Corporate Governance and Institutionalizing Ethics*, Proceedings of the Fifth National Conference on Business Ethics (Lanham, MD: Lexington Books).

Haigh, Martin (2013) "The Roles of an Engineer," in Rob Lawlor, ed., *Engineering in Society* (Royal Society of Engineering) https://www.raeng.org.uk/publications/reports/engineering-in-society

Hanson, Kirk (1983) "Institutionalizing Ethics in the Corporation," in W. Michael Hoffman et al., eds., *Corporate Governance and Institutionalizing Ethics*, Proceedings of the Fifth National Conference on Business Ethics (Lanham, MD: Lexington Books), 185–191.

Jones, Donald G. (1982) *Doing Ethics in Business* (Cambridge, MA: Oelgeschlager, Gunn & Hain).

Kapuska, Sergio (2020) "Why is Ethics Important for Engineering Students," *Rice Center for Engineering Leadership* http://www.rcelconnect.org/ethics-for-engineering-students/

Loui, Michael and Miller, Keith (2008) "Ethics and Professional Responsibility in Computing" in Benjamin W. Wah, ed., *Wiley Encyclopedia of Computer Science and Engineering* (Hoboken, NJ: Wiley).

Ottoson, Gerald E. (1982) "Essentials of an Ethical Corporate Climate," in Donald Jones, ed., *Doing Ethics in Business* (Cambridge, MA: Oelgeschlagen, Gunn & Hain), 155–163.

Perlman, Bruce J. and Varma, Roli (2002) "Improving Ethical Engineering Practice," IEEE Technology and Society Magazine 21:1, 40–47.

Peterson, Martin (2020) *Ethics for Engineers* (New York: Oxford University Press).

Royal Academy of Engineering (2011) *Engineering Ethics in Practice: A Guide for Engineers* (London: Royal Academy of Engineering).

Schlossberger, Eugene (2016) "Dual-Investor Theory and the Case for Benefit Corporations," *Business and Professional Ethics Journal* 35:1, 51–72.

Sherefkin, Robert (2008) "Ford 100: Defective Pinto almost Took Ford's Reputation with It," *Autoweek.* https://www.autoweek.com/news/a2099001/ford-100-defective-pinto-almost-took-fords-reputation-it/

Veach, Coy (2006) "There's No Such Thing as Engineering Ethics," *Leadership and Management in Engineering* 6:3, 97–101.

Webley, Simon and More, Elise (2003) *Does Business Ethics Pay: Ethics and Financial Performance* (London: Institute of Business Ethics).

2

Ethical Decision-Making

So far, we've talked about why ethics is important for engineers and we've thought about how to resolve one particular case. It is time to talk more generally about how to go about making ethical decisions.

A typical approach, used by many professional ethics textbooks (especially older ones), is to explain briefly a few ethical theories, usually including at least the "big three": utilitarianism, Kantian deontology, and virtue theory. Students are then expected to choose a theory and apply it to various issues. (See Appendix II for a brief overview of ethical theories.) Schlossberger (1993) took a different approach, suggesting that such theories are best seen as providing one among many ethical considerations to be used in making decisions. Making an ethical decision about a particular case consists in engineering a solution by identifying the relevant ethical considerations and giving reasons for weighing them in that particular case. Since then, many authors in diverse fields have suggested models somewhat along those lines. Another approach counsels creating an "Ethical Matrix," with stakeholders along the y-axis, ethical principles along the x-axis, and cells containing "value demands." Thus, inside the cell defined by "people" and "autonomy" might be the value "freedom of choice" (Biasetti and de Mori 2021). This value map is used to outline conflicts and lay out what must be weighted. Drawing such a matrix may be useful, especially in straightforward cases, though in complex cases, it may not reflect the nuances and complexities an engineer must ponder.

DOI: 10.1201/9781003242574-3

THE NATURE OF ETHICAL DECISION-MAKING

Case 1 in the previous chapter reveals much about the nature of engineering ethics. Ethics is about how to live, about what makes for a good person and for a good life. Ethical thinking is deciding what really matters in life. So, every significant choice you make is an ethical decision. For example, if you choose not to take a promotion in a distant city in order to remain close to your aging parents, you are deciding that some values, such as family ties, or helping those you love, are more important than others, such as career satisfaction or the pleasure of the sauna you would be able to purchase if you took the promotion. This is a decision about what matters in life, about what is really good and worthwhile, and that is what ethics is about.

Because values can't be precisely measured like steel rods, ethical decisions always involve a lot of individual judgment. Ethical decisions are *not algorithmic*.[1] However, ethical choices are *not arbitrary*: deciding what is right is not like choosing between non-dairy hot fudge and strawberry topping. Making an ethical judgment is more like buying a car. There is no simple answer, usually, to which is the best car for you. But there are good decisions and bad ones, and good decisions are the result of carefully and reasonably weighing the relevant factors. Safety, cost, cargo capacity, aesthetics, and reliability are all considerations. Individuals' needs vary and there is no mathematical formula for weighing safety against reliability. For example, there are three kinds of cost: initial, upkeep, and amortized. The initial cost is the purchase price and fees. The upkeep cost is what it costs to keep the car running: maintenance, repairs, fuel, insurance, etc. The amortized cost is the initial price minus the resale price divided by the number of years you have kept the car. For example, if you buy a $30,000 car (option A), keep it for ten years, and then sell it for $5,000, the amortized cost is $2,500 a year. If you buy an $18,000 car (option B), keep it for two years, and sell it for $8,000, your amortized cost is $5,000 a year. If your income is relatively stable, the amortized cost is often more important than the initial cost. At the end of ten years, you would have $25,000 more in the bank if you choose option A instead of choosing option B five times (to cover the 10 years). While it is generally cost-efficient to spend somewhat more for a car that will last a lot longer (car A), it may make more sense to buy the low-cost, short-lived car (car B) if you expect your income to rise dramatically in the next few years, since, in two years, when you are ready to sell, you can better afford a new car. Reliability may be less important if you live on a bus route to work. Cargo capacity is more important for some people than for others. For example, if this is the only car for a family of five, it must seat at least five (you can't tie a family member to the roof). However, if this is the family's second car, used for commuting to work, and their first car seats five, a four-seater or two-seater is feasible. So, while you can give advice to a friend that will help her make a wise choice, there is no simple rulebook you can give her that will make the right choice for her. There probably isn't even a single "right" decision for her. However, suppose someone has a large family and a job for which being a few hours late might prove quite costly. If she buys, as the family's only car, an unsafe, expensive, unreliable two-seater car because it "looks good," she is clearly making a bad decision.

Ethical decision-making is also much like legal reasoning. A judge who must decide about a difficult question of law draws on a variety of sources (such as statutes, precedents, and legal maxims) in employing the tools of legal reasoning. If they are a good judge, their decision, based on the tools and sources, will be reasonable and defensible, though there is still room for legitimate disagreement. (Supreme Court judges do not always agree about what is constitutional, and even the best judges' decisions are sometimes overturned on appeal.) The good judge can give strong reasons for thinking that precedent N is more central to the case before them than is statute M. Even though their reasons may not convince everyone, any good judge can see the force of their reasons.

In engineering design, there are absolute requirements that must be met, desirable features, and constraints of various sorts. Some designs are simply unacceptable. In many cases, there is no single right design making all others wrong. There are different ways of balancing desirable features while meeting absolute requirements within constraints, some of which are good and some of which are not, and there may not be universal agreement about which of the good solutions is the best.

The same is true of ethics. No simple rules will cover all cases. Just like the judge, the ethical person has to use her judgment. Of course, it doesn't help much to be told "you must use your judgment," when it is not clear what is to be judged, what factors must be weighed, and how to go about weighing those factors. Fortunately, the ethical person is able to draw on a variety of sources, such as specific rules, general principles, guiding ideas, values, and moral factors. The ethical person, facing a difficult moral choice, must reason about which rules, values, principles, and so forth are relevant, analyze the problem in terms of those values and principles, and weigh the different factors in the light of guiding considerations.

Although experts generally agree about most of the ordinary, day-to-day ethical decisions engineers must make, there are areas where experts disagree. Ethical decisions are sometimes difficult. For example, although telling the truth is generally the right thing to do, in rare cases, an engineer might have to choose between loyalty to her employer and telling the truth. Not every ethical person will make the same choice. However, an ethical person will make the decision reasonably, understanding the issues and principles involved, and be able to explain why she thinks that, in this particular case, the value of honesty is more central than loyalty to her employer (or vice versa).

These materials are meant to provide you with some of the tools, sources, and guiding considerations and to give you some general assistance in reasoning, analyzing, and weighing. They may not tell you what to do in every case but they will help you decide in an ethical, fair, and rational way; explain your decision to others; and understand the reasons behind the ethical decisions that others make.

MAKING ETHICAL DECISIONS

The materials in this volume present you with a variety of source materials and techniques to draw upon in making moral decisions. You will be given specific rules to be followed (e.g., do not falsify tests or records), general principles that should be respected (e.g., you should try to promote good consequences), values you should strive for (e.g., you should put a high priority on establishing and maintaining a partnership with nature), and factors and guiding ideas that influence the way these rules, principles, and values apply to particular situations. For example, one factor in weighing competing rules is: a rule that reflects an institutional duty generally, other things being equal, has greater weight than a rule that does not. A guiding idea in deciding whether a given solution is too risky is: am I being a good trustee of the public safety?

Sometimes, the rules, values, and principles pull in opposite directions and you will have to decide how to weigh them. Generally, specific rules have first priority. However, the specific rules are not arbitrary requirements. Rather, they are ways that engineers can put into practice the moral principles and values that apply to engineering. Rules, in other words, derive from the principles and values of engineering. Why, for example, is there a specific rule that engineers must not violate the demands of safety? Because safety is a central value of engineering and because acting unsafely usually violates the general principle that we should try to promote good consequences and generally violates the public trust without which engineers could not practice their professions. (No one would allow an engineer to design a skyscraper or nuclear power plant if engineers cannot be trusted to prioritize safety.) Now, in some cases, following a specific rule might undermine rather than promote the value or principle the rule is based on. When that happens, the rule has to be modified. People are not machines and should not blindly follow rules regardless of the reason for the rule. This is what makes ethics both difficult and fascinating. Moreover, how safe is safe enough? How should we balance a small gain in safety against a large gain in something else, such as cost or consumer autonomy? There is no mechanical way of determining the right thing to do. So, I have tried to give you as many tools as possible. For example, I suggest that in thinking about whether an engineering solution is sufficiently safe, you ask yourself, "would I want my family to undergo this risk for this benefit?" This is often a helpful way of thinking about safety. Sometimes, it won't help at all because you don't know whether you would want your family to take this risk. In that case, you must try something else. Use other factors or rules or values to reach a decision.

Because general moral understanding is as important as specific advice about tough issues, the materials are organized to facilitate understanding, to make clear the reasons for rules, values, and principles, to make clear why certain factors are important, and so forth. For example, since one key moral idea in engineering ethics is that of community, I have devoted one section to the idea of community. The idea of community gives rise to several different rules and values, which are discussed in that section. Another key moral concept is that of fairness (justice), and so another section is devoted to treating others fairly and well. The rules and moral considerations related to justice are discussed in that section. Moral ideas are related in complex ways. While the text sometimes refers readers to a related section, readers should be thinking about how each new consideration introduced relates to others. The realm of the ethical, what I like to call "the moral edifice," contains many ideas that hang together in numerous complex and subtle ways.

Ethics, like most things, requires practice. To help you practice making ethical decisions, I have included a fair number of cases. Some cases I discuss, while others I leave for you to think about and, perhaps, discuss with friends or colleagues. Generally, the cases are brief sketches of situations and do not include all possible relevant features. After all, ethical thinking often begins with asking yourself what else you need to know about a situation. A new piece of information may radically change your solution to an ethical problem. Normally, for example, you should not be a supplier of the company that employs you. Suppose, for example, you are both the owner of a small company that manufactures special coolants and a chemical engineer employed by the Z corporation, which manufactures vitamins. Generally, you should not sell coolants to the Z corporation. However, the fact that your company is the sole manufacturer of the only coolant that meets Z corporation's needs changes the situation dramatically. While you should avoid conflict of interest situations, you need not deny your employer a vital supply. Similarly, a relevant fact may make one of my recommendations inapplicable to your case.

It cannot be stressed enough that the materials and recommendations made in this book are but the considered thoughts of one individual. No one has a monopoly on moral truth. This book is meant to be helpful to you in your thinking, not to replace your own moral judgment. You may disagree with some of the principles and recommendations. You may think of other moral factors or principles not mentioned in these pages. *In ethics, as in almost everything, there is no substitute for your own, carefully considered judgment.*

A TEMPLATE FOR MORAL DECISION-MAKING

One way to learn how to make good decisions is to look carefully at good examples of moral thinking. One example for you to think about is the discussion of Case 1. In analyzing the case, we started by listing our options and then determining which steps we could take and how we might go about taking them. We then identified several relevant moral sources, such as the value of autonomy (the importance of letting people take charge of their own lives), the principles of fair dealing (such as making sure we hear Smith's side of things before taking action against him), and the principle of promoting good consequences (such as making sure our actions do not hurt Jones' career). We thought about how these sources apply to our problem. Then, we gave reasons for a particular weighing of the different factors in this particular case. Finally, we reached a decision about what to do.

This is the general formula for ethical decision-making. Making tough decisions can be thought of as a five-step process. But in order to use that five-step process, one must develop one's ethical toolkit, which is a lifelong, continuing process. So, there are two stages or aspects of good ethical decision-making: preparation (constructing an ethics toolkit) and making particular decisions (the five-step process).

Preparation

Preparation means developing ethical awareness, pre-planning, and constructing an ethics toolkit. These are processes that continue throughout one's life. In the light of new ideas, new experiences, and new arguments, ethical people continue to refine their tools (adding new ones and perhaps discarding old ones), broaden their ethical sensitivity, and re-examine what is right. Preparation, thus, consists in three ongoing (lifelong) activities.

1. Become aware of ethically sensitive issues and situations, so that you can recognize an ethical decision when it arises.

    Often people do wrong not because they do not care about ethics but because they do not even recognize that what they are doing presents an ethical challenge. For example, a secondary school math teacher might say to a student "if you bring these materials to the next room (or set up a technical teaching tool) I will give you extra credit." Now, grades are supposed to be awarded on the basis of demonstrated mastery of the subject matter. Bringing books to a classroom, however helpful, does not show mastery of calculus. The teacher is giving the student a higher grade than earned in exchange for helping the teacher. Technically, that is a bribe. I am sure the teacher did not see it that way – few teachers are willing to throw out their integrity and accept a bribe, especially for such a small gain. But it is an ethically problematic practice if one thinks about it clearly and I suspect teachers would stop doing this if they realized they are giving a higher grade than earned in exchange for a personal favor. Thus, part of preparation is becoming aware of ethically challenging situations (developing ethical awareness).

    Questions to ask:

    - Which situations you might face have ethical dimensions that must be considered?

- What are the relevant issues, laws, and/or professional codes and regulations?
- What are standardly accepted responses?

While you may come to conclude that professional codes or standard responses are faulty, either in general or in a particular case, they are at least a good starting point in your ethical deliberations and should certainly be in your list of options. Even if you ultimately decide to reject or amend them, you should at least know what they are. Engineers have an especially strong general duty to obey the law (see Schlossberger 1995). It is not impossible that a law so strongly violates justice that civil disobedience is called for (e.g., in Nazi Germany). This will be a very rare occurrence. In any case, an engineer who decides to break the law should at least know that they are doing so.

How can an engineer go about doing this?

- Read this book, as well as other books and articles about ethics.
- Attend conferences, seminars, classes, and training sessions.
- Discuss issues with colleagues and friends.

2. Think out in advance (pre-plan) your response to difficult situations – when the situation actually arises, time may be short.

Life can blindside us but it is often possible to anticipate that one might have to make certain sorts of ethical decisions. Give attention to which sorts of decisions may arise and give thought in advance to what one might do. Since a good decision often depends on the circumstances of the particular case, it may not be feasible to reach a firm decision but the more one thinks in advance about the situation, the faster, in general, one will be able to come to a decision.

How can an engineer go about doing this?

- Read this book, as well as other books and articles about ethics.
- Attend conferences, seminars, classes, and training sessions.
- Discuss issues with colleagues and friends.

3. Construct an ethics toolkit.

Your ethics toolkit will include at least three sorts of items: models, ethical factors, and weighing factors. Thus, you need to:

- (a) Select a model of engineering.

Chapter 3 explains what models of a profession are and some ethical consequences of them. It also discusses different models of the engineering profession.

- (b) Determine your values. Identify rules, principles, and key ethical ideas that are frequently applicable.

  Values, rules, and principles can be found throughout this volume. Examples are the value of autonomy, the principle of promoting good consequences, the rule that an engineering firm should not use, in advertising, the name of an engineer no longer with the firm, and the ethical idea that it is helpful to think about how you would feel were the situation reversed (see "The Golden Rule" in Chapter 5).
- (c) Identify factors that help you weigh the importance of rules, principles, etc., when they conflict in a particular case.

  Many such factors are provided throughout this book. For example, it is more important to follow the rules in a competitive situation than in a non-competitive situation.

  How can an engineer go about doing this?

- Read books and articles.
- Discuss with friends and colleagues.
- Attend relevant conference presentations.
- Take classes or attend training sessions.

It is important to select values, models, principles, rules, and weighing factors rationally. An essential part of preparation is looking at the arguments, the rational reasons for believing something to be true. After all, many things that whole societies once felt were obvious and certain may strike us as absurd today. Throughout much of human history, slavery was a common practice across much of the globe (Africa, Europe, Asia, the Middle East, and the Americas). Indeed, slavery was, in many places, considered the obvious natural order of things. Today, it is obvious to most people across most of the globe that slavery is not just wrong but abhorrent. In Victorian England, it was just obvious that a woman who would show her ankles in public was grossly immoral. Although some cultures today share that outlook, billions of others would see that belief as odd, insisting "it's just skin." (Of course, butt skin is also just skin.) Throughout most of human history, marriage and sexual adulthood began with puberty. (We are, after all, evolutionarily designed to seek sex and start producing babies at puberty.) Today, many regard sex with or marriage to a 13-year-old as intolerable abuse. History shows that one's "gut," one's feelings, or one's sense of moral repulsion are not necessarily reliable guides. Either Victorian or contemporary English guts are wrong. That is why it is important to turn to reason and arguments.

Creating a Toolkit Rationally: Looking at the Arguments

While it is obvious to engineers that one can test a material in the lab to determine its cold flow under specified conditions, and what sort of evidence would be helpful in deciding whether it is safe to call for a given material in a design project, it may not be clear how to take an ethical claim into the philosophy laboratory and evaluate it rationally. Who is to say what is right and wrong? The answer is: the arguments say.

Three Rational Responses to an Argument

An argument consists of assumptions (the premises) that are meant to offer a rational reason to believe something (the conclusion). So, the two key questions to ask about an argument are: (1) Are the premises (assumptions) correct? (2) Do the conclusions follow from the assumptions?

Consider this argument:

1. All dogs are mammals (premise).
2. Fido is a dog (premise).
3. Therefore, Fido is a mammal (conclusion).

If the premises are true, then the conclusion absolutely must be true. (After all, if Fido is a dog but not a mammal, then there must be at least one dog that isn't a mammal, namely Fido.)

So, if you want to disagree with the conclusion, you must either insist that all biologists are wrong and that not all dogs are mammals, or you must insist that Fido is not a dog (e.g., Fido is really a goldfish wearing a dog costume). Thus, an argument shows you the price you must pay for disagreeing with the conclusion.

As a result, there are only *three rational responses to an argument:*

- Deny (show false) a premise.
- Show a flaw in the reasoning.
- Accept the conclusion.

The reasoning in this particular argument is unassailable. (If the premises are true, the conclusion has to be true also.) So, if you say that Fido is not a mammal, you must pay a price: you must say that all biology books are wrong and dogs are not mammals or that four-legged, furry Fido is not a dog. Those are your only reasonable options. That is the power of arguments. If I just say "Fido is a mammal," you can say "that's just your opinion." But once I give you that argument, you must agree or pay the price. After all, if you say "yes, all dogs are mammals, and yes, Fido is a dog, but Fido is not a mammal," then you are plain wrong – what you are saying is simply impossible.

Not all arguments are what logicians call "valid arguments" like this one. For example, if I notice that my neighbor runs out of the house at odd hours carrying a black bag and drives an expensive sports car with a hospital parking sticker and a license plate that reads "DOCTOR 4112," I might conclude that they are a physician. This is not a valid argument because there are other possible explanations. They could be a plumber on call for a hospital plumbing emergency whose mother's name was Sarah Doctor. But it is much more likely that they are a physician. The conclusion "my neighbor is probably a physician" is warranted. Of course, this is normally not sufficient evidence to allow your neighbor to perform surgery on you – how strong the evidence must be depends on what is at stake. There are many kinds of arguments you can draw upon in constructing your ethics toolkit. In addition, even with a valid (deductive) argument you still have to determine if the premises are true. Of course, there is an enormous literature about this. What follows just touches the surface of the subject, but, hopefully, it will prove useful.

Four Rational Responses to a Counterexample

Ethical arguments will often contain a general ethical claim, like "killing is always wrong," as a premise or conclusion. For example, someone might argue that abortion is wrong (conclusion) because killing is always wrong (premise) and abortion is killing (premise). [Note: This discussion is NOT about the rights and wrongs of abortion but only about the logic of one particular argument.] One way to test an ethical claim, like "killing is always wrong," is to look for counterexamples to it. A *counterexample* is an example that counters (shows wrong) a claim.

For example:

Claim 1: All US Presidents were Republicans.

Counterexamples to that claim include Joseph Biden, Barak Obama, John Fitzgerald Kennedy (JFK), and Woodrow Wilson, all of whom were US Presidents but were not Republicans.

Claim 2: Killing is always wrong.

Counterexample: Breathing, for when a person breathes they kill large numbers of microorganisms that are taken into the body and destroyed. (Note that we physically cannot refrain from breathing except by dying, and that killing oneself to keep from killing microorganisms is also an instance of killing.)

Of course, finding a possible counterexample to a claim is not the end of the story. There are *four rational responses to a counterexample.*

First, one can *modify the claim to avoid the counterexample.* For instance, the new, improved versions of claims 1 and 2 might be:

Claim 1a: All US Presidents before 1900 were Republicans.

Claim 2a: Killing a human being is always wrong.

One then has to modify the argument in which the claim appears in order to make it follow. The revised argument would be that abortion is wrong because killing a human being is always wrong and abortion is killing a human being. (Note that the second premise, that abortion is killing a human being, is less universally accepted than the original premise that abortion is killing something or other.) If someone still insists that breathing is a counterexample, the second response comes into play: *Show the example does not fall within the scope of the claim.* When I breathe I do not normally kill a human being, and the revised claim talks only about killing human beings – it says nothing about killing microorganisms. Thus, breathing is no longer a legitimate counterexample to the claim. Similarly, the Presidents mentioned in the counterexamples to Claim 1 were all elected after 1900, and so are not counterexamples to claim 1a.

Again, this is not the end of the story. One should now look for counterexamples to the revised claim. For example, George Washington was elected before 1900 and was not a Republican. Many hold that killing a human being in self-defense is not wrong. Some hold that executing a murderer is not wrong. Both are examples of killing a human being that, some people think, are not wrong.

A possible response to these new counterexamples is to employ the third possible response to a counterexample: *Bite the bullet,* that is, insist the predicate of the claim applies to the example. For example, one might insist that capital punishment is indeed wrong (as many do) and that killing a human being in self-defense is wrong (as some pacifists hold). More likely, however, the person giving or considering this argument will use response one and modify the claim again:

Claim 2b: Killing an innocent human being is always wrong.

This revised version avoids the counterexamples because the murderer and the person trying to kill you are not innocent. This requires adjusting the other premise appropriately: abortion is always wrong because killing an innocent human being is always wrong and abortion is killing an innocent human being.

Now, however, a new counterexample arises. Suppose that a mad scientist has planted the trigger to a bomb that would destroy the Earth under some leaves in a park. The Lone Ranger races to the park, only to find that Herbert, out for an innocent stroll in the park, is unknowingly about to step on the leaves hiding the trigger. Given the distance involved and the fact that Herbert's foot will trigger the bomb in less than a second, the Lone Ranger has only two choices. (Let the story be constructed to rule out any other possibility.) He can do nothing and let the Earth explode, killing everyone, or he can shoot Herbert with a necessarily fatal bullet. Most would agree that the right thing to do is (reluctantly) to shoot Herbert. Shooting Herbert, then, is an example of killing an innocent human being that is not wrong: it is a counterexample to claim 2b.

At this point, one may employ any of the three strategies mentioned: revise the claim, argue that it would be wrong to kill Herbert, or insist that Herbert is not innocent. Quite possibly, however, one might employ the fourth possible response to a counterexample: *Give up the claim.*

Of course, giving up the claim, and thus abandoning the argument, does not necessarily mean that the conclusion was wrong. There are plenty of bad arguments for true conclusions. This is an important point. *Showing that an argument is flawed does not prove the conclusion is false.* It just shows that one argument for the conclusion won't work. Thus, the proponent of the original argument might still think that abortion is always wrong. But they do have to find a different argument. Similarly, a proponent of abortion rights would still need to provide an argument that abortion is not always wrong (showing that the argument above is flawed is not enough).

Picture Building and the Moral Edifice

Another way to assess a value or an ethical claim is to ask whether it is an isolated prejudice or part of a comprehensive picture of an important part of human life. (Another way to put this is asking how the value or claim fits into the moral edifice, discussed below.) When deciding whether to add a value to your ethics toolkit, ask whether it fits in with your general picture of a good human life or whether it is arbitrary. For example, in some contemporary Western societies, it is considered immoral to show a woman's nipples in public but acceptable to show the rest of her breast. If those societies changed their minds about this particular belief, deciding with the Himba of Namibia that it was not immoral to appear bare breasted in public, not much would change, apart from minor modifications of fashion. In those societies, it is a relatively isolated belief, no matter how strongly it might be held. While that does not prove that displaying nipples is morally acceptable, it should weaken the confidence in that belief on the part of those who hold it. By contrast, were Portugal or Japan to relinquish entirely the importance of truth telling, major social and legal changes would ensue, for the value of honesty and truth is deeply embedded in a large array of laws, customs, values, and social relationships. That honesty matters is not an isolated belief but an integral part of how we understand human life. It would be hard to function in a world in which we cannot put any trust at all in what people tell us. (There is no point, to take one small example, in asking for directions, since the answer one receives is not particularly likely to be true.) That alone does not prove that honesty is a moral virtue. Perhaps we are grossly mistaken about so much of our lives. But it should strengthen our confidence in that belief because it would take a powerful reason to overthrow so much of our way of life.

It may help to see this idea in action in a moral argument. Consider the following argument:

1. People's projects matter.
2. Killing a person generally puts an end to his or her projects.
3. Therefore: It is wrong to kill people in the absence of a strong justifying reason.

While most people would accept this conclusion, it is not universally held. A famous story about the American Wild West relates how a gunslinger shot a stranger through a hotel wall because he was snoring. The gunslinger did not believe that he needed a strong reason to kill the stranger – he felt that mild annoyance was sufficient. So, there is some point to giving the argument. Of course, the conclusion doesn't tell us what counts as a good enough reason for killing, but it at least tells us that the gunslinger was wrong to do what he did.

Suppose the gunslinger is not inclined to agree and insists that he did nothing wrong. Recall that anyone who wants to deny the conclusion must deny a premise or show a flaw in the reasoning. The second premise seems fairly obvious – it is usually hard to pursue your projects from the grave. So, our gunslinger, if he is inclined to be rational, will most likely deny the first premise. What reason is there to believe premise 1?

Picture building can provide a strong reason. To deny that people's projects matter is to give up as irrational most of our social life and much of our way of regarding things. For example, the reason it is important to be able to form reasonable expectations is because without them we can't feasibly pursue our projects. Much of our way of life, such as contract law and the institution of marriage, exists in order to make it feasible to form reasonable expectations. So, if people's projects do not matter, then neither do reasonable expectations, and so contract law and marriage make little sense. A few minutes thought shows that much of our way of life collapses if we give no importance to people's projects. Thus, the cost of denying premise 1 is giving up contract law, marriage, and so much else. Maybe we ought to throw out our entire way of life. But it is much more likely that we should acknowledge that people's projects matter. There is a good reason to believe that people's projects do matter, and so a good reason to believe that one shouldn't kill someone unless one has a strong reason for doing so. In particular, if the gunslinger is not willing to discard most of our way of life, than he cannot reasonably deny that people's projects matter.

By the way, this argument shows something else very important. People sometimes say "you can't argue about morals." Well, we just did – we gave a strong argument about ethics. Some people may ask "who is to say what's right and wrong?" We just saw that it is the argument that says that what the gunslinger did was wrong.

Relativism

I want to say just a bit more here because many people think that ethics is all in the eye of the beholder. There are no absolute moral duties, they would say, because whatever you think is right is, in fact, right for you. This view is called "individual relativism." A related view, that whatever a society believes is right is, in fact, right in that society, is called "social relativism." Individual relativism is both widely held and deeply flawed. There are many arguments against it. I will mention, briefly, just one. Note first that relativism is different from amoralism.

The amoralist says that nothing is right or wrong. (See Schlossberger 1992 for an argument against amoralism.) Relativism says that if I believe something is right, then it genuinely is right for me. Similarly, this kind of "normative" relativism is different from what philosophers call "descriptive relativism," which simply holds that people have different moral beliefs. When a descriptive relativist says "slavery was not wrong in ancient Sumeria," they mean that Bronze Age Sumerians did not believe slavery was wrong. If a normative relativist says this, they mean that Sumerians who had slaves were actually not wrong to do so. By saying "x is wrong for me," the normative relativist does not mean just that I believe x is wrong, but that it is truly (objectively) wrong for me to do x.

The appeal of (normative) individual relativism is that it appears to avoid judgmentalism and the need for moral justification. If A thinks it is wrong to withhold treatment from someone dying a painful death and B thinks it is wrong and inhumane not to do so, an individual relativist would say that withholding treatment is wrong for A but right for B. So, both are morally fine if A withholds treatment and B does not – judgmentalism avoided. Moreover, I don't need to justify my own view – it is right for me if I think it is. But this is an illusion. If whatever anyone believes is wrong for them is, indeed, wrong for them, then everyone, without exception, has a duty to be sincere (true to their own convictions), because if they think it wrong and do it anyway, what they do would be objectively wrong. So, the relativist contradicts himself when he says that there are no absolute duties – sincerity is the one absolute commandment for everyone. But people often do not follow their own beliefs. If A does withhold treatment, for example, then A is morally bad – judgmentalism is not avoided. Moreover, most people (including me) feel strongly that it is better to be an insincere Nazi who does no harm than a sincere Nazi sending people to the gas chamber. But the individualist relativist must say the opposite (because the sincere Nazi is doing what he thinks is right, while the insincere Nazi is not). So, the individual relativist has to give an argument that we are wrong, that is, the relativist must justify the claim that sincerity is an absolute obligation. So, the need for moral justification is not avoided. Indeed, the relativist's claim is hard to justify. I feel confident that an individual relativist being sent to the gas chamber would not agree that the sincere Nazi is to be morally congratulated because he is doing what is right. The appeal of individual relativism, thus, vanishes. A similar argument can be mounted against social relativism, as it appears to entail that everyone, everywhere, has a moral duty to conform to their society's moral code.[2] (See also Note 2 in Chapter 4 about how to evaluate ends.)

How can an engineer go about improving and developing an ethics toolkit?

- Read books and articles.
- Discuss with friends and colleagues.
- Attend relevant conference presentations.
- Take classes or attend training sessions.

Making Particular Decisions: The Five-Step Process

Making a difficult ethical decision is a five-step process. You can use basically the same process to make any important decision (such as buying a car, as discussed above).

Step 1. *Clarify the moral decision to be made. What must be decided? What are your options? What are the relevant facts?*

Sometimes, your options are clear and limited. Sometimes, you will need to be creative about formulating options to make a good decision. Sometimes, the facts are obvious and sometimes it is not clear what information you need to make a good decision or how to obtain it.

Step 2. *Identify the moral considerations that are pertinent to the particular situation you are thinking about. Which specific rules, general principles, values, factors, and guiding ideas apply to the case? How do those considerations apply to the situation – what does each suggest about each of your options? Which aspects of the situation limit or modify the relevant rules, principles, values, and factors?*

Here you draw upon a wide range of factors. Some will be very general, like a model of engineering (Chapter 3) or universality (Chapter 5), while some will be specific to the field, such as, in complex information systems, it is important that operators be aware of what is happening in other parts of the system (Chapter 13). Because so many factors might potentially apply, you should limit consideration to those likely to have a significant impact on your choice. But the more that rides on your choice, the more comprehensive you should be, within your time constraints. Obviously, if a decision must be made quickly, you are limited to the few most salient factors.

Step 3. Weigh the factors (determine how these moral considerations should be weighed for this situation). Engage in an open-ended process of giving reasons for thinking that some factors are more central, relevant, or important to this particular case.

This is often the most challenging part of making a particular ethical decision. The key phrases here are "reasons" and "this particular case." You are not just picking one factor over another. You are giving a rational reason. In case 1, that Smith is more able to harm Jones than Jones is to harm Smith is a rational reason (R1) for thinking Jones' autonomy might be more important in this case than fairness to Smith. We are not saying that autonomy is always more important than fairness. It depends on the case. A reason for thinking reason R1 less compelling is the fact that Jones has close relationships with Smith's superiors. In general, if you have one reason, R3, to think factor x more important than y for this case and another reason, R4, to think factor y more important than factor x, you must try to give a reason (R5) to think R3 is more compelling in this case than R4 (or vice versa).

Step 4. Calculate the result (design a solution). What follows from Steps 1 to 4? Which of the options listed in Step 1 is best supported by the weighing in Step 4? What is the best way of implementing that option (engineering a solution)? Decide and act.

Usually in Step 1, you are considering somewhat broad options, such as, in case 1, "talk to Jones first." If that appears to be the best option, you should use your thinking in the prior steps to help figure out the best way of implementing that option, a way that is sensitive, as much as possible, to all the ethical considerations. Fairness to Smith, for example, suggests that one should emphasize strongly when talking to Jones that Smith may not be serious, that there may just be a misunderstanding, and so forth. So, in Step 4, you are often not just picking an option but trying to accommodate, as much as feasible, the ethical factors that apply, engineering your decision to best balance your ethical goals. An ethical problem is like a design problem (Whitbeck 1996). In designing a car, an engineer strives to achieve the best compromise between safety, reliability, economy of use, cost, and several other factors. Few can afford a super-safe car that costs $5,000,000 and no responsible engineer would approve a supercheap death trap. Step 4 often requires the same kind of sensitivity to all the competing ethical factors.

Step 5. As feasible, monitor the situation and make adjustments as necessary.

In some cases, making a decision removes the matter from your hands. In other cases, you are still able to influence or rethink matters as events unfold. For example, in case 1, if you had decided to do nothing, you may continue to observe the situation and later, if circumstances warrant, speak to Jones.

This five-step process is the standard procedure for dealing with a difficult moral decision. But you don't necessarily have to go through this procedure every time you make an ethical decision. In some cases, a different strategy may prove more useful. Ethical thinking is a craft and a good craftsperson will use different techniques for different situations. Moreover, in many cases, it is easy to know what is right, and we don't have to go through a complicated decision process to know what we should do. If a situation is clearly covered by a company rule, you must follow the rule unless there is a very strong moral reason to do otherwise. Only in very rare and special cases would the ethical engineer even consider breaking the law.[3] It is virtually never right for an engineer to lie in a professional context. Still, it is important for the ethical engineer to understand the process of ethical decision-making, for two reasons. It is always good to know why something is the right thing to do, and there are many situations in which it is not immediately clear what is right.

You may need to go back and forth between steps. Thinking about Step 2, identifying relevant ethical factors, may make you realize that you need additional information forcing you to return to Step 1 (finding out the relevant facts). If what you come up with in Step 4 (decide and act) makes you too uncomfortable, you may need to go back to Step 1 and be more creative in formulating options. Of course, in an imperfect world, the best decision may not be a perfect decision. Sometimes, every good option has at least some ethical downside. Try to engineer the best available solution. Some situations are just difficult and all options are less than satisfactory. You just have to pick the least bad option. Note that these situations can sometimes, but not always, be avoided by steps taken before the choice arises. Someone who made two incompatible promises, and hence cannot help but break at least one of them, could, in some cases, have avoided the situation by being more careful about what they promise.

Since the process is open-ended at several points, one may wonder how one knows one is finished. How much time one should devote depends on the importance of the decision, the other demands on your time, and any decision deadlines you might have. You don't need to spend hours agonizing over a 50-cent library fine, while a decision in which lives are at stake should have priority on your time and attention. Some decisions must be made quickly, while others give you more time to decide. Some decisions have built-in deadlines. If you are Roger Boisjoly contemplating what to do before the launch of the Challenger (see Case 27 in Chapter 7), you can't wait until after the shuttle launches. The basic answer is that one is done when either one can't think of anything else or one runs out of time. *In making an ethical decision, you are trying to do the best you reasonably can, in the circumstances.* For example, talking to others may help clarify your thinking and suggest options or factors you hadn't considered. Do you have time to talk your decision through with others? Does confidentiality prevent it? Is the decision important enough?

Sometimes, it is helpful to construct a specific flow chart for making the decision. Which flow chart is most useful for a given decision depends upon the nature of the decision. One example of a flow chart for ethical decision-making is given below. Only rarely will you need to go through such an elaborate procedure, but moral thinking often uses some simplified version of the decision strategies elaborated in the flow chart below.

Part II of this volume gives you tools and sources for making ethical decisions. They are often illustrated with cases and examples. Further and more comprehensive applications of these tools and sources to specific issues and problems in engineering ethics come in Part III. Together, they are meant to give you a good foundation in ethical engineering.

Case 2: Hurricane Katrina and New Orleans' Levee System

On August 29 2005, when Hurricane Katrina hit the city of New Orleans, 50 failures occurred of the levee and flood wall system designed to protect New Orleans and surrounding suburbs. "Of the 284 miles of federal levees and floodwalls—there are approximately 350 miles in total—169 miles were damaged" (ASCE 2007). Eighty percent of the city was inundated, with water reaching 15 feet in some areas. Floodwater was toxic with solvents and gasoline from flooded cars and buildings as well as from seven flood-related oil spills (ASCE 2007), and 70% of the city's housing was severely damaged or destroyed. One hundred twenty-five billion in damages occurred and over 1,800 lives were lost. Although 1.5 million individuals were evacuated before the storm, some chose to remain and some were unable to leave. More than 150,000 people did not evacuate before the storm. Thirty thousand people packed into the Superdome, seeking shelter. They endured hellish conditions, including five days without water, before being evacuated. Others were stranded on rooftops.

Failures occurred either from overtopping of the walls by floodwaters, improper design, or walls being set in ground containing sand rather than clay. At various locations, the Army Corps of Engineers, not adhering fully to its guidelines, overestimated the strength of the soil, such as the subsurface, water-filled peat at 17th Street, partly because of taking too few samples and ignoring the variability of the soil in different parts of the area. They did not account for natural sinkage gradually lowering the effective size of the walls, and in some cases used incorrect information about land elevation. Walls were built to withstand 100 mpg winds, at the low end of the US Weather Bureau's estimate of typical winds of 100–111 mph and well below both Katrina's 125 mph and the National Weather Service's revised estimate of possible 151–160 mph winds (Hoke, 2015).

The evacuation and rescue efforts have been widely criticized. Some speculated (and others deny) that evacuation instructions were delayed from fear of losing face if the hurricane altered its path. The Mayor of New Orleans was criticized by some (and defended by others) for not using school buses for evacuation, due to liability concerns and a shortage of drivers. Rescue efforts were delayed due to coordination issues between local and federal authorities, both of which have been criticized and defended. FEMA was surprised by the large number of people seeking emergency water and supplies, and the original supply was quickly exhausted. Official requests for federal help were sometimes delayed before reaching FEMA. FEMA reportedly sent volunteer firefighters to Atlanta for two days of training. Vice President Cheney purportedly asked crews to be diverted to support a gas pipeline. According to the White House in 2006, "Our current system for homeland security does not provide the necessary framework to manage the challenges posed by 21st Century catastrophic threats" (Townsend 2006). Hoke (2015) claims:

> Additional problems brought to light in postdisaster assessments of the hurricane protection system include failures of coordination on the part of federal, state, and local agencies; the absence of a central authority with responsibility for the system; a poor funding mechanism

and pressures from government to lower design standards to increase affordability; and the failure of city disaster planners to mitigate the risk with more effective evacuation procedures.

Discussion: The levee, evacuation, and rescue raise several ethical issues. Multiple individuals at various points in the story had to make ethical decisions. Think about identifying the various ethical decisions that had to be made at different points, the key ethical issues and factors for each one, and what a rational person using the five-step process might decide to do.

Sample Flow Chart

S1.   List options.

S2.   See if any option is *Required*

Goto subroutine R1

S3.   Are any options marked "R"?

Yes		No	
Check for conflict Go to subroutine E1 for *those options*. Is there an option marked both X and R, or are two or more incompatible options marked R?		Go to subroutine E1 for *all* options. Eliminate all options marked "X." Is more than one option left?	
yes	No	yes	no
Go to subroutine C1. Proceed to S4	*Decision*! Select the option. End.	Proceed to S4.	*Decision*! Select the option. End

S4.   Consult values and principles

We are now at the point at which the rules do not help us. All we can do is ask ourselves which values and principles tell against each remaining option, and which values and principles speak in favor of each remaining option. Then we must weigh these values and principles in the light of relevant factors and guiding ideas, to try to make the best choice possible.

Decision!
Select option
End

Subroutine R1
For each option:
Do any specific rules require
selecting the option?

yes does following the specific rule violate a general principle or value?			no Next option
yes Does the value or principle justify breaking the rule in this case?		no Mark the Option "R." Next option	
yes Next option	no Mark the option "R." Next option.		

Return

Subroutine E1
For each option:
Does the option violate a specific rule?
A.

yes Does following the specific rule in this case violate a general principle?			no Proceed to next option.
Yes Is the general principle of sufficient importance in this case to justify breaking the specific rule? (See "When to Break the Rules.")		no Place a question mark under the option. Proceed to B.	
yes Next option.	No Put a question mark under the option. Proceed to B.		

B.
Does following the specific rule in this case
violate a value?

yes Is the value of sufficient importance in this case to justify breaking the specific rule?		no Mark the option "X." Next option.
yes Next option.	No Put a question mark under the option. If there are TWO question marks under the option, mark the option "X." Next option.	

Return
Subroutine C1
You have a moral dilemma: one rule requires an action that
conflicts with another rule. You must resolve the dilemma.

F. Is there a way to modify the option so it still follows one rule without violating the other? If not, proceed to G.

G. Consider the relative importance of the two rules, and the extent to which the act violates the rule (degree of violation). Do any values, principles, factors or guiding ideas mitigate or strengthen the force of the two rules in this case? (Example: if the rule requiring an option is an institutional duty, and the rule forbidding the action is not an institutional duty, this suggests one should perform the action). How direct is your responsibility in this case? How much harm would be caused in each case? Etc. This can only be done on a case-by-case basis.

H. Is there a clear preference?

Yes *Decision!* Select that option. End.	no Eliminate all options not marked "R." Return.

*

NOTES

1. Cf. Pavlovic (1983, 230): "If you think you can nicely circumscribe a problem area, the usual next step is either to bring in or manufacture the 'experts,' professionals who will generate the specialized knowledge and then stand ready to supply professional solutions....This approach...makes no sense in ethics."
2. There are quite a few additional and related arguments against relativism. For example, an individual relativist would have to agree that someone who believes relativism is mistaken would be morally wrong to follow relativism. For more about relativism, see Rachels (1999) and Gowans (2021).
3. For example, an engineer given a legal order by the Nazis to engineer more efficient extermination facilities ought to defy the law.

REFERENCES

ASCE (2007) *The New Orleans Hurricane Protection System: What Went Wrong and Why (A Report by The Hurricane Katrina External Review Panel)* (Reston, VA: ASCE) https://doi.org/10.1061/9780784408933

Biasetti, Pierfranceso and de Mori, Barbara (2021) "The Ethical Matrix as a Tool for Decision-Making Process in Conservation," Frontiers in Environmental Science https://www.frontiersin.org/articles/10.3389/fenvs.2021.584636/full

Gowans, Chris (2021) "Moral Relativism," *Stanford Encyclopedia of Philosophy* https://plato.stanford.edu/entries/moral-relativism/

Hoke, Tara (2015) "The Lessons of Katrina," *Civil Engineering* 85:7, 46–47 https://ascelibrary.org/doi/pdf/10.1061/ciegag.0001021

Pavlovic, K.R. (1983) "Autonomy and Obligation: Is There an Engineering Ethics?" in James Schaub and Karl Pavlovic, eds., *Engineering Professionalism and Ethics* (Hoboken, NJ: John Wiley & Sons).

Rachels, J. (1999) "The Challenge of Cultural Relativism," *The Elements of Moral Philosophy*, 3rd ed. (New York: Random House) 20–36.

Schlossberger, Eugene (1992) *Moral Responsibility and Persons* (Philadelphia, PA: Temple University Press).

Schlossberger, Eugene (1993) *The Ethical Engineer* (Philadelphia, PA: Temple University Press).

Schlossberger, Eugene, (1995) "Technology and Civil Disobedience: Why Engineers Have a Special Duty to Obey the Law," *Science and Engineering Ethics* 1, 163–168.

Townsend, Frances F. [Review Leader] (2006) "The Federal Response to Hurricane Katrina: Lessons Learned, Chapter 5" *The White House* https://georgewbush-whitehouse.archives.gov/reports/katrina-lessons-learned/chapter5.html

Whitbeck, Caroline (1996) "Ethics as Design: Doing Justice to Moral Problems," *Hasting Center Report* 26:3, 9–16.

Part II

Sources of Ethical Decision-Making

3

The Engineering Way

Moral thinking consists of using reason-guided judgment to make particular decisions by drawing upon moral sources, factors, and guiding ideas. Part II provides you with many of the tools you will need to make ethical decisions and illustrates them with cases and examples. Part II also deals with some of the important issues in engineering ethics, such as product safety, and prepares you to think about the further issues, such as whistleblowing and GMOs, that are dealt with in Part III. A good place to start is the nature of engineering itself.

WHAT IS ENGINEERING

Engineering is not just a way of making a living. It is a *profession*, a "calling," in which individuals are personally committed to using their skills and abilities to achieve a high social goal. If you see engineering as "just a job," you will not obtain the satisfaction and sense of purpose that engineering offers,[1] and you will not be an ethical engineer. You will be a better and happier engineer if you see your work as dedicated to a worthwhile ideal, just as the best and happiest physicians are those devoted to healing and to the ideal of health. Engineering is not just a way of making money. It is also a moral commitment. (See also "Values of the Engineering Profession" in Chapter 4.)

What is so important about engineering? There are many different types of engineers, and each serves society in a different way. Nonetheless, every engineer, no matter how small or big her job, helps to expand the frontiers of human knowledge and create a better life. Philosophers as diverse as John Locke and Karl Marx have argued that we are by nature creative

DOI: 10.1201/9781003242574-5

producers: the distinctive human attribute is the ability to reshape the world in line with our dreams and visions. If this is indeed the essential human characteristic, then engineering is the essential human science. Engineers, after all, give shape to dreams: they are the sculptors of society, for together, in the countless projects completed in companies and firms throughout the world, they determine the shape of the world we all live in.

What ties engineers together into a profession is not just that engineers make things but that they make things in a special way we might call "the engineering way." Engineers don't guess haphazardly or sloppily: the engineering way is precise, rational, and careful. Engineers don't take wild risks: the engineering way is to be responsible about safety. While painters often tend to be solitary, each working alone on his own painting, the engineering way is to work as a team. More generally, engineering is the science of technology, and technology is practical wisdom. A widely quoted definition suggests that engineering is "the art or science of making practical application of the knowledge of pure sciences." ABET (Accreditation Board for Engineering and Technology) defines engineering as "the profession in which a knowledge of the mathematical and natural sciences gained by study, experience, and practice is applied with judgment to develop ways to utilize economically the materials and forces of nature for the benefit of mankind" (Kemper and Sanders 2001, 104). In Canada, the Professional Engineers Act (R.S.O. 1990 c. P.28) defines the practice of engineering as "any act of planning, designing, composing, evaluating, advising, reporting, directing or supervising that requires the application of engineering principles and concerns the safeguarding of life, health, property, economic interests, the public welfare or the environment, or the managing of any such act." Malaysia's Board of Engineers explains that "'engineering works' means all works which include any publicly or privately owned public utilities, buildings, machines, equipment, processes, works or projects that require the application of engineering principles and data" (BEM 2015). Ross and Athanassoulis (2010) emphasize the social nature of engineering.[2]

Very broadly, engineering consists in designing and implementing solutions that meet design objectives to achieve a goal. The vocabulary writers use in talking about objectives varies. Jack (2013) says, "Design objectives can be clarified as a set of basic requirements and hopeful outcomes, subject to realistic constraints." I will use the terms "design objectives" and "design criteria" in this very broad way. *Design objectives* include the following more specific elements:

1. *Design goals* articulate the general outcome aimed at. An example might be: a secure system of employee record data storage for company X.
2. *Design specifications/requirements* are the specific limits and outcomes that must be met for the design to be successful. For example, Pinto engineers were told that the Pinto must cost less than $2,000 and weigh less than 2,000 lbs.
3. *Design desiderata* are the things a design should strive to optimize (or at least accommodate). Ease of use is an example of a design desideratum. Good designs accomplish a good balance between different desiderata (a balance that is appropriate for the goal).

Thus, engineering designs strive to achieve the design goals by meeting the design specifications and optimizing the best balance of design desiderata.

Because engineering is understood both descriptively and normatively, that is, in a neutral way and as an ideal, it is helpful to have both a neutral definition of engineering and a definition of engineering as an ideal.

I would offer the following *neutral definition of engineering*:

Engineering is the practice of innovatively drawing on empirical and scientific theory to help develop and/or implement a process or object or system that a) articulates goals to address efficiently a problem or need by b) meeting criteria of adequacy (specifications) and c) well balances a framework of desiderata.

This neutral definition allows us to talk about Nazi engineering, for example, because it doesn't characterize the kind of problem or need (whether feeding the hungry or killing millions) or the kind of criteria and desiderata (safety or gratifying the ego of the CEO).

But engineering is also a calling, an ideal, in a way that makes designing Nazi death chambers a betrayal of engineering. To capture this sense of engineering, we need to add a few words and phrases.

I would offer the following *definition of engineering as an ideal*:

Engineering is the practice of safely advancing the progress of the human community, in partnership with nature, through know-how that innovatively and honorably employs clear, clean, and precise practical deci-sion-making, drawing on empirical and scientific theory, to help develop and/or implement a process or object or system that a) articulates goals to address efficiently a genuine problem or need, b) meets criteria of adequacy (specifications), c) well balances a framework of desiderata, and d) overall, helps make the world a better place.

DESIGN ETHICS AND CULTURAL APPROPRIATION

In a seminal article, Van Gorp and van de Poel (2001) call attention to two aspects of design in which ethical issues may arise: in drawing up design criteria and in evaluating and balancing trade-offs between them. For example, because "the design of products often invites certain forms of use and discourages others ... the way a technology is designed is also relevant to how it will be used and what kinds of effects it will produce." A gun with an elaborate safety release mechanism is suited for target prac-tice but less likely to be used in a gang shooting, where ready access to firing is critical. In addition, engineer's responsibility for safety and other engineering values (see Chapter 4) may exceed legal or business require-ments, and design problems are often ill-defined; for example, there may not be a definitive solution and the process of working through and envi-sioning possible solutions may help sharpen and define a problem that was not initially clearly understood or well-defined. Van Gorp and van de Poel suggest that engineers should be actively involved in formulat-ing design goals, requirements, and criteria. When desired design features conflict, engineers often seek novel technical options that eliminate or lessen the conflict. Despite the ingenuity of engineers, however, stubborn conflicts are sometimes inevitable. A common approach to this situation is satisficing, that is, setting minimal levels for each design goal that must be met for a solution to be acceptable (good enough). Engineers may pick a preferred approach and then simply tweak their approach to reach those thresholds. Van Gorp and van de Poel point out, however, that this may create pressure to lower the thresholds to make that solution work. Using multiple-criteria analysis instead of satisficing can be problematic, they point out, when there is no common scale to measure different features, such as safety, effect on the environment, and convenience. (How may "convenience points" equal one "environmental sustainability point"?)

In addition, ethics also enters the design process in terms of who is included and whose needs are considered. Are consumers and members of the community given a chance to be heard in formulating design criteria and evaluating possible solutions? Are broader factors like sustainability or effect on climate change included?

Finally, design ethics must address the issue of *cultural appropriation*. Cultural appropriation is typically defined as the adoption by a dominant culture of symbols, artifacts, designs, and so forth of an oppressed culture without the latter's permission that may cause offense to members of the oppressed culture. "Appropriation is defined as taking something from a less-dominant culture in a way its members find undesirable and offensive – so that its heritage is misused by those in a position of privilege" (Chesterton 2020). Reasons that have been given to object to cultural appropriation include the following: it can cause what Joel Feinberg calls "profound offense," that is, "an offense to one's moral sensibilities" that "strikes at a person's core values or sense of self" (Young 2005, 135)[3]; it doesn't give groups credit for their own culture; it creates/maintains stereotypes; it reduces something of cultural significance to mere fashion, design accessory, or entertainment; it proclaims the power of the dominant group over the oppressed group by taking an element of their culture; it erodes the identity of the oppressed group – "Identity is precious and those who feel marginalised may not wish to see theirs dissolve in the deluge of globalisation" (Chesterton 2020); the appropriator is pretending to be of a different race or ethnicity; and it misrepresents the original culture. Against this, some argue that cultural exchange is among the major drivers of human progress and that the distinction sometimes drawn between cultural appropriation and cultural appreciation is spurious. Since all modern humans originated in Africa, there is, in some sense, only one human culture to which we all belong. In one sense, we are all part of humanity: everything any human being has achieved is part of the common history of humanity. Beethoven, Confucius, Neruda, Dostoevsky, Averroes, Joy Harjo, Maya Angelou, Nelson Mandela, Gandhiji, and countless others are all part of our common spiritual heritage. "If we can only exist in and guard the cultures we emerged from, from those we resemble, we will shrink into the superficiality of newly contrived tribes" (Chesterton 2020).

As Meghan Gallagher notes, "there is a big debate about this. There is also a large spectrum of what is considered to be cultural appropriation. For some people, white Americans walking around in traditional east African garb, for example, is considered cultural appropriation. Others would argue it is a form of self-expression and shows appreciation and respect of another culture" (Gallagher 2017). Conscientious engineers need to think about these issues with thoughtful sensitivity.

One area in which the issue of cultural appropriation arises in the practice of engineering is in the use of indigenous technologies and ways of knowing or "*indigeneering.*" Watson (2020) describes indigenous ways of living with nature that may be fruitfully applied to contemporary problems, but stresses the importance of obtaining permission before using or publicizing these ancient technologies.

In sum:

1. Engineers need to be involved in careful weighing of the ethical ramifications of design problem formulation and trade-offs.
2. Users and others affected need to be involved in the design process.
3. Ethical engineers should (a) at least be aware of when they are incorporating in a design or product a significant element of a culture with a history of oppression, (b) think about whether their use of the element may cause offense, and (c) address and evaluate the moral issues raised.
4. In difficult cases, there is no substitute for the kind of careful ethical thinking described in Chapter 2.

THINKING LIKE AN ENGINEER

To this somewhat formal definition of engineering should be added two more things: a characterization of the engineering process and a more robust picture of the engineer as a professional, that is, as a participating member of an institution carrying out the institutional practice of engineering as we know it. Engineering is not only a job but also a way of life. And so the ethical engineer's commitment to safety does not end when she leaves the office: a concern for safety is something she brings to yoga, dancing, balap karung, going to the opera, tejo, soccer/football, or building furniture in her workshop. Of course, each of us is different and we all have limits. No one can expect every engineer to give full weight to every value in her off-hours. However, each engineer should strive to come as close to the "ideal" engineer as she comfortably can. So, it is useful to talk about an ideal or model engineer, "the compleat engineer," who not only makes correct ethical decisions but instantiates all the ideals of engineering as a profession. When the definition of engineering as an ideal is supplemented by pictures of the engineering process and the compleat engineer, what emerges is "the engineering way."

Characteristics of the Compleat Engineer

1. *Precision, clarity, rationality*: Engineers don't guess haphazardly or sloppily: the engineering way is precise, rational, and careful. Even outside of engineering contests, such as buying an appliance where there is limited space, the complete engineer tends to prefer measuring to eyeballing.
2. *Passion for and appreciation of excellence*: Outside of work, the compleat engineer has an appreciation of fine craft and excellence, in areas as diverse as poetry writing, woodworking, teaching, and clock-repairing. The compleat engineer prefers neatness of dress and speech to lazy, vague, ungrammatical speech and a slovenly appearance. She attempts to be fair and precise in all things, including personal relationships and decisions about which car to purchase. The compleat engineer does not vote for political candidates because of how they look on TV, but on the basis of careful and rigorous attention to the facts and issues. She keeps accurate income tax records and is meticulous about her personal finances, paying her bills on time, balancing her checkbook, and so forth. However, engineers also show
3. An understanding that *good is good enough*. Tolerance is a key engineering concept. For a mathematician, in most contexts, "approximately" is another word for "wrong." Engineers think within tolerance, getting it right enough. It is often said that engineering seeks the optimal solution to a problem, but this is not quite right. Engineers seek a good enough solution. True, a better solution is preferable and an optimal solution is optimal. Better is generally preferable. But the goal is not to keep working until the optimal solution is found. The goal is to find a good enough solution. Since safety is a key value (see below), you often go beyond the minimum (make it safer than good enough). But because the goal is getting it to work and it will work if it is within tolerance, success is usually not getting it exactly right, but getting it right enough (which is usually pretty close to exact, which is why there is no contradiction in saying that engineers love excellence and good is good enough). Perhaps this is

why engineers took so naturally to the digital world, where, unlike the analogue world in which there are infinitely many values, things are either close enough to one or close enough to zero. The digital world is built on the notion of tolerance.

Closely related to tolerance is the idea of *heuristic thinking*, widely equated with the engineering method (see Koen 1985 and 2003). A heuristic is a rule of thumb or strategy or trick. Koen (2003) defines a heuristic thusly:

A heuristic is anything that provides a plausible aid or direction in the solution of a problem but is in the final analysis unjustified, incapable of justification, and potentially fallible [p. 28] .... The engineer's best solution to a problem is found by trade-offs in a multivariant space in which criteria and weighting coefficients are the context that determines the optimal solution. There is never an implication that a true, rational answer even exists. The answer the engineer gives is never *the* answer to a problem, but it is his engineering best answer to the problem he is given[,] all things considered [p. 61].

4. *Love of innovation that builds on the past*: Engineers are generally unhappy doing exactly what was done before. However, engineers tend not to set off alone into uncharted waters. They look at what has been done before and look for some new aspect, some new idea, some new approach or conception. Engineering innovation is often different in this respect from artistic innovation, which may celebrate radical departures. A natural concomitant of engineering innovation is

5. *The urge to tinker*: This can be understood as the confluence of two things: an urge to improve and an urge to see "what happens when …." The particular way in which engineers love to play with things, coupled with characteristic 1 (clarity and rationality), makes characteristic 6 natural.

6. *Love for* (and appreciation of the logic and art of) *puzzles and games*. One study suggested that, in engineering classrooms, "both student learning and attitudes are improved by game-based activities" (Bodnar et al., 2016). But even far away from the classroom or the drawing board, engineers tend to enjoy the process of working out a solution to a puzzle.

7. Engineers are *practical* in the sense that they aim for what works in response to a felt need (a solution that works). Abstract questions

with no application to human or environmental needs tend to hold greater interest to physicists and philosophers than to engineers. While the compleat engineer respects knowledge in all its forms, she is most drawn to questions that address a need. This practicality is partly fed by

8. *Dedication to advancing human and environmental welfare*: This dedication is shown, of course, not only in the compleat engineer's work but also in the way the compleat engineer is likely to help out a stranger, donate to charity, or volunteer to help clean up a river.

9. *Commitment to safety*: While all professionals should care about safety, for engineers, safety is a core, almost defining value. Why is engineering different from other professions? Medical providers are always choosing between risky paths, such as starting surgery or doing nothing, weighing the risk of death from the operation against the risk of death without the operation. In engineering, doing nothing is rarely very risky. If the bridge collapses, people die. If the bridge is not built, the absence of the bridge is not, as a rule, fatal to anyone. If a food additive causes cancer, people die. Generally, no one dies because the additive was not manufactured. Thus, while the physicians' activities generally modify pre-existing risks, the engineer's activities generally create risks where none existed. There are exceptions. No one dies without cosmetic surgery. Some things engineers do prevent natural deaths or calamities. More often, engineers make changes in an existing technology already posing some risk (imposed, hence, by other engineers). Still, part of the reason engineering as a field is more safety-conscious and more risk-averse than many fields is that engineering as a practice tends to create serious risks where none existed before, risks that can generally be lessened or eliminated by better engineering. And, unlike mountain climbers or skiers (who also undergo risks where none existed before), those killed or injured by engineering projects often have no choice about being subject to the risk. After all, most people do not climb mountains, in part because they are free to choose not to take the risk. It is easy enough not to climb K2. So, those who do climb naturally expect some risk as part of the activity. However, we often can't feasibly avoid the risks posed by engineers, so we generally expect not to be exposed to more risk than necessary. Finally, risks created by engineers often affect a broad range of people, including those who have no involvement in the process. The risk of surgery usually affects

only the patient. The risks involved in pursuing a legal strategy generally involve the client and perhaps those close to or allied with the client, such as a spouse or business partner. An engineer who takes a risk in designing a drug may harm thousands of consumers and an engineer who takes a risk about building a skyscraper may kill pedestrians who had nothing to do with the skyscraper apart from walking on the pavement next to it.

Partly because of these features, there is a culture of engineering that is particularly safety-conscious.

10. Penchant for *Teamwork and collaboration*: While some projects may involve only a single engineer, quite often projects call for, at various stages, the participation of two or more engineers working together as a team. More generally, engineering as a profession is a team enterprise, sharing and advancing not only techniques and methods but ideals. The compleat engineer welcomes feedback and the participation of others.

Some of these may seem like odd items to place among the much grander sounding entries above. Puzzles and games, in particular, may sound trifling. But puzzles need not be trivial. Games need not be irrelevant distractions or waste of time. The concept of play and the concept of puzzle are fundamental to human nature. The term "Lila" in Sanskrit means play or sport, but it is used in Hinduism to describe the entire cosmos, which is regarded as the creative play of the Brahman, a spontaneous drama expressing the nature of divinity. Sir Arthur Conan Doyle chose the title, "The Final Problem" for the story in which he killed off Sherlock Holmes not only because it was meant to be Holmes' last case but also to suggest that death itself is humanity's final problem. Theologians speak of the great game of good and evil and the riddle of existence. But almost any problem or puzzle or game can catch the eye of the imagination. Wittgenstein described a philosophical puzzle as an itch that must be scratched, and I think virtually every engineer must understand exactly what he meant. The urge to tinker is the inability to leave things alone, as they are. It is the need to improve, to take things apart and see how they work, and/or to try things another way, just because it's cool, or just to see if it will work, or just because things can always be made better.

All human beings are fallible and have limits, and no one could expect anyone to be "Wonder Woman" in all ten ways. Moreover, people are different. An engineer who loves abstract questions or who dislikes games is no less excellent an engineer. The idea of the compleat engineer is meant to capture the engineering way while respecting and even celebrating individual differences. *It is meant to inspire, not limit, engineers.*

The Engineering Process

While every problem is different, it can be helpful to think in general terms about the engineering process. It is important to remember, of course, that this is a flexible general template. An experienced engineer will know when to do things a bit differently.

Searching the library and the internet will uncover many articulations of the engineering process. Typically, they include several of the following:

Identify the need and define the problem; do background research; brainstorm (at several stages); specify requirements; generate a bouquet of possible solutions; test and evaluate the solutions; design a prototype; test the prototype; create a detailed design/solution; get feedback/approval; implement a revised solution; and test the final solution. Koen (1985) holds that "the engineering method is the use of heuristics to cause the best change in a poorly understood situation within the available resources."

Here is one formulation of the process:

1. Receive (or create) and clarify the assignment. What is the problem to be solved or need to be met? What constraints, outcome goals are given? What further factors does the engineer believe should be considered?
2. Conduct background research.
3. Create outcome goals and design criteria:
   - Specific goals
   - Requirements
   - Desiderata and preferences
   - Balancing factors
4. Construct a process (flow chart or equivalent).
   - For each stage:
     - Identify problems
     - Design solutions

Typically, this involves examining and comparing several approaches, each of which must be developed, tested, and compared. Generally, this involves constructing models. Often, in the course of this process, the need for new information becomes clear. That information can be obtained either through further research or by innovative testing and model creation.

Obtain feedback at every stage, as feasible.
  - Conduct performance reviews-improvement analysis.
  - Adopt a solution.
  - Further test and research the solution.
  - Create a prototype.
  - Beta test and finalize.
5. Obtain any approval necessary.
6. Implement and monitor.

Typical engineering strategies in carrying out the process include:

Consider the worst-case scenario.
Look for patterns.
Build on previous results.

This process and these strategies can be applied fruitfully outside of traditional engineering projects. In a broad sense, making good ethical decisions amounts to defining a problem and engineering a solution. In addition, there are many sorts of tasks, from fixing your media system to repairing a troubled friendship, that can sometimes benefit from thinking like an engineer.

FUN ACTIVITY TO STRETCH YOUR BRAIN:

This is something I had my engineering students do, but it would also work for any group of engineers (brainstorming session, corporate training, or informal game). You can do it alone, but it is more fun in a group.

Thinking like an engineer, figure out how you would go about one or more of these tasks – that is, what is the engineering way of doing this? How would you approach this as an engineer?

Writing a novel (you are a novelist planning to write a novel).

Conducting marital therapy (you're a therapist helping a couple).

Doing historical research (you're a historian planning to research the causes of the American Civil War).

Diagnosing a patient (you're a physician diagnosing a difficult case).

MODELS OF THE PROFESSION

Moral conclusions about professional conduct can be drawn from the model one gives of a profession generally, and of engineering in particular.

Definitions of a "profession" are plentiful.[4] Loui and Miller (2008) state that "the defining characteristics of a strongly differentiated profession are specialized knowledge and skills, systematic research, professional autonomy, a robust professional association, and a well-defined social good associated with the profession." Bayles (1981, 7–8) identifies three features necessary for an occupation to count as a profession: *(1) extensive training, (2) a significant intellectual component,* and *(3) providing an important [I would say "critical"] service to society.* By "critical" here I mean that, without it, lives would be lost or endangered, society would collapse, or other drastic consequences ensue. Thus, nurses and physicians, teachers, judges, and engineers provide a critical service, in this sense, while ballplayers and hair stylists do not (however much people may enjoy and value what they do). Bayles also gives three common features of professions: *(4) there is a process of certification or licensure, (5) there is an organization of members,* to promote the goals of the profession and the economic well-being of its members, and *(6) there is room for autonomy in one's work.* Certification can range from medicine and law, where it is a crime to practice without a license, to certification that certain standards are met (such as a licensed daycare in the US), to a general understanding (e.g., tenured faculty need a terminal degree from a reputable institution). Harris et al. (2019) add that "there is a monopoly or near-monopoly on the provision of professional services" (p. 3) and that a profession claims to adhere to ethical standards and fosters public welfare. Davis (1997, 417) adds that professions serve "a moral ideal in a morally permissible way beyond what law, market, morality, and public opinion would otherwise require." McGinn (2018, 273) says an engineering professional "integrates abiding primary concern for protecting and promoting societal well-being into his or her everyday engineering practice." Those three ideas are captured by a seventh feature: *(7) the field constitutes an ideal-guided institution that functions because of public trust.* (See "Institutional Duties" in Chapter 6.) An institution (such as engineering) has a publicly understood role and set of ideals and standards, generally accepted, both within the field and by the public, as partly defining the nature of the institution, such that the public trusts members of the institution (engineers) to uphold those ideals and standards, and it is precisely because of that trust that members of the institution are allowed room to carry out their practice, such as building skyscrapers that could kill passersby and inhabitants if improperly designed and constructed. In the case of engineering, those ideals include the values of the engineering profession: safety, human progress, clean, clear decision-making, community, and partnership with nature (see Chapter 4).

The purpose of our definition is to point out key features that have moral significance. What lessons can we draw from this definition?

First, extensive training, intellectual component, providing a crucial service, licensure, and, to a lesser extent, organization of members together mean that professionals have a coercive bargaining position. In a free bargaining situation, people are free to make whatever agreement they wish. For example, I don't owe it to my neighbor to mow his lawn for him. So, I may offer to mow his lawn in exchange for the deed to his house. This is an exorbitant demand, of course, but, as he is quite free to turn down my offer, it is up to him to decide whether having his lawn mown by me is worth giving up his house. However, suppose he is drowning in a secluded lake, and I am standing on the edge of the lake with a life preserver. It is unethical to refuse to throw him the life preserver unless he promises to give me his house, and no court would uphold such a "contract." This is not a free bargaining position, but a coercive bargaining position, and I am not legally or morally entitled to exploit him this way. (Similarly, contract law refuses to uphold "contracts of adhesion."[5]) Our definition of a profession reveals that professionals are generally in a coercive rather than a free bargaining position: the critical importance of the service combined with legal licensure mean that the professional has a monopoly on a service that society cannot really do without, and (because of the extensive training and intellectual component) society cannot easily replace the professional who "strikes." Moreover, clients lack the organization that membership in professional societies provides. Thus, unrestricted rules of bargaining do not apply to the professional as they might to a beautician. So, professionals are morally required to exercise restraint in what they demand in exchange for their services. One consequence of this conclusion is that physicians and attorneys either may not be morally justified in demanding exorbitantly high fees or, as a profession, must make sure that no-cost or low-cost services are available (e.g., pro bono work, public defenders, and free clinics). For the same reasons, there are moral limitations on the conditions of service that might not apply to non-professionals. I may offer to cut your lawn on a take-it-as-is basis, where, as long as it is cut, you have little recourse if you are dissatisfied, while it is unethical to offer surgery or design a bridge on that basis. The work of professionals must meet certain standards. Another is that engineers have a moral obligation to look out for the public interest, since those adversely affected by engineering projects are not always in a position to bargain effectively.

Second, features (6) and (1)–(3) indicate the need for professional self-regulation: because the service is critical, and so quality is crucial and the public has a legitimate interest in the exercise of the profession, and because the professional is largely autonomous and evaluation of professional work requires difficult to obtain expertise, outside regulation is less feasible than it is in other areas. By contrast, no one dies from a bad haircut and people can generally tell whether they received a good or a bad haircut. Thus, engineering as a profession has a special duty to regulate the competence and ethics of practicing engineers. For the individual engineer, this means participating in professional societies. It means taking some responsibility to see that incompetent or unethical colleagues do no harm. This may mean reporting them, not "covering up" for them, or simply speaking to them.

Third, (2)–(4) indicate that professionals owe a debt of gratitude to society: the granting of a monopoly, the training social institutions such as universities provide,[6] and the social institutions of the professions themselves mean that professionals' skills are obtained and employed only by virtue of considerable assistance from their society. How this debt of gratitude should be repaid depends on the profession. In general, professions have a special duty to be good public citizens, and to use their skills for the public good. (See also the discussion of dual-investor theory in "Engineering and Business," below.)

Finally, (7) and (3) together mean that professionals have a special responsibility toward society, which has given them public trust concerning an important service.[7] The institutional duties of professionals have special moral weight. Thus, for engineers, safety, advancing human welfare, and environmental responsibility, for example, are powerfully important.

Models of the Engineering Profession

Another source of ethical insight is the way one views the engineering profession. Two complementary models of the engineering profession may be drawn upon in making ethical decisions.

First, Martin and Schinzinger (1983, 242) mention six possible models of the engineer: (1) a "savior," who will create utopia and rescue society from the ills of poverty, inefficiency, etc., (2) a "guardian" who can best guide society in its development, (3) a "bureaucratic servant" who simply "translates the directives of management into concrete achievements," (4) a "social servant" who turns society's commands into concrete achievements, (5) a "social enabler and catalyst" who, ceding ultimate authority to management and/or society, must help them to "understand their own needs and to make informed decisions about desirable ends ... and means ...," and (6) a "game player" who simply plays by the corporate rules to win and move ahead.

Of these, (5) is clearly the most acceptable. Model (1) is ruled out by the value of autonomy, Models (3) and (4) are ruled out by the duty to leave the world no worse, and model (6) is ruled out by the value of community. Model (2) simply becomes Model (5) when the importance of autonomy is fully appreciated. So, Model (5), *the engineer as social catalyst and enabler*, seems the most appropriate. Two important things follow from this model. (See also "Codes of Ethics" in Chapter 6 for more about these two points.)

Consequence (1) is that *engineers as a group have a duty to offer their professional perceptions about the needs of society and technological problems facing the community*. Within the company, this means speaking up to supervisors and management about public policy, to the extent one knowledgeably can. In addition, engineers may have a special perspective on non-company matters (such as national energy policy and what kind of publicly funded research is likely to prove useful). Engineers as a group have a duty to testify before legislative committees, write articles and/or give interviews to the press about technologically oriented public issues, etc. In short, engineers who, because of their expertise, have a special perspective on public issues should make their voices heard. Consequence (2) is that *the public must be involved in important engineering decisions that affect the life of the community*. Companies can help achieve this result by establishing an Environmental and Community Issues Advisory Board. (See Appendix IV.)

In addition, the definition of engineering as an ideal given in the first section may be called the *ideal model of engineering*. It emphasizes the importance of safety; innovation; clean, clear decision-making based on science and empirical knowledge; partnership with nature; community; integrity; advancing human progress; and making the world a better place.

ENGINEERING AND BUSINESS

Engineering is a field of its own, but engineers who work for a corporation or an engineering firm are also part of a business, and so considerations of business ethics also apply to their decisions. Moreover, engineers in any society with a free market component, whether they are part of a business operation or not, often work with businesses (suppliers, contractors, users, etc.). Thus, it is helpful for engineers to have some understanding of the ethics of business.

It is often helpful to begin with a case.

Case 3: Delta Company Relocation

The Delta Company is considering closing or relocating its Smallville plant. The plant has been the major employer in Smallville for over 40 years, and the economy of Smallville is thus highly dependent on the Delta plant. The plant is fairly profitable: return on investment (ROI) is 10% and will remain so for the foreseeable future. (Industry average is 9%.) If Delta relocates to Bigville, 500 miles away, projected ROI, after all costs of moving have been factored in, is 11%. Is it ethical for Delta to relocate the plant?

Discussion: Why is this an ethical question? Because closing the plant would start a cascade of serious consequence for the residents of Smallville. A large percentage of residents would be thrown out of work. Their financial difficulties mean that they will have relatively little money to spend at local businesses, forcing those businesses to close or lay off employees, further raising unemployment. The growing unemployment decreases the tax base of the community, requiring cuts to services, creating more unemployment. Crime often increases with unemployment, and the curtailment of services, lack of jobs, and rising crime induce people to leave Smallville, thus further reducing the tax base and money available to be spent at local businesses, resulting in further cuts (e.g., to schools, as there are fewer students), resulting in further unemployment and more people leaving, etc. A further result of this downward spiral is that property values plummet, hurting the retired as well as those trying to leave. (Note also that it would be difficult for Smallville to compete with Nearville in attracting new companies, since Nearville can offer better tax breaks, better schools and parks, less crime, etc.)

An economist might say that the increase in efficiency gained by closing the plant means that the losses in Smallville are more than offset by the gains in Bigville. But this response overlooks the transition costs of moving a job from Smallville to Bigville. When A loses the job they have held for 20 years, they will need to relocate to Bigville or elsewhere. A's spouse, B, who has put blood and tears and hard work into building a restaurant, must either close the restaurant or stop living with A. The once prosperous restaurant has little value, since Smallville is dying. B must start over from scratch in a new city, providing that they can gather the capital needed to start a new restaurant. It takes years to make connections, find and establish relationships with vendors and staff, learn what B knew about Smallville (such as where to find the best artichokes on Tuesday, that one must hire extra help when the Shriner meeting ends each third Friday of the month, and which dishes are successful with the new clientele), attract and establish regular diners, and so forth. Most restaurants fail in their first half-year, so it is far from clear that B's new restaurant will survive. Personal and professional networks established over the years (a friend who will babysit or take you to the doctor, a podiatrist who knows and understands your particular problems, and so forth) must be built anew. Pensions may be lost, and decisions made in the old environment may make no sense in the new. Transition costs are why it is generally much worse to lose a long-held job than it is not to get a particular new job one for which one is applying. In both cases, one does not have a job and there are other jobs to apply for, but it is not the same.

Since closing the plant makes money for the company but hurts employees and residents of Smallville, one way to decide whether to relocate is to think about who should count when executives are making their decision. In general, there are four models of business, based on the purpose of a business venture.[8]

The *shareholder model*, famously supported by Milton Friedman,[9] says that the purpose of business is to maximize shareholder value[10] by any legal means.[11] The shareholders of a business venture are the stockholders of a corporation, the partners in a limited partnership, the sole proprietor of a store, or the photographer in the case of a freelance photographer. The shareholder model thinks of business as akin to a game of monopoly: players shouldn't cheat – they must abide by the rules (the law) – but within the rules their goal is simply to win, not to be nice or help out other players. One of Friedman's main arguments is that the market knows best, and so, in the long run, utility is best served by everyone trying to maximize profits (within the limits established by law). This argument, which dates back to Adam Smith, is known as the "invisible hand argument" – when the market governs, the claim is that everything works out for the best as if guided by an invisible hand.[12]

The *stakeholder model*, the modern version of which was formulated by R. Edward Freeman in 1984, suggests that businesses must serve all those who have a major stake in (are importantly affected by) its operations: shareholders, employees, customers, vendors, the local community, and even, in the case of a multinational logging company, every denizen of planet earth.

The *executive model* is virtually never advocated or even mentioned (I had to make up the name), but it is what in fact governs most large corporations: the reality is that most corporations are run primarily for the interests of the executives of the company, and only secondarily for the benefit of shareholders or stakeholders, when serving them also benefits executives (e.g., when executives are given stock options, although sometimes the price is manipulated to benefit executives more than other shareholders).

Finally, *dual-investor theory*, first proposed in Schlossberger (1994),[13] suggests that society is a kind of shareholder, and so the purpose of business is to increase shareholder value by serving society. Society, by many generations of hard work, creates the "opportunity capital" without which no business venture can flourish. No modern business could operate without roads and transportation, water mains, sewer lines, currency, electrical power, computers, telephones, and a large body of knowledge, skills, and techniques, all of which (and much else) are provided by society. Henry Ford did not begin by inventing the wheel or discovering how to smelt ore. No physician re-invents the history of medicine from her fertile brain: her skills are borrowed from the treasure trove of human knowledge, knowledge gained by generations of people who dedicated their lives to the battle against ignorance and disease. There is, in the modern world, virtually no such thing as a person who "did it all myself." (This is what Barak Obama meant in 2012 when he said, referring to the roads and bridges businesses rely on, "if you have a business–you didn't build that.") Business ventures also need the "specific capital" individual investors supply, which includes, in the case of B's restaurant, the initial funds to buy/rent a location; buy food to cook, plates, silverware, cookware, and tables and chairs; and B's hard work and knowledge (e.g., developing the recipes). The restaurant relies essentially upon the contributions of both society and B, upon the opportunity capital and the specific capital. Thus, the business should provide a good return to both categories of investors. It should return value to B, who provided the specific capital, in a way that serves society, which provided the opportunity capital. For further discussion of dual-investor theory, shareholder theory, and stakeholder theory, see Schlossberger (1994 and 1998), Freeman et al. (2010), and Friedman (2016).

In this case, Delta is already doing well by the stockholders and, providing Delta produces a good product, does not pollute or abuse workers, Delta is serving society by helping support the community of Smallville. Were Delta to relocate, it would only marginally better serve stockholders but significantly make society, overall, worse off. Thus, both dual-investor theory and the duty we have to leave the world no worse than we found it (see Chapter 5) suggest that Delta should remain in Smallville.

Obviously, the situation becomes less clear if the Smallville plant is not profitable, in which case Delta is not providing a good return for its shareholders, or if the extra productivity resulting from closing the Smallville plant is large, since these factors change the balance between specific investors and society and make it less clear that closing the Smallville plant creates more overall hardship and ill-serves society. In that case, Delta should make efforts to make the Smallville plant more profitable, if feasible. If that is not feasible and Delta closes the plant, it should take steps to ameliorate the harsh effects of the plant closure. For example, when Stroh's Brewery acquired Schlitz in 1982, Stroh's, a major Detroit employer for 70 years, decided to close its own Detroit breweries, which were inefficient compared to the newly acquired Schlitz breweries. Detroit's unemployment rate stood at 9%. Stroh's spent 1.5 million, supplemented by $600,000 of government funds, in a program to find new jobs for the displaced employees. The program involved lobbying of employers by Peter Stroh, as well as "orientation, counseling, job skills workshops, skills testing, and training in resume preparation and interviewing." Virtually, all of Stroh's employees found new jobs (at a cost of $2,000 per worker). Other ameliorative strategies include finding a purchaser for the old facilities and adaptive re-use of the old facilities (Desjardins and McCall 1990, 447).

In general, engineers who are part of a business venture should keep firmly in mind the obligation that business has not only to make money but to give society a good return on its investment (providing the opportunity capital without which the business could not exist). When designing a plant that may adversely affect the local community (e.g., untreated pollution from the plant would adversely affect the health of and bring down the property values of local residents), engineers must consider not only the financial effects of design decisions on their firm or client but also the obligation to serve society, to give society a good return on its investment. That obligation can supplement and/or strengthen other relevant factors such as safety and partnership with nature.

NOTES

1. Florman (1976, 94) cites a number of relevant remarks. The engineer has "a responsibility to help society" by working "for the use and convenience of man" (James R. Killian, Jr.), and the engineer's "opportunity for productive contributions is almost without limit; his obligation to judge wisely and imaginatively is profound" (Newman A. Hall). Engineering "can provide a life of genuine satisfaction" (Vannevar Bush), namely "the deep satisfaction that stems from an understanding of the world in which we live" (George E. Holbrook).

2. For the reader who seeks a few more definitions: in 1828, Thomas Tredgold defined engineering as "the art of directing the great sources of power in nature for the use and convenience of man" (Florman 1976, 19). Martin and Schinzinger view engineering as "social experimentation." From this, they conclude that:

   The general features of morally responsible engineers [are] ... a conscientious commitment to live by moral values, a comprehensive perspective, autonomy and accountability (Haydon, Graham, "On Being Responsible," *Philosophical Quarterly* 28 (1978), pp. 46–57). Or stated in greater detail ... 1) A primary obligation to protect the safety of and respect the right of consent of human subjects. 2) A constant awareness of the experimental nature of any project, imaginative forecasting of its possible side effects, and a reasonable effort to monitor them. 3) Autonomous, personal involvement in all steps of a project. 4) Accepting accountability for the results of a project (Martin and Schinzinger 1983, 63).

3. Profound offense "is offensive even when unwitnessed .... The knowledge that artworks are being produced by means of cultural appropriation may be offensive even to people who do not experience the works themselves" (Young 2005, 135).

4. Cf. (A) Cogan (1983, 319), summarizing Abraham Flexner: professions are characterized by "1) intellectual operations coupled with large individual responsibilities 2) raw materials drawn from science and learning 3) practical application 4) an educationally communicable technique 5) tendency toward self-organization, and 6) increasingly altruistic motivation." (B) Goland (1983, 286): "The virtues of professionalism are in independent thought, the creative dedication to the dual and compatible goals of advancing the employer's interests while serving the just needs of the society." (C) The ASCE amplification of its definition of a "profession," cited by Nelson and Peterson (1983, 331): "A profession is a calling in which special knowledge and skill are used in a distinctly intellectual plane in the service of mankind, and in which the successful expression of creative ability and application of professional knowledge are the primary rewards. There is implied the application of the highest standards of excellence in the educational fields prerequisite to the calling, in the performance of services, and in the ethical conduct of its members." (D) Newton (1988, 48): Professionals "possess a specialized art, skill or capacity, requiring long and difficult education and extended practice; they are employed full-time in the practice of that art...they render a service to individual clients and to society in the practice of the art." (E) Newton (1989, 50–51): Two features of professionalism are "maximal competence in a certain area of knowledge or skill, and a moral commitment to the public good in that area," which often results in "social award of a legitimate monopoly of practice." A third "prominent feature" is "commitment on the part of individual professionals to the welfare and interests of individuals in their charge." (F) Sullivan (1988, 41): The original meaning of a "profession" is a calling, "a promise to enter on a distinct way of life .... a free act in response to a belief that one had received a 'call,' ... and it meant a commitment to embody virtues central to the community's highest purposes." (G) Layton (1989, 483): the values of professionalism are "autonomy, collegial control of professional work, and social responsibility," where the last means "special responsibility to see that [one's] knowledge is used for the benefit of the community," that is, the use of "esoteric knowledge...guided by a sense of public duty" (p. 484). (H) Epstein and Hundert (2002): "Professional competence is the habitual and judicious use of communication, knowledge, technical skills, clinical reasoning, emotions, values, and reflection in daily practice for the benefit of the individual and community being served. Competence builds on a foundation of basic clinical skills, scientific knowledge, and moral development." (I) Australian Council of Professions (2003): "A Profession is a disciplined group of individuals who adhere to ethical standards and who hold themselves out as, and are accepted by the public as possessing special knowledge and skills in a widely recognised body of learning derived from research, education and training at a high level, and who are prepared to apply this knowledge and exercise these skills in the interest of others."

5. For example, suppose all the owners of rental properties within a hundred miles of Midville offer tenants the same "boilerplate" five-year lease, according to which the landlord may raise the rent at will, and the tenant, who may not inspect the premises before signing the lease, must pay for any repairs or improvements the landlord decides upon. Although this is a grossly unfair contract, Midville renters have no real choice but to sign it. Such a contract is called a "contract of adhesion," and is not enforceable.

6. Most universities are subsidized, by grants, private donors, or direct government funding, so that the education even of those who pay full tuition is partly a gift from society.

7. Cf. Newton, Lisa (1989, 55): "two candidates for the essence of 'professionalism,' two possible *raisons d'etre* for a profession: it can exist for the sake of excellence in the practice, or it can exist for the sake of profit for the practitioners. A third possible motivating principle...might be the direct service rendered to identifiable others."

8. Mixed economies, in which some business concerns are privately owned and others are state-owned, or in which all, while state-owned, are independent competitors, present variants of these four possibilities. Most "socialist" countries are examples of mixed economies. A true alternative is a pure command economy in which all business entities are entirely state-owned and centrally controlled. There are problems with the way such a system concentrates power.

9. Since Friedman (2016), originally published in 1970, is often reprinted, it is worth pointing out a few flaws in his argument. Friedman argues that executives are agents of the stockholders hired for the express purpose of making a profit. An executive who uses stockholders' money to support the arts is imposing, says Friedman, a tax on the stockholders. Similarly, an executive who spends stockholders' money on, for example, a stack scrubber not required by law is imposing environmental restrictions on the stockholders. But, says Friedman, only the legislature has the right to impose a tax or to impose environmental restrictions. Thus, it is improper for an executive to spend corporate money supporting the arts or instituting non-mandatory environmental safeguards, unless those actions, ultimately, maximize profits. In short, executives have a duty to maximize profits by any legal means. I will mention two problems with Friedman's argument. First, corporate executives are in fact elected by the shareholders, generally via a board, while executives of a private concern are hired by the owners. If the owner is dissatisfied with executive decisions, the owner can fire those executives. If shareholders are unhappy with the executives' decisions, they can, in principle, elect a new board, and, de facto, vote with their feet (by selling their shares) more easily than, for example, citizens of New Hampshire or Uttar Pradesh can pick up and move to California or Tamil Nadu. Second, Friedman's argument is based on the importance of freedom. But unless corporations can be trusted to be good citizens, society must restrict and regulate every aspect of corporate behavior. For example, it seems to follow from Friedman's position that if a company can save a small amount of money by dumping poison in a reservoir, it ought to do so, provided the dumping is not illegal, runs no risk of litigation, and will not hurt sales. The only way, if executives really thought as Friedman suggests they should, that society can prevent its members from being poisoned in this way is to generate a long list of restrictions, perform constant inspections to make sure that no new safety hazards are emerging, and so forth. In short, rather than being a recipe for greater freedom, Friedman's position is a recipe for more governmental interference than any socialist would wish for. Put another way, the precondition for freedom is responsibility. If executives cannot be trusted to use common sense in being good citizens, citizens cannot afford to allow them much free reign. Thus, Friedman's position is self-defeating: his recommendation, meant to respect freedom, would actually decrease freedom. In any case, my argument that society provides the opportunity capital without which corporations could not function shows that executives' duty to make profits for stockholders is counterbalanced by their duty to society to be good citizens. For references to other discussions of this argument, see Schlossberger (1994), note 17.

10. I will sometimes say "profit" for convenience but economists say "shareholder value" because, for example, a startup or expanding company that actually loses money but whose stock (or resale value) shoots up is still increasing shareholder value.

11. Friedman also adds customary morals as a constraint on business, but it plays little role in most of his actual discussions.

12. Problems with the invisible hand argument include the existence of externalities; the power of advertising; the difficulty consumers face in making informed decisions in a world in which a large number of competing products are constantly changing and complex products require considerable information and expertise to evaluate; the presence of collusion; and the complex ways in which, in the contemporary world, individuals' choices are constrained by social structures and the actions of others (see Schlossberger 1994).

13. Sometimes known as the "stakeholder as shareholder" approach, subsequent writers who have taken a related stance include Blair (1995) and Etzioni (1998).

REFERENCES

Australian Council of Professions (2003) "What Is a Profession," https://www.professions.org.au/what-is-a-professional/

Bayles, Michael (1981) *Professional Ethics* (Belmont, CA: Wadsworth).

BEM (Board of Engineers Malaysia) Registration of Engineers Act 1967 (Revised 2015) http://bem.org.my/web/guest/registration-of-engineers-act-1967-revised-2015-

Blair, Margaret (1995) *Ownership and Control: Rethinking Corporate Governance for the Twenty-First Century* (Washington, D.C.: Brookings Institution Press).

Bodnar, Cheryl, Anastasio, Daniel, Enszer, Joshua and Burkey, Daniel (2016) *Journal of Education Engineering* 105:1, 147–200.

Chesterton, George (2020) "Cultural Appropriation: Everything Is Culture and It's All Appropriated," *GQ Magazine* https://www.gq-magazine.co.uk/article/the-trouble-with-cultural-appropriation

Cogan, Morris (1983) "The Problem of Defining a Profession," in James Schaub and Karl Pavlovic, eds., *Engineering Professionalism and Ethics* (New York: John Wiley and Sons).

Davis, Michael (1977) "Is There a Profession of Engineering," *Science and Engineering Ethics* 3:4, 407–428.

Desjardins, Joseph R. and McCall, John J. (1990) *Contemporary Issues in Business Ethics* 2nd ed. (Belmont, CA: Wadsworth).

Epstein, Ronald and Hundert, Edward (2002) "Defining and Assessing Professional Competence," *Journal of the American Medical Association* 287, 226–235.

Etzioni, Amitai (1998) "A Communitarian Note on Stakeholder Theory," *Business Ethics Quarterly* 8:4, 679–691.

Florman, Samuel C. (1976) *The Existential Pleasures of Engineering* (New York: St. Martin's).

Freeman, R. Edward, Harrison, Jeffrey S., Wicks, Andrew C., Parmar, Bidhan and de Colle, Simone (2010) *Stakeholder Theory: The State of the Art* (Cambridge, MA: Cambridge University Press).

Friedman, Milton (2016) "The Social Responsibility of Business Is to Increase Its Profits," in Fritz Allhoff, Alexander Sager and Anand J. Vainya, eds., *Business Ethics in Focus: An Anthology*, 2nd ed. (Peterborough, Ontario: Broadview).

Gallagher, Meghan (2017) "The Debate about Cultural Appropriation," *O'Neill Institute for National and Global Health Law* (Georgetown University) https://oneill.law.georgetown.edu/the-debate-about-cultural-appropriation/

Goland, Martin (1983) "Can Professionalism Be Attained within the Corporate Structure?" in James Schaub and Karl Pavlovic, eds., *Engineering Professionalism and Ethics* (Washington, D.C.: John Wiley and Sons).

Harris, Charles E. Jr., Pritchard, Michael S.E., James, Ray W., Englehardt, Elaine E., and Rabins, Michael J. (2019) *Engineering Ethics: Concepts and Cases* (Boston, MA: Cengage).

Jack, Hugh (2013) *Engineering Design, Planning, and Management* (London: Academic Press).

Kemper, John D. and Sanders, Billy R. (2001) *Engineers and Their Profession*, 5th ed. (New York: Oxford).

Koen, Billy V. (1985) *Definition of the Engineering Method* (Washington, DC: American Society for Engineering Education).

Koen, B.V. (2003) *Discussion of the Method: Conducting the Engineer's Approach to Problem Solving* (Oxford: Oxford University Press).

Layton, Edwin T. (1989) "The Engineer and Business," in Peter Windt et al., eds., *Ethical Issues in the Professions* (Hoboken, NJ: Prentice Hall), 481–489.

Loui, Michael and Miller, Keith (2008) "Ethics and Professional Responsibility in Computing" in Benjamin W. Wah, ed., *Wiley Encyclopedia of Computer Science and Engineering* (New York: Wiley).

Martin, Mike and Schinzinger, Roland (1983) *Ethics in Engineering* (McGraw-Hill).

McGinn, Robert (2018) *The Ethical Engineer: Contemporary Concepts and Cases* (Princeton University Press).

Nelson, Carl and Peterson, Susan (1983) "Ethical Decisions for Engineers: Systematic Avoidance the Need for Confrontation," in James Schaub and Karl Pavlovic, eds., *Engineering Professionalism and Ethics* (John Wiley and Sons).

Newton, Lisa (1988) "Lawgiving for Professional Life: Reflections on the Place of the Professional Code," in Albert Flores, ed., *Professional Ideals* (Wadsworth), 47–55.

Newton, Lisa (1989) "Professionalization: The Intractable Plurality of Values," in Peter Windt et al., eds., *Ethical Issues in the Professions* (Prentice-Hall), 49–59.

Ross, Allison and Athanassoulis, Nafsika (2010) "The Social Nature of Engineering and Its Implications for Risk Taking," *Science and Engineering Ethics* 16, 147–168.

Schlossberger, Eugene (1994) "A New Model of Business: Dual-Investor Theory," *Business Ethics Quarterly* 4, 459–474.

Schlossberger, Eugene (1998) "The Middle Path: Using Dual-Investor Theory in Teaching Business Ethics," *Teaching Business Ethics* 2:2, 127–136.

Sullivan, William (1988) "Calling or Career: The Tension of Modern Professional Life," in Albert Flores, ed., *Professional Ideals* (Wadsworth), 40–46.

Van Gorp, Anke and van de Poel, Ibo (2001) "Ethical Considerations in Engineering Design Process," *IEEE Technology and Society Magazine*. DOI: 10.1109/44.952761

Watson, Julia (2020) *Lo-Tek: Design by Radical Indigenism* (Taschen).

Young, James O. (2005) "Profound Offense and Cultural Appropriation," *Journal of Aesthetics and Art Criticism*, 63:2, 135–146. DOI: 10.1111/J.0021-8529.2005.00190.X

4

Values of the Engineering Profession

One important source engineers can draw upon in making ethical decisions is the set of values that are central to engineering as a profession. Remember the definition of engineering as an ideal. If this definition correctly describes the essence of engineering as a profession, then engineering, by its nature, is dedicated to five key values: safety; human progress; clean, clear decision-making; community; and partnership with nature.

In addition, engineering is the science of technology. So, it is worth taking a moment to think about the nature of technology. The topic is, of course, an important one. And it is worth noting that we can get guidance in making particular tough decisions by reflecting on theoretical matters such as human nature and the nature of technology.

TECHNOLOGY AS PRACTICAL WISDOM

A common view is that technology as such is neither good nor bad: technology is value-neutral and it is the way technology is used that is good or bad. I want to suggest to you that this is not correct. A look at the technology of television shows why. ("Television" here generally includes other methods of viewing programs, such as phone streaming, and other forms of video content. Much of what is said applies equally well to content created for *TikTok/Douyin*, *YouTube*, and so forth.)

What characterizes TV (TikTok, YouTube, Netflix, etc.) as a technology is: there are a large number of choices (multiple channels and platforms), there are (often) advertisements or other promotions, it is available right in our home, and, even if viewers pay for a subscription, there is usually no charge for turning it on (except for pay-per-view events, rentals, etc.).

DOI: 10.1201/9781003242574-6

By contrast, in going to a movie or a play or playing scrabble, one makes a commitment to being there for a while: in the case of movies and theater plays, we ante up something (going out there, paying money for our tickets, etc.), while in the case of scrabble, we have a commitment to a person. We usually don't walk into the middle of a movie, theater play, or game of chess, while people often turn on a TV program or live-streamed event in the middle. With TV, there is no commitment: we can change channels, turn it off, or go make a sandwich with complete ease, without losing anything, and without hurting anyone's feelings. We can usually pause a program if we wish or record it for later viewing. As a result, a television program is like a conversation with someone walking out the door, one hand already on the doorknob. It has to keep "hooking" the viewer, or she won't be there after the commercial or if the pace slows for a moment. No one walks out of a Chekhov play or K-pop concert if the pace momentarily slows, but restless channel/content searching is common. Moreover, the cost of television means that, in most cases, programs must attract large numbers of viewers to survive. A book that sells 10,000 copies in a week is a bestseller, while a weekly television show with 10,000 viewers is, in most cases, a dismal failure. To keep grabbing the attention of large numbers of viewers, who may tune in at any point during the program, TV shows must keep coming hot and fast things that appeal to the lowest common denominator. It must cater to short attention spans and must have continual highs. This makes it difficult to develop long-term, subtle relationships, to explore the subtleties and complexities of an issue, etc. Any show that tries must balance those aims with frequent attention grabbers. In addition, because TV is usually viewed at home, it has a capacity for replacing life that theater and movies do not. When people watch talk shows, they typically feel as if they had a conversation – TV provides an ersatz life. (Note that while viewers do not tend to confuse movie actors with their roles, they do tend to confuse TV actors with their roles. Anthony Hopkins is not assailed by people in the street reprimanding him for his misdeeds as the villains he plays in the movies, but actors who play TV villains report such experiences routinely, and soap opera actors are often given advice about which character in the show to marry, etc. Viewers fully understand that movies are fiction, while people often tend to respond to TV characters as real neighbors.) Movies, plays, and scrabble are social – they bring people together. TV is fragmenting, since TV watching is usually something done home, behind closed doors. Moreover, since many homes have multiple viewing devices (including smartphones), TV may break apart parents from children, sibling from sibling, spouse from spouse. The dominance of networks and big providers like Netflix and Disney means that TV is also national rather than regional. One result is that regional communities tend to give way to homogeneity. People no longer live in Peoria but in the land of *Real Housewives*, *Lang ya bang*, *La Desalmada*, or *The Bachelor*. Television has also become the great educator. Before mass media, people used to learn how to respond to a death in the family by going to a funeral. Now they have seen a thousand responses to grief on TV before they see a real one. Thus, responses to grief, which once were as diverse as the ethnic background of the world's many neighborhoods and families, are now homogenous. TV, in short, tends to replace the local communities, which are social (depend on person-to-person interaction and involvement), with a national or even international virtual world in which people participate in a solitary and passive way (watching alone). TV produces social fragmentation, isolation, homogeneity, and passivity. As a result, we may identify what I call the TV sensibility: short attention span, need for constant highs, passivity, lack of commitment, and homogeneity. A generation raised on TV will import this sensibility to other areas of their lives. (For example, it is not uncommon to see people conducting romantic relationships with these same features.) Moreover, when most movie viewers spend 20 hours a week or more watching TV, they bring TV sensibility to movies. Thus, movies increasingly come to resemble TV shows. TV also affects the way political debate is conducted. While *The New York Times* may run a President's entire speech, TV newscasts and internet highlights, responding to the "hand on the door" syndrome, typically run 5–10 seconds of a speech. Thus, political debate tends to be conducted entirely in pithy slogans (responding to short attention spans and the need for highs). Slogans always played an important role in politics, of course; what is new is that they now constitute almost the entirety of political discourse. Television panelists are rarely allowed even 2 minutes to state and develop a view.[1] What is worth noting is that all of these features flow from the nature of the technology itself. Given national or even international networks and providers, multiple channels/platforms, home viewing, and that there is no cost to turning off the tube or switching channels, the effects I have mentioned are almost inevitable. Of course, if people in our society were excited by books rather than sex, we'd see more and more books instead of more and more sex. To this extent, TV just reflects public taste. But given the fact that people are excited by sex and violence, it is almost inevitable that TV will show increasing amounts of increasingly explicit sex and violence. TV is not the only factor that accounts for increasing homogeneity, and we are not entirely helpless in the face of TV. But a tendency to increasing homogeneity is built into the technology itself. Technologies like television are not value-neutral.

I suggest instead that *technology is practical wisdom*,[2] where practical wisdom is engaging wisely in the enterprise of building a human world. Now a human world is a world that reflects human nature. Moreover, as Aristotle points out, wisdom is the habit of acting in ways that lead to excellence in expressing our nature. So, if technology is practical wisdom, then technology reflects human nature. Three examples will serve. First, again citing Aristotle, we are rational animals. And so technology must reflect both our rationality and our animality. Because we are animals, we are part of nature, part of the beauty and wonder of the natural world. Because we are rational, because we have visions and standards, we must shape our lives and our world. As Marx pointed out, human beings are by nature creative producers. As producers, as Homo Faber (man the maker), we remake the world. We are not alone in this respect: apes and crows make tools and beavers build dams. But human beings do so innovatively, rationally, and collectively. Beavers and humans both build dams, but beavers never go to conferences on new techniques for dam building. They build, if I may use this expression, the same dam thing over and over. Human beings, unlike beavers, can envision a novel future that better reflects our values and then remake the world in accordance with that vision. These two sides of human nature mean that we must live in partnership with nature. We should be neither passive nor oblivious to the natural characteristics of the world as we find it. Just as the good woodworker respects and works with the individual characteristics of her wood, its particular grain and color, while shaping it into a box or chair that serves human needs and expresses her own sense of beauty, so the good engineer respects the peculiar beauties of nature while reshaping the world in a way that is conducive to human welfare, rationally conceived. This is wisdom in being a rational animal. Second, because we are *rational* animals, we should reshape the world rationally: technology requires the thoughtful and innovative use of precise, clearly articulated knowledge. Finally, because it is our nature, as Aristotle says, to be social animals, technology is a community activity, in two ways: technology is conducted communally, and it takes account of the communal character of human welfare. A process that isolates people is, in at least that one respect, not a technological advance but a technological step back.

There are some important consequences of these ruminations on technology as practical wisdom. For example, it follows that technology is not value-neutral. Technology is not the ability to manipulate the environment as such, but the search for human excellence. Moreover, technology is committed, by its very nature, to community, to rationality, and to partnership with nature. So, technology is value-laden both in its defining goal and in its defining methods.

SAFETY

Engineering as a profession is committed to safety, as every engineer knows.[3] But what exactly is safety and how safe is safe enough?[4]

Safety does not mean the absence of all risk. Risk is the necessary concomitant of progress, indeed, of life itself. Some risks are worth taking because of the benefits obtained by taking the risk. For example, we accept 1.3 million deaths a year, worldwide, as the price of the convenience of the automobile.[5] The value of safety does not require that engineers must sacrifice everything else to eliminate the risk of a single life being lost, and so engineers needn't (on grounds of safety alone) demand an end to automobile manufacture.[6] Again, were we willing to spend $500,000 on safety devices, we could make cars much safer than they are. This would mean, of course, that very few people could afford to drive. The value of safety does not require automotive engineers to design cars that almost no one can afford.

In short, every risk must be balanced against the benefits of the product or process. In making this risk/benefit assessment, it is important to consider alternatives. For example, if an alternative to the automobile would provide all the same benefits but at a cost of $10 more per person, then the effective benefit of the automobile over the alternative is just saving $10 a person.

Extent of a Risk

There are three aspects to determining the extent of a risk:

    a. *How severe is the possible harm to each individual?*
    b. *How widespread is the danger?*
    c. *How likely is the danger?*

The greater the risk, the more imperative it is to reveal the risk fully and promptly. The degree of risk depends on three related factors. Factor (c) concerns the *likelihood* of harm. For each risk, there is a threshold of significance. If there is only a one in a hundred fifty million chance of an accident that might mildly harm a single person, that is below the threshold of significance. (It is orders of magnitude below the general death rate, for example). However, a one in a million chance of a major catastrophe, such as nuclear war, is a significant risk. In general, the higher the stakes, the lower the threshold of significance. Once the threshold of significance is passed, the obligation to publicize the risk increases with the likelihood of harm resulting. Factors (a) and (b) concern risk distribution. A million-dollar risk to one person is more troublesome than a one-dollar risk to a million people. Of first importance are devastating harms to individuals, such as death or severe bodily injury. If a company's proposed action poses a significantly likely possibility of devastating harm, then only very strong reasons (such as an imperative need for national security) could justify keeping the risk a secret. The importance of widespread risks depends upon two factors: the average possible harm and the total possible harm. For example, a one-dollar harm to a million people is a low average harm (one dollar) but a high total harm (a million dollars). A total harm of a million dollars may be a significant loss for a community because of its cumulative effect on commerce, government revenues, etc. Obviously, a small, poor community will be more affected by a given total loss than will a large, wealthy community, so the total harm must be assessed in the context of the particular community and its ability to withstand such a loss. Both average and total harm must be considered: the greater the average and the total harm, the greater the obligation to avoid the risk, or at least publicize the risk.

Balancing Risks against Benefits

Of course, the engineer would prefer to take no risks at all. Unfortunately, this is not feasible. Risk is inescapable: even breathing incurs the risk of a respiratory-borne infection. So, the engineer must decide whether the benefit is worth the risk involved. There is no simple way to do this. However, there are a few factors worth mentioning. In some cases, one of these factors may not be particularly helpful, but usually at least one of these factors will be useful in thinking about a particular risk.

First, when weighing risks against benefits, the responsible engineer *places a high value on minimizing risk.* (Other things being equal, she prefers the lowest feasible value of LIKELIHOOD OF HARM X TOTAL POSSIBLE HARM). The responsible designer of automobiles goes to considerable lengths to make automobile travel as safe as feasible, even if that means slightly higher costs or requires compromises in style and image. *When in doubt, be conservative.*

Second, a rough but useful test an engineer may use is to ask herself, *"would I be willing to have my family undergo this risk for these benefits?"*

A third guideline comes from the fact that an engineer is a *trustee* of the public welfare and should act like the administrator of a trust fund or the conservator of an estate. A trustee does not perform their task well if they are so risk-averse that the fund they manage does overwhelmingly worse than the market average but they ought to be more cautious than the average investor, since their primary directive is to protect (preserve) the fund or estate. Thus, the engineer should *exercise somewhat more caution on behalf of others than they would exercise for themselves.*

Finally, safety costs more in the short term but saves money in the long run. As a rule, the more reliable a machine is, the more it costs to make it, though more reliable machines save money in down-time and repairs. The same is true of safety. The initial or primary costs of a safer product or process tend to be higher. Riskier products and processes, however, tend to have higher long-term or secondary costs, such as damage caused by accidents, liability costs (lawsuits and fines), recall costs, lost sales because of unfavorable publicity, and time lost when the product or process has to be modified. It is a poor business decision to fail to replace or install a hundred-dollar part when that results in the destruction of the plant by fire or explosion. As Martin and Schinzinger (1983, 107) point out, the low primary costs of high-risk products and processes must be balanced against the high secondary costs, and the high primary costs of low-risk products must be balanced against the low secondary costs. Minimal total cost to the manufacturer is thus secured somewhere in the middle of the safety scale, at a point M where "incremental savings in primary cost ... are offset by an equal incremental increase in secondary cost ...." Although point M is ideal from a purely financial point of view, the ethical engineer has a tendency to *opt for solutions somewhat less risky than M.*

Nature of Risks

The extent of the risk is not the only factor to be weighed against the benefit: the *nature* of the risk is also important. For example, those who use automobiles are aware of the risks and the benefits, and have some choice in the matter. Drivers and passengers voluntarily take the risk. Furthermore, the same people who take the risk also get the benefit. There is a big difference between *my* voluntarily taking a risk in order to get something *I* want, and *your* putting me at risk, without my knowledge or consent, to get something *you* want. So, there is a special reason to avoid imposing risks on the unwary, or on those who have no choice, or on those who don't receive the benefit.[7]

This suggests three key factors in assessing safety:

1. *Is the risk voluntarily taken?*
2. *Are the potential risks known to those at risk?*
3. *Do those at risk reap the benefits of the risk?*

These factors are useful in three ways. First, when the engineer has to choose between two risks, she will prefer (other things being equal) a voluntary to an involuntary risk, and she will prefer a risk borne by those who benefit to a risk borne by those who do not benefit. For example, in a nuclear power plant, a risk to workers is, other things being equal, less bad than a risk to residents, since workers have more choice than residents, and workers get more of the benefits of the plant than do residents. So, the three factors help us decide which risk to take when we have a choice. Second, the factors are useful in deciding whether to impose a particular risk. A given risk is worse to the extent that it is involuntary, to the extent that the risks are not known to those at risk, and to the extent that those taking the risk do not benefit. So, the three factors help us decide how bad a given risk is. Third, these factors make clear the importance of publicizing risks.

Publicizing Risks

Since voluntarily taken risks are, other things being equal, better than involuntary ones, it might seem that companies should always publicize every risk its operations pose to the public. This is too stringent a standard, however. If even minor and extremely remote risks are publicized, the dangerous risks wind up buried in a long list of remote risks. Newspapers looking for stories that sell papers, politicians seeking votes, and posters seeking likes or retweets are likely to give undue attention to very remote risks. Public fanfare about an extremely remote risk serves well neither the public nor the company. Moreover, publicizing a risk may compromise trade secrets. So, although the default value is to publicize, there are times when a company need not advertise a risk.

In determining when and whether to publicize a risk posed by a contemplated company action, the engineer or executive should ask herself five key questions:

1. *Is there a legal duty to publicize the risk?*

   This question takes priority over the other three. A company *must* meet its legal obligations. If there is no legal duty to publicize the risk, the executive or engineer should consider the next three questions.

2. *Can the community/user/others at risk take feasible action to reduce the risk?*

   Depending on the nature of the risk, the community/user may choose from a wide range of options designed to mitigate or guard against the risk posed by the company's or firm's activities. To give but a few examples, the community might make emergency preparations (such as formulating an evacuation plan and instituting periodic evacuation drills, arranging for extra fire-fighting assistance in case of an industrial accident, installing extra mains in the vicinity of the plant, and giving special training to local emergency crews), construct retaining walls, alter flood planning and management, institute periodic state inspections, oversee construction, impose restrictions on plant operations, or even relocate homeowners. Users of a battery that may leak can take care not to touch the battery with bare hands.

   When such options are feasible and significantly lower the risk of potential harm, the community must be given the opportunity to decide whether implementing one or more of these options is warranted. This decision cannot be made intelligently if the community is unaware of the nature of the risk. Thus, when there are feasible and appropriate measures the community might take, the company or firm has a duty to inform the community of the risk and co-operate with the community in exploring options, sufficiently far in advance that the community can make intelligent decisions. Thus, how far in advance of implementation the company must publicize the risk depends on the nature of the community action available. A simple retaining wall can be built quickly. Altering complex flood plans requires considerably more time for study, pricing, legislative action, and construction of facilities such as dams.

The same logic applies to individual consumers/users or others, such as pedestrians passing by a building, and their ability to take steps to avoid or mitigate the risks.[8]

3. *How great is the risk?*

For example, because the risks posed by nuclear power plants are generally extremely severe and very widespread, it is imperative that virtually *any* significant risks a nuclear power plant poses to either workers or residents be well publicized, rather than hidden or down-played.

4. *Are there legitimate reasons for withholding the information?*

National security, legitimate business and trade secrets, and preventing panic are legitimate factors that tell in favor of not publicizing a risk. These factors must be balanced against the first three.

5. *How much will the company be hurt by publicizing the risk?*

Firms and companies must balance public stewardship against the economic facts of survival. At one extreme, a firm or company must be prepared to go out of business rather than pose a significant risk of world disaster. At the other extreme, a company need not publicize a trivial risk (such as a highly remote risk of trivial harm to one or two individuals) if publicizing the risk would destroy the company. Most cases lie between these extremes. Because innovation is the soul of engineering, the company must do its best to find ways of publicizing the risk that minimize damage to the company. For example, the company can propose community action to minimize risk at the time it publicizes the risk, thus helping the community protect itself.

In sum, the ethical engineer places a high value on avoiding risks, and on publicizing those that are worth taking.

The value engineers place on safety extends beyond the workplace: it includes placing a premium on health as well as avoiding accidents, and a commitment to due care in all things. So the compleat engineer will not drive recklessly, will not drive an unsafe car, will not leave tools lying about where others may trip over them, will maintain a healthy diet, etc.

Case 4: Chemical N in a Lipstick

Large doses of chemical N have been shown to cause cancer in laboratory rats. In addition, studies show that workers who have had long-term exposure to N have a significantly higher than average rate of cancer. Using N in a lipstick under study by your company would give that lipstick a slightly brighter shade than any lipstick now in the market. Is it ethical to use N in a lipstick?

Discussion: The mere fact that large doses of N cause cancer in rats does not mean that N will cause cancer to human beings when applied to the lips in small quantities: human beings differ from rats, large doses differ from small doses, and application to the lips differs from injection or ingestion. However, if N is used in a lipstick, customers, unlike the rats in the study, would be subject to long-term exposure. For these reasons, it is hard to assess the likelihood of the risk. It is clear, however, that the risk would be both severe and widespread. In sum, the lab results are enough to raise a significant possibility of harm to customers. Now a significant possibility is not the same as a certainty. We must balance the benefits of using N in a lipstick against the risks. To a large extent, this is guesswork, since we don't really know how likely the risk is. Fortunately, we are all used to operating under uncertainty. When you choose between two films for your Saturday night outing, you don't really know which film would be more enjoyable or worthwhile to see. It could even be that neither film is worth the time or money. You make the best decision you can, taking into consideration any relevant factors (such as reviews, friend's recommendations, and how much difference the cost of attending the film will make to your budget).

In this case, the value of safety suggests that N should not be used in a commercial lipstick. The benefits and advancement of human welfare that a slightly brighter lipstick provides are slight. There are already many satisfying lipsticks in the market, and no one's health and happiness depend on how bright their lipstick is. However, the risk and likely suffering (setback of human welfare) are significant. You would not, after all, want yourself, your son, your daughter, or your mother to risk getting cancer in order to have slightly brighter lipstick.

Moreover, while the benefit does go to those at risk, buyers of the lipstick are unlikely to be fully aware of the risk. Clearly, the potential costs of lawsuits and decreased sales of all the company's products as a result of adverse publicity make point M fall on the side of forbearance. And an engineer who used N in a lipstick would not be acting as a faithful trustee of the public welfare. Thus, using N in a commercial lipstick would violate the engineer's commitment to safety. In addition, the strong possibility of producing a significant number of carcinomas clearly sets back human welfare more than the slight advantage of a brighter lipstick advances it. So, using N in the lipstick violates the engineer's commitment to advancing human welfare. (See also "The Duty to Leave the World No Worse" in Chapter 5.) Of course, the engineer might monitor the situation by keeping an eye out for further studies. If the engineer did, ultimately, decide to use N, part of engineering a solution would be publicizing the risk so that the risk is borne voluntarily.

Notice what we have done so far. We began with somewhat theoretical conceptions of human nature and of technology as practical wisdom. We then derived as a consequence a definition of engineering from which we extracted a value, safety. We developed some guiding ideas in applying that value and used our guidelines to help decide an actual case by giving reasons for the particular case. This illustrates the kind of ethical thinking at the heart of sound decision-making.

Case 5: Drug P and Pancreatic Cancer

Drug P shows some promise of both ameliorating the symptoms of and lengthening the lives of late-stage pancreatic cancer patients. However, laboratory tests indicate that P may cause liver cancer and ulcers. Is it ethical to market P?

Discussion: Drug P is a worthwhile risk. First, since P would be a prescription drug, patients can be informed of the possible dangers of P and have a choice about whether to take the drug. So, the risk is known and voluntarily borne by those who would receive the benefit. Moreover, it is not irrational for a late-stage pancreatic cancer patient, who may not live very long without P, to choose to take the risk of developing ulcers and liver cancer some years later, especially if P improves the quality of their remaining time. If a member of your family were a late-stage pancreatic cancer patient, you would want them to have the choice of taking P. Finally, marketing P may also advance human knowledge and is more likely to advance human welfare than to cause overall suffering. So, the value of safety is not compromised by marketing P, and the commitment to human progress suggests that one should market P (making sure to publicize the risks).

Case 6: Amtrak/Conrail Collision

On January 4, 1987, 16 people died and 174 were injured when an Amtrak passenger train, traveling at 125 mph, collided with a Conrail freight train, traveling about 60 mph. The speed limits for the two trains are 103 and 30 mph, respectively. (Speeding is often due to commercial pressure to keep on schedule.) The blood and urine of the Conrail train's engineer tested positive for marijuana. He also "failed to stop at a warning signal or properly test his cab signals as required before departure" and "was charged with manslaughter by locomotive. He was sentenced to five years in state prison and three years on federal charges of lying to the NTSB" (Robinson 2015). He had previously had his driver's license suspended and had been indicted for drunk driving. Amtrack trains have automatic brakes that are designed to slow a speeding train, while the Conrail train did not. The radio on the Conrail train was broken, temporarily replaced with a hand-held radio. The whistle that was supposed to serve as a warning had been taped over. The Federal Railroad Commission discovered 75 other trains with taped whistles. Separate tracks for freight and passenger trains are considered too expensive (Westrum 1991, 255–256; Flynn and Kaye 1987).

Case 7: The Dalkon Shield

In September of 1968, tests were begun on a new type of contraceptive device, the Dalkon Shield. By summer of 1969, 640 women had been tested. The Dalkon Shield showed a pregnancy rate of 1.1%, as low or lower than that of birth control pills. By June of 1970, the Shield was being marketed on a small scale. Several changes had been made to the shield, after the tests were run, making it thinner and more flexible.

Moreover, copper had been added to the shield. The A. H. Robins Corporation was aware of these problems. It was also aware that there were questions about the actual pregnancy rate of the original study. (For example, Jack Freund, Vice-President of Robins' medical department, was concerned that the studies had not been conducted long enough. He was told, he said, that the actual pregnancy rate, with a follow-up period, had been 2.3%.) Nonetheless, Robins decided to purchase rights to the device on June 12. Two weeks later, Robins became aware that the tail of the device might wick material into the uterus. In July, further changes were made to the width of the device, and the legs were tear-dropped. These changes were not tested (though Robins later claimed to have consulted experts and claimed that "there is no evidence that these changes affected the safety of the Shield"). Evidence of septic abortions caused by the Shield, effectiveness rates below those of the pill, deteriorating Shields, and other problems, continued to mount (Buchholz 1989). In particular, studies indicated that the risk of pelvic inflammatory disease was more than twice the risk for users of other IUDs (Henig 1985). Another study found the risk to be five times greater (Kolata 1987). Lawsuits against Robins were filed. Eventually, on June 28, 1974, two days after a letter from the FDA recommending removal, Robins removed the Shield from the market. However, the company did not issue a formal recall and claimed in a press release that "performance has clearly been satisfactory" (Buchholz 1989). A.H. Robins eventually sued for bankruptcy and established a 2.4 billion trust fund to compensate users of the shield.

Case 8: Narbitrol Emulsifier

The FDA ran a series of tests in which Narbitrol, an emulsifier used in certain foods, produced mutations in laboratory animals. The FDA required a warning be printed on the labels of all food products containing Narbitrol. This requirement does not apply to imported foods. A significant part of XYZ Company's sales come from Dark Dreams, an expensive chocolate sauce containing Narbitrol. Marketing research indicates that a warning would seriously dampen sales of Dark Dreams. XYZ has several options. (A) Discontinue producing Dark Dreams until a safer emulsifier can feasibly be used. This will take considerable time and research, as the most commonly used safe emulsifiers are not compatible with Dark Dreams, causing clotting and discoloration. (B) Use another emulsifier, Starbitol, about which no FDA regulation as yet exists, but which early evidence shows likely to be carcinogenic. (C) Sell Dark Dreams through foreign import markets, thereby circumventing the FDA order. (D) Sell Dark Dreams with the warning label.

Case 9: Fukushima Daiishi

Fukushima Daiichi Nuclear Accident, 2011, Honshu, Japan. The Fukushima plant consisted of six light water reactors. In March 2011, units 1–3 were operational. Reactor 4 was used to store spent rods, while 5 and 6 were shut down for planned maintenance. Responding to a 9.0 magnitude earthquake on March 11, 2011, the reactors were immediately shut down. When the electric grid went down, emergency generators powered cooling of decay heat in the cores. Forty minutes later, a tsunami with waves well over 40-feet high surged over the plant's 19-foot seawall, damaging the emergency diesel generators (as well as the battery backup systems) powering the cooling pumps. As a result, although reactors 1–3 were shut down, the cooling system failed. Residual decay heat built up in the cores, creating a partial meltdown of the fuel rods. Melting material falling to the floor of containment vessels burned holes, exposing the core. Many of the internal components were made of zircaloy (zirconium alloy), since it does not absorb neutrons. However, steam from evaporated water produced hydrogen and oxygen resulting in oxidation of zirconium at high temperatures, producing large amounts of hydrogen gas. Explosions of the hydrogen gas on March 12, 14, and 15 damaged the outer containment buildings of reactors 1, 2, and 3, respectively. Large amounts of radioactive material were released into the air and the Pacific Ocean. The disaster was classified as a level 7 on the International Nuclear Event Scale, a dubious honor shared only by Chernobyl. More than 154,000 people were eventually evacuated. The cleanup is estimated to require 30–40 years.

National Diet of Japan Fukushima Nuclear Accident Independent Investigation Commission (NAIIC 2012) identified numerous problems in planning and in response to the event. The seawall was too low, there was insufficient redundancy in the power supply, and adequate containment plans were lacking. Tsunami studies in 2000 and 2008 were ignored. The 2008 study said that the site could be subjected to waves over 33 feet. The report stated that NISA, a government regulatory body, did not keep records nor release to the public information "on their evaluations or their instructions to reconsider the assumptions used in designing the plant's tsunami defenses" (NAIIC 2012, 27). Inadequately protected turbine buildings housed the switching stations from the backup generators. Moving them to a secure location would have maintained power to the cooling system. "TEPCO [the plant operator] had not prepared any measures to lessen or eliminate the risk, but failed to provide specific instructions to remedy the situation" (NAIIC 2012, 16), leaving plant workers unprepared to deal with the failure of the emergency generators. According to Amory Lovins, Japan's "rigid bureaucratic structures, reluctance to send bad news upwards, need to save face, weak development of policy alternatives, eagerness to preserve nuclear power's public acceptance, and politically fragile government, along with TEPCO's very hierarchical management culture, also contributed to the way the accident unfolded. Moreover, the information Japanese people receive about nuclear energy and its alternatives has long been tightly controlled by both TEPCO and the government" (Lovins 2011).

Discussion: The 19-foot wall was clearly too low. The actual wave, well over 40 feet, was the result of a very rare event (estimated by some to be a once-in-a-thousand-year event). How much should engineers prepare for extremely rare events? Clearly, there is a trade-off between the rarity of the event (likelihood) and the cost of the extra precautions, on the one hand, and the severity of the consequences on the other. What principles or guiding ideas should govern that trade-off?

HUMAN PROGRESS

Human progress has two basic components, knowledge and welfare (improving the lives of fellow human beings both materially and psychologically). The ethical engineer is dedicated to the advancement of knowledge, both practical and theoretical. This means that the compleat engineer is drawn to innovative solutions that advance human knowledge. He keeps up with his field, always eager to learn new techniques and approaches. He is intellectually curious and loves to learn about all sorts of things. And he seeks appropriate means of sharing his knowledge (in ways that do not compromise his loyalty to his company). The ethical engineer is also dedicated to the advancement of human welfare. He wants to make people's lives better. This means that he seeks to use technology to help others rather than to hurt them. In his work, he is mindful of the human costs and potential benefits of engineering and employment decisions. In general, the compleat engineer is civic-minded, compassionate, and helpful to others. He may volunteer his services for community and charitable projects, acquaint himself with social problems such as world hunger and future energy needs, and speak out publicly on those issues about which he has special knowledge. Kindness to a lost child or stranger, helping a neighbor with a problem, giving to charity and writing a letter to the editor about nuclear energy are all ways the compleat engineer might express his commitment to human progress.

CLEAN, CLEAR DECISION-MAKING

There are many ways in which people make decisions. Engineering as a profession is dedicated to making decisions by clean, clear, rigorous thinking. The engineer analyzes the problem, gathers and records precise and reliable information, uses care, concentration and ingenuity in solving the problem, and tests their solution in order to achieve excellence. Sloppy, haphazard, vague, and half-baked thinking are contrary to the engineering way.[9] So the engineer is dedicated to values such as:

Precision and clarity.
Ingenuity and creativity, concentration and care.
The value of excellence generally.

In the workplace, this means meticulous precision and concentration, careful and accurate record-keeping, and striving for excellence in all facets of one's work. As Cropley (2015, 155) says, "Creativity is a fundamental element of engineering." The ethical engineer welcomes new ideas and creativity, though always striving to make innovations precise and safe.

Case 10: Quebec Bridge

Quebec Bridge collapse, Quebec, Canada. The Quebec Bridge has the dubious distinction of collapsing twice, nine years apart. From the mid-1800s onward, building a bridge over the St. Lawrence River was a topic of discussion. In 1887, the Quebec Bridge Company was formed to make the plan a reality. It took years to raise financing, choose a site, and settle on team and a design. In 1900, the Consulting Engineer, Theodore Cooper, lengthened the planned bridge span from 1,600 feet to 1,800 feet to save costs and time on building piers as well as decrease negative effects of ice. When a government engineer questioned the extra stress caused by the longer length, Cooper ignored the criticism because he refused to be put, in his words, "in the position of a subordinate." Ill health prevented Cooper from supervising the construction. According to Griggs (2015), "almost immediately after the beginning of construction of the suspended span in July, problems with member 8L started, setting into motion one of the most bizarre set of miscalculations, mis-communications and plain incompetence in the history of bridge building." As a result of miscommunications between Cooper, the on-site bridge inspector, and the Chief Engineer regarding bent cords, work simply continued. (See also Marsh 2015.) Moreover, Cooper discovered in 1906 that the actual weight of the steel parts exceeded the pre-manufacture estimates but decided to proceed anyway (Whalen 2000). In 1907, a rivet snapped and a cantilever dropped into the St. Laurence River, killing 75. During construction of the second incarnation of the bridge in 1916, the failure of a casting in the erection equipment caused the 5,100-ton center span, being hoisted to join the two cantilever arms, to plunge into the river. Thirteen workers died (Whalen 2000).

COMMUNITY

Community is a central value in engineering ethics. Co-operation between engineers is essential, of course, but "community" denotes much more than sharing information and ideas with colleagues and co-workers. Engineers are part of an important community at four levels: the *workplace community* (co-workers and perhaps clients), the *local community* (from family and friends to the broader region), the *engineering community* (being part of the institution of engineering), and the *world community* of fellow human beings and ecosystems throughout the globe. All four are sources of bonds and ethical concern.

A community is a group of persons working out a joint moral vision through common institutions, practices, relationships, etc., all of which are dedicated to that vision. Perhaps the clearest example would be a small space colony, in which everyone sees herself as a partner in a common project. Each person plays a role in this project, whether as a gardener or science officer, and each is committed to the others' success and welfare in this remote region of space. The value of their enterprise, as well as its precariousness, gives each member of the colony a sense of self-worth, and leads her to view her team-mates with respect and concern.

It is important to employers and employees alike that the workplace be a community of this sort.

Being part of a community is important to the individual engineer because it makes them a happier and better engineer. Most people spend a third of their waking lives at work and derive a good part of their self-esteem and sense of who they are from their jobs. An engineer's life will be enormously more pleasant and rewarding if they see themselves as part of worthwhile and socially important project. Not only does this give their life as a whole a sense of meaning but it also makes rewarding what might otherwise be less interesting tasks. So, it is important that each engineer understand the way in which their contribution fits into the bigger picture and has a sense of the importance and value of the larger project of which their work is a part. In addition, people enjoy going to work if the workplace is supportive, caring, and demanding in ways that make sense. All people need to feel that they belong somewhere, that they are valued, and that they are among friends. If engineers feel that the company is fair and cares about them, that their colleagues are out to help them rather than stab them in the back, and that they can trust others in the workplace to be moral and sincere, then they will love their work. If, however, engineers feel that the company is just using them, that their colleagues are deceitful, cruel or underhanded, and that they have to look out for themselves instead of concentrating on their work, then they will be unhappy and will not flourish and develop as engineers. If the workplace has a community atmosphere, engineers will be highly motivated, and will reach their personal best in skill and performance. A good workplace makes for a good and a happy engineer.

The same facts make a community atmosphere attractive to an employer. Engineers who are part of a community make better employees. If the company is loyal to its employees, treats them fairly and supportively, and makes them feel valued, then the employees will give their best to the company. Loyalty begets loyalty and a company has no greater asset than the loyalty, skills, and motivation of its employees. However, a company that deceives its employees can expect dishonesty in return. If the company is not ethical, neither will its employees be ethical. Stealing, featherbedding, fudging, coasting, attrition, and not caring about the company's interests are rare in companies that truly function as a community. (See also "Cut-throat v. Community Workplaces" in Chapter 1.)

The same considerations apply, in different ways, to the three other sorts of communities. Local communities are perhaps the most diverse, as they may include a close circle of friends, a religious community, a village, the community of chess players, a State or Province, and much more. The world community unites not only geographically dispersed human beings in Zimbabwe, Brazil, India, Italy, the US, and everywhere else human beings may be found, but other inhabitants of earth, such as dolphins and pandas, and the wondrously varied ecosystems that support them. All of these call for mutual concern and support and instill a sense of belonging to something greater than oneself.

It is important to keep this all this firmly in mind. Often both employees and employers can get *short-term* gains by violating the demands of community, within the workplace or without (such as unsafe or environmentally harmful practices). So, it may be tempting to gain a short-term advantage by lying, exploiting, etc. But if you do this, you will pay for it in the long run. Fear is an effective short-term motivator of employees, but a poor long-term motivator. You may get a promotion by lying or fudging, but, in the long run, you will make coming to work miserable. It is worthwhile to make short-term sacrifices or give up an opportunity, for the sake of community. In the long run, *everyone* benefits.

An interesting illustration of this is game theory's "prisoner's dilemma." Two prisoners, A and B, are told that if no one confesses, each will get two years in jail. If one confesses and the other does not, the one who confesses will get one year in jail, while the other will get ten years. If both confess, both will get six years in jail. Clearly, they should both refuse to confess. In a community atmosphere, where people see themselves as part of a team, that is what will happen. But in a self-seeking atmosphere, each prisoner will confess because he cannot trust the other not to try to get an advantage for himself (to "defect"). Thus, in a self-seeking atmosphere, both lose, while in a community atmosphere, both gain. Of course, this is an artificial situation, but it is important because many real-life situations have the same logic as the prisoner's dilemma. For example, companies A and B, whose plants are close to each other in Airsville, both produce almost identical regulators, used in the process of making rubber for sneakers. The manufacturing process produces sulfur dioxide which, reacting with water and oxygen in the air, corrode the factories. Installing scrubbers at both factories to remove the pollutant would be cost-effective within a year. However, air pollution is what economists call an externality, in that the cost falls on everyone equally, not just the polluter. If A installs scrubbers and B does not, B receives the same benefit as A does but avoids the cost of the scrubber. Other things being equal, this means that B can produce the regulator more cheaply, and will thus put A out of business. A cannot compensate by advertising that they are a more environmentally friendly company, since the regulators are too far removed from the public: no one will buy a more expensive sneaker because of pollution from the factory that makes the regulator used in the manufacture of the rubber that goes into the less expensive but otherwise identical sneaker. Thus, neither A nor B can take the risk of installing scrubbers, even though both would benefit if both installed them.

The compleat engineer views her job, her family, her town, her country, humanity as a whole, and nature as a community of this sort. She does her part in the human enterprise, takes pleasure in the success of others, and feels a sense of kinship with all human beings and with nature.

Thinking of a company as a community affects the way companies operate. A company that operates as a community will not overcompartmentalize its processes, with the result that each group does not know what the other is doing. For example, if design and manufacture are completely separated, manufacturing engineers will not feel a close sense of teamship with design engineers. This not only impairs the attitudes of engineers but also hurts efficiency:

> We learned this lesson at Hewlett-Packard a decade ago .... Our manufacturing engineers used to play a somewhat passive role in the innovation process. They assumed that whatever the design engineers threw over the fence, manufacturing would build. Today, manufacturing engineers are part of the product design team from day one of a project .... The collaboration between R&D and manufacturing has changed both functions for

the better …. In order to make day-to-day adaptations and mid-course corrections in production, there must be continual communication between engineers and workers and between the design and manufacturing arms of the company.

Young (1990, 306)

Case 11: Sandbagging a Rival

You and Smith are the two major candidates for a promotion to the head of department N. Smith's chances for the promotion depend largely, you think, upon the outcome of the project he is now directing. You can "sandbag" Smith's project by getting a superior to reassign some of the key people in Smith's project to your project, although you don't really need any more help. Should you do it?

Discussion: To do this would violate the value of community. You would be seeking your own advantage in an unfair way that would hurt the company. You might benefit in the short run, but in the long run this kind of underhanded back-stabbing hurts both you and the company. If you do this kind of thing, others will treat you the same way and you cannot expect the company to be loyal to you. Do you really want to work in a place in which other people treat you this way? (In addition, to the extent that Smith's project would advance human progress, you are setting back rather than promoting human knowledge and welfare.)

PARTNERSHIP WITH NATURE[10]

As we said earlier, the essence of being human is our urge and ability to build and make things that change the environment in which we live. Thus, engineering is the distinctively human science. And the way engineers regard their task has profound moral significance. Engineering is the way we relate to the world in which we find ourselves, and how we do this defines our role in the cosmos. Engineering, in other words, is one way to answer the question "what is humanity?"

This is why partnership with nature is such an important part of engineering as a calling. Human beings are rational animals and we are untrue to ourselves if we deny either the "rational" or the "animal." Because we are animals, we are part of nature, part of the beauty and wonder of the natural world. Because we are rational, because we have visions and standards, we must shape our lives and our world. These two sides of human nature mean that we must live in partnership with nature.

There are two aspects to environmental awareness. The first aspect concerns direct harm to the quality of human life. This may be called *"environmentalism as prudence."* For example, we have to be concerned about the effect on climate of atmospheric CO_2 and other greenhouse gasses. Well-established effects of unrestrained global warming include the inundation of seacoasts by the rising ocean, loss of many species, including many of the food crops we rely upon, and more violent extremes of temperature and storms, resulting in many deaths through natural disasters. More speculative effects include halting of ocean currents resulting in a new ice age and "runaway" warming turning the earth into a hellish inferno like Venus. A second type of example is the Aral Sea project. The former Soviet Union began a giant engineering project to take water from the Aral Sea to make the arid regions around it arable. The project was a disaster: changing the riverbed patterns resulted in an irreversible drying of the once scenic Aral (which was so lovely there were once plans to make it a "Russian Riviera"). As the Aral dried out, vast salt fields were left behind. The wind scattered the salt, making the land unfit for farming. The result is that thriving villages have become ghost towns, and a once scenic region became an ugly moonscape (Kotlyankov 1991). Today, what remains of the Aral Sea is less than a tenth of its original size. The Aral Sea is one of many examples. In 2019, nitrate pollution from fertilizer runoff and untreated sewage affected 387 districts in India and one study revealed "the Harike Wetland and Sutlej River are critically polluted with heavy metals" (Kumar et al. 2020, 1). "Across the U.S., drinking water systems serving millions of people fail to meet state and federal safety standards" (Peeples 2020). According to WHO/UNICEF (2021), 771 million people worldwide lack clean water near their homes.

The second aspect of environmentalism concerns the preservation of nature for its own sake. This may be called *"environmentalism as an ethic."* Perhaps the most famous argument for environmentalism as an ethic is Routley's "Last Man Argument" (Routley 1973). ("Last Person Argument" would be a better name.) Would the last human being, knowing they were about to die, have done something wrong were they to eliminate all nonhuman life (or, in another version, all non-sentient life)? If it would be wrong to do this, the argument goes, it must be because nature has an intrinsic value of its own, since, in this scenario, there will be no human beings left to be adversely affected. (We have to assume that humanoid organisms or other organisms with advanced intelligence could not re-evolve and that the remaining organisms would not be observed by alien species of advanced intelligence.) It is this second aspect of environmentalism that requires an understanding of our own place in the natural world.

One view about our place in nature is called the *Gaia Theory*.[11] According to Gaia Theory, we should understand the entire earth as if it were a single organism. More precisely, it claims that through complex mutual interactions between the biosphere and the geophysical realm, the earth maintains itself much as a living organism does. In short, the earth is a homeostatic system. Although much that has been written about Gaia theory is premature or even silly, there is nothing particularly mystical about Gaia theory, properly understood. For example, it assumes "no plan or intention," and ascribes to the earth "no 'soul' or other mystical power" (Joseph 1990, 6). The mechanisms it describes are purely physical. For example, "dormant microbes ... come out of their spore states and remove gas from the atmosphere when conditions are right, as though they were built-in safety mechanisms for correcting environmental imbalances" (Joseph 1990, 115).

There are several important consequences of Gaia theory. One consequence is that the "climatic system is in many ways more robust and resilient than has been generally believed" (Joseph 1990, 2). However, this resiliency depends on biodiversity, and so it is crucial to the health of the planet as a whole to maintain biological diversity. Thus, Gaia theorists are disturbed by large-scale agriculture, which produces massive destruction of forests, wild lands, natural shelters and windbreaks, pollutes the watershed, erodes the soil, and replaces a highly diverse field or forest with acres of a single plant. In short, problems such as the greenhouse effect, depletion of the ozone layer, and acid rain are not necessarily catastrophic, so long as we leave intact the earth's mechanisms for dealing with such imbalances. However, since natural mechanisms can take thousands of years, "to count on the biota collectively to restore comfortable conditions is ... foolhardy" (Joseph 1990, 151). What Gaia theory counsels, then, is a balanced strategy of minimizing environmental deterioration while maintaining diversity in the biorealm.

Another approach is ecofeminism. *Ecofeminism* refers to a range of views that emphasize the "woman-nature connection," and articulate ways in which the oppression and domination of women parallel and/or stem from the same outlook as the abuse of nature. For more on ecofeminism, see Warren (2015).

Several authors have advocated for *environmental virtue theory*, that is taking a virtue ethics approach to environmentalism. Thomas Hill (1983) suggests that we need to ask, about a person who cuts down and paves pristine woodland in order to escape paying for maintaining it, "what kind of person would do a thing like that?" That is often a useful question for engineers balancing environmental against financial concerns. Sandler (2013, 1667) lists as environmental virtues "care for living things, appreciativeness of natural beauty, and moderation in use of natural resources."[12] Schlossberger (2001) argues that: "Practical wisdom consists in acting in ways that habitually tend to promote human excellence," which stem from the appropriate way human organisms function as a unit over time. Similarly, nature's interests can be understood in terms of the appropriate way ecosystems and nature function over time. "Human excellence is defined in terms of the sorts of lives that enduring cultures ideally interacting over time would find to be good." Since partnership with nature "is implicit in any adequate conception of human excellence … practical wisdom requires a partnership with nature. A partnership, in this sense, is a pattern of mutually interested interactions to advance the interests of the partners. Partnerships entail an obligation on the part of each partner to strive to obtain a reasonable balance between the (objective) interests of the partners." For example, "if complex diversity is a virtue of nature, a partnership with nature demands that we not systematically replace, on a global level, the complex diversity of a rainforest or plains with the single crop and paucity of organisms produced by large-scale agriculture" (Schlossberger 2001). Environmental virtues, thus, include partnership virtues, especially those pertaining to a partnership with nature, such as respect for and appreciation of natural diversity.

Other views range from the overlord view to the view that nature is sacrosanct (nothing in nature may be changed). One way to classify views on the environment is where they draw the line of moral significance, that is, what counts (has *moral considerability*). One way to think about this is to ask what it would be wrong for Routley's last human being to eliminate. For example, would it be wrong for the last person to eradicate all plant life if all sentient life were about to be wiped out? Would it be wrong to destroy the Grand Canyon or the Puerto Princesa Subterranean River in Palawan (Philippines) if all life on the planet were about to cease? *Anthropocentric* views draw the line after human beings: human beings are solely or primary the objects of moral concern. The *overlord view*, for example, holds that human beings alone are of moral significance and the only legitimate environmental ethical questions concern effects on human beings. John Passmore claimed that nature has only extrinsic, utilitarian value to people, and that proper aesthetic enjoyment of nature derives from seeing human improvements upon natural forms (such as the English Garden). Eugene Hargrove describes the early 18th-century view of nature as either picturesque (pretty) or sublime (vast). Hargrove says that this view gave way to nature as interesting, and the interest of nature centered on the properties of complexity, diversity, variety, individuality, and geological time. *Sentiocentric* views extend moral considerability to sentient organisms. Closely related is the view that all animals (including non-sentient animals, if there are any) are of moral significance. Some vegans hold this view. Yet broader are biocentric views, extending moral considerability to other life forms, such as plants and microbes. *Ecocentric* views place primary importance on ecosystems: entire ecosystems, more than individual living things, are of moral importance. Leopold (1949) famously said, "A thing is right when it tends to preserve the integrity, stability and beauty of the biotic community. It is wrong when it tends otherwise." Leopold was an avid hunter, since, for him, moral concern centered on ecosystems, not individuals. "*Deep Ecology*," a term coined by Arne Naess, sees value in all life and insists that a major restructuring of human activity is needed to minimize a human impact on the environment and promote biodiversity. Finally, a *geocentric view* regards all of nature, from human beings to geodes, as having an intrinsic value apart from all human perceptions, a value to which human values are, ultimately, subordinate.[13] One kind of geocentric view is the *caretaker view*, which suggests that the role of human beings is to leave nature as we found it. (Someone who hired a caretaker for their home would be shocked to return and find that the caretaker had remodeled their home.)

Some writers suggest that *trees and other natural objects have rights*.[14] I would argue, however, that natural objects do not have rights, because in order to have rights a thing must have interests, and to have interests a thing must have at least a primitive worldview, that is, standards or values in terms of which it measures some futures preferable to others. To have interests, a thing must have rudimentary goals and commitments. By contrast, whatever happens to the Grand Canyon is all the Grand

Canyon could ask for. Consider this: if there were never any sentient creatures in the world, if the universe were forever empty of thinking, feeling organisms, if there were never anything to see the Grand Canyon, then it wouldn't much matter whether there was a Grand Canyon in Colorado or a mudflat. So, I would argue, the Grand Canyon and the Puerto Princesa underground river do not have rights. Only people and, to a lesser extent, animals have rights. But with rights come obligations, and people do have obligations to the natural world. Human beings have a two-sided nature: we are animals that create. We reshape nature to our vision, but we are a part of that nature. We are thus true to our nature only if we are good partners with nature, working with nature in fulfilling our dreams.[15]

So, I suggest, we are neither lords of the planet nor caretakers of the planet, but partners.[16] A partnership involves mutual growth. This means respecting nature even as we change the face of the globe. If trees have rights, it is wrong to cut down any trees. If we have no obligation toward nature, we could cut down all the trees if we choose. A partnership with nature means that we may log selectively, preserving some virgin forests intact, clear-cutting some selected areas (though we've already done too much of that), and partially harvesting other regions in ways that minimize damage to the ecosystem. It is *not* ok, however, to cut down all the hardwoods and replace them with fast-growing pine, as some logging companies do, for this diminishes the diversity of nature.

There are no simple rules for establishing a partnership with nature. It is not even easy to be clear about what "nature" and "natural" mean. In one sense, since we are a part of nature, everything we do is "natural": ants and human beings are both animals, and so what we make is as much a part of nature as an ant-hill.[17] Bridge-builders and beavers obey the same laws of physics. In this sense, human processes are also natural processes. Perhaps some people mean by "natural" that which occurs without human intervention. But in this sense shaving is un-natural, since, without human intervention, many human chins and underarms would not be hairless. Three-year-old children are also "unnatural," in this sense, since without human intervention in the form of care and support, an infant would quickly perish. (Indeed, in some sense, human procreation itself is a form of human intervention.) And, in this sense, there is no such thing as "natural" bread, since, however free of additives and preservatives, bread has to be baked by people.

Still, we all understand that a parking lot is quite different from a "natural setting" such as a wildlife refuge. What, then, is "nature"?

It is helpful to realize that while shaving my chin does not seriously interfere with the growth and development of other plants and animals, paving the ground with asphalt prevents grass and other plants from growing, deer from grazing, and so forth. This thought suggests a characterization of partnership with nature. When human beings work with nature, the product of human ingenuity leaves room for other natural elements, such as trees, flowers, animals, ecosystems, and geological formations to realize their own special beauty and nature. A partnership with nature preserves the richness of the natural world. It is important that there be *some* wilderness unchanged by human effort, or a special and lovely feature of our world would be forever lost. We cannot and should not leave the whole world as virgin wilderness. But when designing a structure to be built on a site with interesting natural features, such as old trees, a waterfall, or an interestingly shaped hill, the design should work with and preserve those features. Frank Lloyd Wright's "Fallingwater" is perhaps the most famous example. But working with a landscape need not be that expensive, nor require the genius of a Frank Lloyd Wright. A shopping-center may "fit into" a hill, using the slopes as a design element. It may incorporate, build around rather than raze, an old tree in the middle of the property, thus preserving a natural element and providing a pleasant environment for the shopper.

Partnership with nature means several things. First, we should make sure that the special qualities of the natural world are not destroyed. We must not destroy all wilderness. We must try not to eradicate a species or destroy unique objects such as the Gran Salar de Uyuni in Bolivia or California's Sequoia forests. In short, we must preserve the diversity and wonder of the natural world. Second, when we build or remake the world, we should be sensitive to the beauty of nature. For example, bridge design should work with the site, not against it. It is perverse to bulldoze the area around a highway in construction and then replant shrubs. The highways on which people most enjoy driving are those that interact with their natural setting. Highways should convey a sense of place. They should not be anonymous pipelines for shipping human cargo. Finally, our remaking of the world should be no more intrusive than necessary. Examples include preserving the flora when installing oil wells, instead of clear-cutting the area to make access to and installation of the wells easier, building bird roosts over power lines, and using netting and/or sonic devices to keep sea life from becoming enmeshed in or killed by offshore facilities. Every day ships throw 450,000 plastic containers in the sea. Five species of sea turtles confuse floating plastic bags and wraps for jellyfish (their food), and die. It's easy enough not to dump plastic garbage in the sea.

While the idea of a partnership with nature is a powerful guiding idea, it is not a neat rule that will settle all disputes. Two key things are needed. First, we must develop legal remedies and guidelines. These will necessarily be somewhat *ad hoc*, but the guiding judicial ideal should be that of a partnership with nature. Legal remedies must be used carefully, since they sometimes have unexpected results. For example, one consequence of the Endangered Species Act, which protected alligators, was that alligators were no longer seen by some as being of commercial value. In some cases, instead of protecting alligators, this fact led land owners to drain alligator habitats for farming (ReVelle and ReVelle 1981). Second, engineers must show sensitivity in their decisions and approaches to the demands of partnership with nature, so that legal problems are less likely to arise. Toward this end, engineers should cultivate an appreciation of nature in all its splendor and diversity. Go camping or hiking, develop an enjoyment and appreciation of flowers and animals, wetlands and forests, the early morning mist fluttering between the sentinels of dark cypress.

In sum, what characterizes the compleat engineer is not just a commitment to particular gestures of this sort, but a frame of mind. The compleat engineer feels a sense of kinship with and appreciation of the complex and multifarious splendors of the world about her, from the dainty lilt of the smallest wildflower to the majesty of the Alps. She enjoys camping and hiking in the Black Forest, canoeing in the bayous, traversing the African savannahs, or sitting by the side of a pebbly brook. She views the earth and its other creatures as a partner, and is as considerate and protective of them as she would be of a friend. Friends are not revered from afar, nor are they kicked about heedlessly. Rather, they are interacted with in a loving, caring, and mutually beneficial way. Indeed, some of the greatest feats of engineering have their roots in a love of nature. (For example, a love of and study of birds was important to early aviationists.)

At the company level, partnership with nature means an increased concern with the environment. Monsanto Chairman Richard J. Mahoney, speaking at a National Wildlife Federation dinner, pointed out that "corporations like ours are experiencing what can only be called a revolution in environmental stewardship" (Siegal and Baden 1990, 45).[18] It is important to remember that a partnership is mutually beneficial. Thus, engineers should look for ways to make taking care of the environment profitable.

As George Pilko points out, "If a company can justify a wellness program because it reduces absenteeism and improves productivity, it can justify environmental expenditures on the same grounds. It not only is the moral and ethical thing to do, it makes sense on a dollars-and-cents basis" (Siegal and Baden 1990, 45–46).

Climate change, sustainability, and the environment are the subject of Chapter 10, which addresses these issues in more detail.

Case 12: Economical Cooling System

You have an idea for an economical cooling method that would solve a major problem for your company. Unfortunately, the method involves heat exchange that would significantly raise the temperature of a river beyond the level tolerable to an endangered species of fish. How do you handle the problem? Should you present your idea for consideration? If you do, does your responsibility end at that point, or is there more you should do?

NOTES

1. In the 1858 Lincoln-Douglas debates, each candidate spoke for 60 minutes. In the 2016 televised US presidential debates, each candidate had 2 minutes to answer a question.
2. Most economists follow the philosopher David Hume in thinking that only means (and not ends) can be evaluated. As a result, they have regarded practical wisdom as wisdom about means, that is, the ability to achieve any given end. Thus, economists have viewed technology simply as the capacity to make outcomes conform to one's will. If, however, as I would argue, ends also are subject to rational scrutiny, then technology applies to the choice of ends as well as the choice of means, and the means must reflect the values inherent in rational ends. Unfortunately, space does not permit me to argue for this important claim. However, because this point helps illustrate the nature of ethical reasoning, I will point out that one reason people have found Hume's view plausible is that they have taken too narrow a view of reason. It is true enough that deduction and induction alone will not serve to show that one end is better than another. But neither will deduction and induction alone serve to show that one scientific theory is better than another, as philosophers of science have pointed out. Rather, rational assessment must draw on a wide assortment of tools, none of which is conclusive, but which, together, give some reason for preferring one end to another. Those tools include do deduction and induction. But they also include what Chapter 2 called "picture building." For example, can the values inherent in a particular end be shown to be part of the structure of values upon which our whole way of life is built? Arguments of recognition also play an

important role in ethical assessment. One way to show the value of X is to describe a form of life based on X in such a way that the hearer recognizes value in that form of life (films and fiction excel at this sort of argument). One can ask whether a given end tends generally to facilitate or impede a good human life, where social convergence helps determine what is a good human life: Would enduring societies, interacting in the right sorts of ways, tend to acknowledge that a life of type Z is among the collection of good possible human lives? (See Chapter 6 as well as Schlossberger 2022.) There are more tools of rational assessment on heaven and earth than are dreamt of in Hume's philosophy.

3. This is a "value-based" reason, stemming from the definition of engineering as a profession. Martin and Schinzinger (1983) mention three "contractual" reasons for thinking that engineers have a special obligation to promote safety: by joining a professional society with a safety-based code, by accepting employment, or by entering a career underwritten by the public, engineers tacitly promise to strive for public safety. Michael Davis has long argued that engineers have a duty to follow their professional codes of ethics, which virtually all emphasize safety. (See, e.g., Davis 2010.)

4. Martin and Schinzinger (1983, 97) assert that "A thing is safe (to a certain degree) with respect to a given person or group at a given time if, were they fully aware of its risks and expressing their most settled values, they would judge those risks to be acceptable (to a certain degree)." This definition is not useful for two reasons. First, the qualifier "to a certain degree" takes the teeth out of the definition – everything, on this definition, would be safe *to a certain degree* (perhaps a very small one) if it yields any benefit whatsoever. Of course, the authors are really trying to define the *degree* of safety. But this leaves untouched the crucial question, namely, how much safety (so defined) the engineer should seek. More importantly, the definition relativizes safety to the values of a group. To a motorcycle gang, which, let us suppose, does not highly value what the rest of us call "safety," racing on a precipice might turn out to be "safe" on this definition. We need to add, in other words, that the groups' "settled values" are rational ones. But the problem of defining safety is precisely the problem of deciding how much risk it is rational to assume.

5. "Road traffic crashes are also the leading killer of children and young people worldwide, aged five to 29. As things stand, they are set to cause a further estimated 13 million deaths and 500 million injuries during the next decade" (UN 2021). Even if the automobile saves some lives a year (e.g., via ambulances), and even if the number of passengers who die in train accidents would increase were automobile travel banned, it is patent that many lives would be saved each year were we to give up the automobile.

6. Whether or not we should stop producing cars because of their environmental and social impacts is quite a different question.

7. Bazelon (1980, 40) says that evaluating acceptable risks involves "critical value choices [that] ultimately are reserved for the public." This may be accomplished through government and/or through citizen advisory boards (see Appendix IV).

8. Cf. Manheim (1983, 119): "There must be full opportunity for timely and constructive involvement of affected interests in the process, such that every interest–individual or group–[that] may potentially be affected by the changes being considered has full and timely access to all relevant information and has full opportunity to influence the process constructively."

9. Florman (1976, 32) points out that "failure [in engineering] results from lack of imagination," from carelessness and human error, and from ignorance.

10. This section contains information on various products and processes, on the effects and amounts of pollution, etc. Such information quickly becomes outdated. It is given primarily for the purpose of illustration.

11. Although the idea is not a new one, the invention of Gaia theory is generally credited to James E. Lovelock and Lynn Margulis. Some further books on Gaia theory include Thompson 1987, Margulis and Sagan 1986, and Lovelock 1995.

12. Which other virtues are environmentally central, Sandler points out, depends partly on where the line for moral considerability is drawn. Respect for persons, compassion, and ecological sensitivity will be central for, respectively, anthropocentric, sentiocentric, and ecocentric views.

13. The Transcendentalists took a transitional, if somewhat fuzzy, view: nature is a mirror image of the human soul, and the attitude toward which human beings should aspire is one of unity with nature, which represents human nature in its "pure" form.

14. Some writers who want the law to recognize the rights of natural objects view their suggestion as a "legal fiction"; treating trees *as if* they had rights is the best legal means of protecting the environment. Since this is really a legal point rather than an ethical one, I won't discuss this view.

15. I would add a somewhat more speculative note. Natural processes in a world without any intelligent beings, such as *homo sapiens*, are a meaningless chatter of subatomic events: one quantum state is as good as another. In this sense, nature has no value apart from human (or human-like) beings. However, human worldviews create a logical space within which mountains have meaning apart from our perception of them. For example, it requires a human worldview to establish the category of the powerful. Once established, however, the concept of power has application to the natural world whether we see it or not. There is something grand and powerful about Mt. Ranier rising above the mists, even if no one ever sees Mt. Ranier, and so Mt. Ranier's beauty and value do not depend upon our seeing it. Once the categories of meaning exist, the world resonates with meaning: the very stones do prate of value. But without human (or human-like) consciousness, there would be no Mt. Ranier. There would just be neutrinos and electrons hopping between quantum states. The relation between natural processes and human thought is also dynamic: our concepts derive from natural processes at the same time as nature acquires meaning by virtue of those concepts. Our concept of human power derives in part from our experience and understanding of Mt. Ranier, just as the thrill of seeing Mt. Ranier derives, in part, from a metaphorical view of the great rock as like, for example, the deaf Beethoven creating the 9th symphony. Moreover, our concepts are reshaped by paying closer attention to natural processes (as happens, e.g., in political philosophy, when the body politic is viewed as a kind of ecosystem). In short, the idea of partnership with nature governs even ethics and human consciousness.

16. By seeing human beings and the rest of nature as two partners, partnership with nature does give more weight to human needs. The needs and interests of the Fruit Dove count, but not equally with human needs and interests. Other things being equal, if you saw a bird and a person lying on the ground bleeding and you can save only one, you should save the person. This does not mean that every human interest counts more than any interest of a bird. A chimpanzee's interest in living outweighs a person's interest in a wall mounting. Saving an entire species may, in some cases, count more than saving an individual human life. Despite these caveats, some

would call what I am saying here is "speciesism," akin to sexism or racism, because I grant more status, other things being equal, to human lives than to the lives of other organisms. I would point out that taking even an herbal antibiotic or giving one to another human being does privilege human life over the lives of streptococci and herbs. That this counts as unfair discrimination is not a widely shared intuition. In any case, those who actually follow that view do not live long.

17. See Florman (1976, 67): "Does nature consist of farms, seashores, lakes and meadows, to use Reich's list?" Must we not also include, asks Florman, hostile environments such as outer space, ice fields, and deserts? "If farms and meadows are considered 'natural' even though they have been made by men...what is 'unnatural'?"

18. Without commenting myself, it is only fair to mention that some would raise an eyebrow at such a statement coming from Monsanto.

19. The article states, "The engineer and 15 passengers were killed," so the headline is incorrect and does not match the article.

REFERENCES

Bazelon, David (1980) "Risk and Democracy," *Professional Engineer* 50, 40–42.

Buchholz, Rogene A. (1989) *Fundamental Concepts and Problems in Business Ethics* (Prentice-Hall).

Cropley, David H. (2015) "Creativity in Engineering," in Giovanni Emanuele Corazza and Sergio Agnoli, eds., *Multidisciplinary Contributions to the Science of Creative Thinking. Creativity in the Twenty First Century* (Singapore: Springer), 155–173. DOI: 10.1007/978-981-287-618-8_10

Davis, Michael (2010) "Professional Responsibility: Just Following the Rules?" in Clancy Martin, Wayne Vaught and Robert C. Solomon, eds., *Ethics Across the Professions* (Oxford University Press), 12–19.

Florman, Samuel C. (1976) *The Existential Pleasures of Engineering* (St. Martin's).

Flynn, Ramsey and Kaye, Steven (1987) "On the Wrong Track," *Baltimore Magazine* November 1987 https://www.baltimoremagazine.com/section/community/special-report-on-the-crash-of-amtrak-colonial-94/

Griggs, Frank Jr. (2015) "The Quebec Bridge Part 1," *Structure* December 2015, 42–44 https://www.structuremag.org/?p=9300

Henig, Robin Marantz (1985) "The Dalkon Shield Tragedy," *Washington Post* November 17, 1985 https://www.washingtonpost.com/archive/entertainment/books/1985/11/17/the-dalkon-shield-disaster/6c58f354-fa50-46e5-877a-10d96e1de610/

Hill, Thomas (1983) "Ideals of Human Excellences and Preserving Natural Environments," *Environmental Ethics*, 5:3, 211–224.

Joseph, Lawrence E. (1990) *Gaia: The Growth of an Idea* (New York: St. Martin's Press).

Kolata, Gina (1987) "The Sad Legacy of the Dalkon Shield," *NY Times* December 6, 1987 https://www.nytimes.com/1987/12/06/magazine/the-sad-legacy-of-the-dalkon-shield.html

Kotlyankov, V. M. (1991) "The Aral Sea: A Critical Environmental Zone," *Environment* 33, 4–7.

Kumar, Vinod, Sharma, Anket, Kumar, Rakesh, Bhardwaj, Renu, Thukral, Ashwani Kumar and Rodrigo-Comino, Jesús (2020) "Assessment of Heavy-Metal Pollution in Three Different Indian Water Bodies by Combination of Multivariate Analysis and

Water Pollution Indices," *Human and Ecological Risk Assessment: An International Journal*, 26:1, 1–16.

Leopold, Aldo (1949) *A Sand County Almanac: And Sketches Here and There* (Oxford).

Lovelock, James (1995) *The Ages of Gaia: A Biography of Our Living Earth* (New York: W.W. Norton and Company).

Lovins, Amory (2011) "Soft energy Paths for the 21st Century," http://healthstudiescollegium.org/wp-content/uploads/2016/06/2011-09_GaikoSoftEnergyPaths_AL.pdf

Manheim, Marvin L. (1983), "Values and Professional Practice," in James Schaub and Karl Pavlovic, eds., *Engineering Professionalism and Ethics* (John Wiley and Sons).

Margulis, Lynn and Sagan, Dorion (1986) *Microcosmos: Four Billion Years of Microbial Evolution* (NY: Summit Books).

Marsh, John (2015) "Quebec Bridge Disaster," *The Canadian Encyclopedia* https://www.thecanadianencyclopedia.ca/en/article/quebec-bridge-disaster-feature

Martin, Mike and Schinzinger, Roland (1983) *Ethics in Engineering* (McGraw-Hill).

NAIIC (The Fukushima Nuclear Accident Independent Investigation Commission) (2012) "Official Report," *National Diet of Japan* https://www.nirs.org/wp-content/uploads/fukushima/naiic_report.pdf

Peeples, Lynne (2020) "Across the U.S. Millions of People Are Drinking Unsafe Water. How Can We Fix That?" PBS (Detroit Public TV) https://www.greatlakesnow.org/2020/09/drinking-unsafe-water-contaminants-solutions/

ReVelle, Penelope and ReVelle, Charles (1981) *The Environment* (Willard Grant).

Robinson, Lisa (2015) "Look Back: Trains Collide in Chase in 1987, Killing 15,"[19] *WBAL* https://www.wbaltv.com/article/look-back-trains-collide-in-chase-in-1987-killing-15/7093564#

Routley, Richard (1973) "Is There a Need for a New, an Environmental, Ethic?" *Proceedings of the XVth World Congress of Philosophy* 1, 205–210.

Sandler, Ronald L. (2013) "Environmental Virtue Ethics," in Hugh LaFollette, ed., *International Encyclopedia of Ethics* (Hoboken, NJ: Wiley-Blackwell), 1665–1674.

Schlossberger, Eugene (2001) "Environmental Ethics: An Aristotelian Approach," *Philosophy in the Contemporary World* 8:2, 15–26.

Schlossberger, Eugene (2022) *Moral Responsibility beyond Our Fingertips: Collective Responsibility, Leaders, and Attributionism* (Lexington).

Siegal, D. and Baden, J.A. (1990) "Business's Green Revolution," U.S. News and World Report (February 19, 1990), 45–48.

Thompson, William Irwin, ed. (1987) *Gaia: A Way of Knowing: Political Implications of the New Biology* (Great Barrington, MA: Lindisfarne Press).

UN (2021) "With 1.3 Million Annual Road Deaths, UN Wants to Halve Number by 2030," *United Nations News* https://news.un.org/en/story/2021/12/1107152

Warren, Karen J. (2015) "Feminist Environmental Philosophy," *Stanford Encyclopedia of Philosophy* https://plato.stanford.edu/archives/sum2015/entries/feminism-environmental/.

Westrum, Ron (1991) *Technologies and Society* (Belmont, CA: Wadsworth).

Whalen, James M. (2000) "A Bridge with Two Tragedies," *Legion Magazine* https://legion-magazine.com/en/2000/11/a-bridge-with-two-tragedies/

WHO/UNICEF (2021) "Progress on Household Drinking Water, Sanitation and Hygiene 2000-2020," *Joint Monitoring Programme* (Geneva: World Health Organization).

Young, John A. (1990) "Technology and Competitiveness: A Key to the Economic Future of the United States," in Albert H. Teich, ed., *Technology and the Future*, 5th ed. (St. Martin's Press).

5

Additional Ethical Sources (Part 1)

Some ethical decisions are difficult and gut-wrenching. Others slip by unnoticed; an engineer may not realize, until it is too late, that something she has done is unethical. To make difficult decisions and avoid overlooking ethical problems, you need a large set of rules, principles, values, and guiding ideas. The more you have, the better decisions you will make. You are less likely to overlook an ethical problem if you ask yourself, before taking any important step, "Must I fight this battle? Does doing this violate the duty to leave the world no worse? Does it treat others fairly and well? Does it respect rights? Does it violate the principle of universality? Am I promoting good consequences and showing respect for persons?" You are better able to balance competing moral demands (such as a conflict between honesty and loyalty to your company) if you ask yourself "are any institutional duties involved? What do the principles of accountability require of me? Should I break the rules in this case? How do the Golden Rule and the value of autonomy apply?" The next two chapters arm you with 20 additional ethical sources you can use in making sound ethical decisions.

WHEN TO FIGHT A BATTLE

Morality is everyone's concern and the engineer has a real stake in ensuring that their colleagues and company are ethical. The principles of accountability, the importance of a community atmosphere, and the ethical engineer's commitment to values such as fairness and safety together mean that no unethical behavior in the workplace is "not my business."

DOI: 10.1201/9781003242574-7

However, there are too many worthwhile battles for any person to fight all of them. For example, the ethical person is concerned about many issues, including combating poverty and hunger; improving education; research and treatment for cancer, infectious disease, heart disease, and birth defects; providing basic legal and medical care for all; and problems faced by refugees, those with disabilities, and victims of discrimination and political repression. None of these problems is "not my business." However, no individual can contribute time and/or money to *all* of these causes. Rather, the ethical person chooses to make a contribution to some of these causes, leaving the others for their fellow citizens. In the same way, the individual engineer must leave some moral battles to others.

There is no easy way to decide which moral battles one should fight; yet, this is a decision we must make every time we encounter wrongdoing. So, it is very useful to identify some *factors* that help determine which moral battles one should fight.

There are some battles one is *obligated* to fight. One may be required to fight a battle:

1. *as a matter of principle*: the evil is so heinous that to remain silent is to ignore one's deepest moral commitments,
2. *as a matter of responsibility*: one is directly responsible for righting the wrong,
3. *as a practical matter*: the harm to oneself or one's goals is much greater than the harm one will incur by fighting, or
4. *because there is no neutral option*.

Example 1

You have documented evidence that your company lied to a regulatory agency, suppressing facts proving that a current project has major design flaws and is being built with substandard materials, thus seriously threatening hundreds of lives. This is a battle you must fight because the evil involved is so great that you cannot in good conscience permit it to go unchallenged. You must, at the very least, speak to your supervisor about your concerns. (See also "Whistleblowing" in Chapter 7.)

Example 2

You are in charge of safety tests for a project, and one of your subordinates is ordered by upper management to falsify test results. This is a battle you must fight because you are directly responsible for overseeing the integrity of the tests.

Example 3

In violation of OSIA regulations, the plant in which you work registers dangerous levels of PCBs. This is a battle you must fight because the PCB level poses a severe danger to you.

Example 4

You are subpoenaed as a witness by a court and asked a question whose answer is harmful to your company. This is a battle you must fight because there is no way to avoid the problem. When asked a competent question as a witness in a court of law, you have no choice but to tell the truth, commit perjury, or refuse to answer and risk contempt of court. It is just impossible to "stay out of it." (It is normally obvious that telling the truth is the correct choice in this situation, though journalists and priests have, as a matter of principle, chosen jail over compromising a source or the sacrament of confession.)

Many of the morally troubling situations engineers face are not like this. In many cases, the ethical engineer has some choice about which moral battles to fight. There is a set of battles, in other words, such that one must fight some but need not fight any given one. Although there is no formula for deciding which of them to fight, there are considerations that make some decisions reasonable and others not.

When deciding which of these battles to fight, ask yourself three crucial questions. First, *how bad is the infraction*? What are the expected consequences to the persons who are hurt by the infraction? (It is more important to fight to save a life than to avoid the imposition of an unfair 50 cent fine.) What is the moral character of the infraction? (For example, suppose the Dean of the School of Engineering tells faculty who deserve to be tenured "pay me ten Euros or I will see that you don't get tenure." Here, the actual consequence is that the faculty member has to pay ten Euros – not very significant. But it is *morally* heinous that the Dean uses her power this way.) Second, *what is the expected cost to me and to other non-guilty parties*? If the infraction is not severe, either morally or consequentially, but the cost to me of fighting would be disastrous, that is a weighty reason for choosing not to fight this battle. Third, *to what extent am I implicated*? I am more responsible for an infraction if I am in a special position to know about or correct it, if I have close ties to the infracting organization (I'm more responsible for what my department does than for what the advertising department of another company does), if I played a role in the infraction itself (e.g., I have to make out the pink slip), or if overlooking the infraction compromises doing my job faithfully.

TREATING OTHERS FAIRLY AND WELL

There are several key elements in treating others in the work environment well and fairly.

Fairness in the workplace requires people to *be straightforward* with each other. It is better to *be direct and honest* about a problem, even if that causes some tension. It is also important to *be impartial.* Whenever possible, *give other people a chance to present their side* of things. This is important not only because it shows basic respect for others but also because, no matter how clear things seem to you, you may have a misleading picture of things. Suppose I see Carla running out of Mel's house with a TV in her arms. Mel yells "Stop!" but she continues running. As she passes, I happen to notice the serial number of the TV. Mel tells me that Carla stole the TV and shows me the original store invoice for the TV. The serial number matches that of the TV Carla was carrying. Things look pretty bad for Carla. But all she has to do to clear herself completely is show me the signed receipt Mel gave her when she bought the TV from him. Perhaps she was running and did not stop because she was in a rush. Perhaps Mel lied to me because he does not want me to be friends with Carla or to gain attention and sympathy. The point is, no matter how bad it looks, I can't be sure I understand the whole picture unless I speak to Carla. I should (if feasible) take no action until I've given Carla a chance to present her side. Moreover, the straightforward approach often resolves problems or misunderstandings that would have remained unresolved had you been less direct. Give yourselves a chance to work things out. In any case, being straightforward with someone shows respect for him as a person. (Remember the point of the Golden Rule: would you prefer to be manipulated or to be treated straightforwardly?) Straightforwardness may cause some discomfort to you or to the other person, but, when you deal with each other honestly, both of you are acting as responsible adults, facing the situation squarely and honestly.

With respect to giving others a chance to present their side, it is worth noting that, in writing this book, when feasible, I sent the relevant case or text to the companies involved (many are now out of business), asking for any clarifications, additions, or corrections they would make to what I wrote. (For example, in cases 33 and 74, I incorporated what GM and Union Carbide, respectively, replied.)

Don't jump to conclusions. Don't let your personal feelings for or against someone cloud your judgment. Instead, *judge on the basis of facts and sound evidence*, not superficial appearances. *Observe the principle of just deserts*: people should be given what they deserve, and the way you treat people must be based on a fair assessment of the merits of case, that is, of the relevant features.

Don't exploit or manipulate others. You do, of course, want to motivate and get good work out of your subordinates. Several things distinguish manipulation and exploitation from honest persuasion and motivation. One is deceit: you want to motivate your subordinates by being honest and straightforward. Another is advantage taking. Don't make demands no reasonable person would agree to if she had any choice at all. While it is hard to give a strict definition of exploitation and manipulation,[1] it is often fairly easy to tell when you're exploiting someone or being manipulative. In general, you want your relationships with others to be *mutually beneficial*.

Help each person to make the most of herself. Helping each person make the most of herself is not only the ethical thing to do, but is good leadership as well. You want your department to be a good team, in which everyone participates according to her unique talents and abilities. This requires that everyone has a stake in what is going on, that everyone's ideas are taken seriously, that people are treated as valuable individuals, not as replaceable and uniform machines, and that everyone has a chance to grow toward becoming the best she can be. In order to help each person to make the most of herself, you need to do five things: *allow scope for individual differences; maximize opportunity for growth and taking responsibility; encourage feedback and thoughtful participation; reward and encourage effort, achievement and commitment; and be patient but firm*.

Finally, you want to *build a community atmosphere*, and you want to *respect the rights of others*, such as the right to privacy.

Many of these ideas are developed further in Chapter 9, "Employee-Employer Relations."

Illustrating Cases

Case 13: The Tardy Employee Chapter

Thorp, a reliable employee for several years, has been late rather frequently in the last month.

Discussion: Being straightforward, not jumping to conclusions, and giving people a chance to present their side require that you speak to

Thorp about this before coming to any conclusions. Building a community atmosphere means that the conversation should be supportive rather than threatening – point out the lateness to Thorp, and ask Thorp if he has a problem you might be able to help with. Invite Thorp to discuss the matter with you, not defensively, but as a colleague on a team. Now suppose, as a result of your discussion, you find that Thorp has been oversleeping because he has had to stay up with his ailing mother every night. Allowing scope for individual differences means that if some adjustment is feasible, it should be suggested. Perhaps Thorp could come in and leave an hour later while his mother is ill. This may not be feasible: if Thorp's work requires him to interact with other engineers, it does no good to have him stay an hour after everyone else has left. If being an hour late now and then for a month or two does not pose a significant problem for the company, Thorp's lateness might be overlooked, provided it is clear that this situation is *temporary*. However, if Thorp's group is working feverishly toward meeting a contractual deadline, Thorp's lateness may mean losing the contract. So, although you should try to make some adjustment, it may not be feasible. If it is not feasible and if Thorp's lateness does cause problems, then the problems caused by Thorp's lateness should be made clear to him, in a sympathetic rather than a hostile way. Building a community atmosphere suggests that personal and corporate gestures of concern for Thorp's mother are appropriate. Finally, being firm but patient suggests that, given the circumstances, you should expect Thorp to make a serious effort, and cut down significantly the number of days in which he shows up late but you should also realize that he may need some time to adjust: expect a serious effort from Thorp, rather than perfection. If in the week following your discussion Thorp is late one day rather than three or four, tell him you are pleased at the progress he is making, and that you have confidence that next week he won't be late at all.

Case 14: Informing Employees of Outside Opportunities

A subordinate of yours, Cleaver, has a good job at which they do well, but there is nowhere for them to advance at your firm, at least for some time. You become aware that firm Y is seeking someone with Cleaver's credentials for a more responsible position than your firm can offer Cleaver. Should you inform Cleaver of this opportunity? Does it matter if your awareness of the opportunity at Y is a result of your position? (For example, Y writes a letter to "Head of N Department" describing the opportunity and asking for a recommendation.)

Discussion: Building a community atmosphere and maximizing opportunity for growth suggest you ought to feel some concern that Cleaver is unable to advance and grow properly within your company. This suggests that you should inform Cleaver of the opportunity, at the same time telling Cleaver that you are happy with their work and hope they will stay. If the information comes to you as a result of your position, the case for informing Cleaver becomes even stronger: straightforwardness and not manipulating suggest that you should not use your position to keep such information from an employee. However, all these considerations must be balanced against your commitment to excellence and the company. How much of an opportunity is this for Cleaver? How hard are they to replace, and how much of a problem would their leaving cause? Is Y a competitor? (No company should be expected to undermine itself by going out of its way to provide its competition with top people.)

Some employers will ask "why should I go out of my way to lose a valuable employee?" The answer is that the long-term benefits of a community atmosphere are worth occasional minor sacrifices. No employer wants its employees to be willing to leave in the middle of an important project if a better-paying job becomes available with another company. Every employer wants its employees to be willing to put in extra hours when needed, to go the extra mile for the company, etc. But an employer cannot expect employees to put loyalty to the company ahead of short-term personal advantage unless the company is willing to show the same loyalty to employees. A company that expects loyalty from employees, but is unwilling to make small sacrifices for the sake of employees, is violating the principle of universality (discussed below). It is also kidding itself.

Case 15: Promoting "My Kind of Person"

You must decide whether to promote Perez or Yi. Perez would do the job slightly better but Yi is more "your kind of person."

Discussion: Promoting someone because you like them violates the principle of just deserts: they have not earned the promotion, since your liking them is not a relevant reason for promotion. However, Yi's personal characteristics could be a job-related factor: if you think Yi would be more supportive of their subordinates, could get more out of them, etc., this is a relevant reason for promoting Yi over Perez. However, it is important to beware of *perceptual bias*. People who would never knowingly discriminate sometimes do so because their tendency to perceive is biased. Moss-Racusin et al. (2012) found that when identical resumes for a lab manager were submitted under male and female names, both male and female science faculty from research universities rated the male-named applicant as more competent and more hireable. Similar experiments have shown that study participants who were given a written description of a superior interacting with a subordinate are more likely to assess the superior as "aggressive" when given a female name in the description. It is important to step back from our perceptions and examine whether they are the result of unconscious bias. (See also "Equality/Equity, Diversity, and Inclusion (EDI)" and "Hiring Practices" in Chapter 9.)

THE DUTY TO LEAVE THE WORLD NO WORSE

Professionals are also persons. Whatever is required of professionals by law and by company policy, they also have personal moral responsibilities that may not be overlooked. One personal responsibility is of special importance in professional life: we are each personally responsible for doing our best to insure that, whatever we do, we do not leave the world worse off because of our activities.

What we do affects others, for better or worse. It is a worthy goal to make the world better than we found it. But even if we don't improve the world in some way, we have a duty not to make the world worse than it would have been without us.[2] Just as campers must leave the forest in the same condition they found it and must not leave their litter scattered about, so we have a general duty not to "trash up" the world as we find it. Obvious as this might seem, it is often overlooked or ignored by professionals. Advocates of shareholder theory think that the duty of an engineer or corporate officer is to maximize profits for the company and its shareholders, and not to look out for the rest of the world. (See "Engineering and Business" in Chapter 3.) Attorneys sometimes think that their only duty is to the welfare of their clients, and politicians sometimes think that any kind of immorality is fine as long as it benefits their nation. I am suggesting that making the world a significantly worse place in order to maximize profits for the shareholder, benefit her nation's foreign policy, or help out a client goes against one's responsibilities as a human being, and so, in many cases, is the wrong thing to do. (As always, relevant factors must be weighed in each particular case.)

Why do we have such a duty? Why not "trash up" the world if you can profit from it? There are (at least) two answers to this question. The first answer is that the duty to leave the world no worse is at the center of living well, of leading a good life. It is a key to being a happy and fulfilled human being. As we saw in Chapter 1, the consumer life is dedicated to the pursuit of feelings that are essentially private and essentially fleeting, and so one who leads the consumer life is condemned to loneliness and frustration. (See "A Revealing Case" in Chapter 1.) In contrast, those who have values, who care about things like justice and excellence for their own sake, have a stake in other people. Since progress toward a value endures, someone with values can look back at his life and take satisfaction in what he has achieved. So, unless I want to doom myself to the unhappiness and isolation of the consumer life, I must see myself as owing loyalty to values. And leading a life committed to values means trying to leave the world no worse. For example, if I value knowledge, then I think the world would be a better place without ignorance, and I am committed to fighting ignorance. Thus, if I do not want to live a life that is trivial, isolated, and unhappy (the consumer life), I must be committed to making the world a better place. At the very least, I must not be a traitor to what I value, whether that be knowledge or excellence or human welfare. And so I have a personal responsibility to do my best not to make the world a worse place than I found it. In short, the only life worth living is the life committed to values, and you betray that commitment if you don't strive to leave the world no worse than one found it.

The second reason for thinking we have a duty to leave the world no worse is an extension of dual-investment theory discussed in Chapter 3. Every engineer, every executive, every physician, every lawyer, and every other professional makes extensive use of their society's educational system. Everyone, not just businesses, draws heavily upon the opportunity capital provided by society through generations of hard work. No one is entirely "self-made." In an important sense, then, society is a stockholder in each of our lives, putting up the opportunity capital we need to function in the modern world. Thus, we each owe society the loyalty that executives of a corporation owe its stockholders, and have a duty to do our best not to put society in the red by our operations. Thus, we have a duty to do our best not to make the world worse for our existence

Given that we have a duty to leave the world no worse than we found it, how does this duty apply to engineering and business decisions? Let's begin by looking at some cases and examples.

Case 16: The Ford Pinto Case

The Ford Pinto Case. Engineers designing the Ford Pinto were told that the car must weigh less than 2,000 pounds and cost less than $2,000. Although the Pinto did not violate laws or regulations of the time (1971–1976), the gas tank of the Pinto was subject to rupture during rear-end collisions of 20 mph (Baura 2006). The Ford Motor Company had to decide whether to employ an improvement, costing $11 per vehicle that would cause gas tanks to rupture less easily. (Other alternatives were also available.) Ford did not do so. According to Mark Dowie (1977), by 1977 the Pinto had been responsible for 500–900 burn deaths of passengers who, had the Pinto not erupted in flames, would have sustained no serious injury. Ford disputes these figures. In any case, an internal report released by Ford's automotive safety director, Echold (1972), entitled "Fatalities Associated with Crash-Induced Fuel Leakage and Fires," estimated that 180 burn deaths and 180 serious burn injuries could be prevented by making the $11 change. Multiplied by 12.5 million vehicles, this comes to a cost of 137 million. Since the report estimated the cost of each death at $200,000, each serious burn at $67,000, and each of the anticipated 21,000 burned vehicles at $700, for a total of 49.5 million, the report concluded that the $11 change was not cost-effective. (Note: Ford eventually paid out $50 million in settlements and lost an untold amount in sales. For more about this case, see Schlossberger 2013.)

Discussion: There is considerable debate about what a company owes its customers but it is clear that the benefit to each purchaser of saving $11 is minimal, while the world is made worse by the pain and suffering of 360 deaths and serious injuries (using Ford's figure). Those who made the decision not to make the $11 improvement violated their personal responsibility not to make the world worse. The responsibility not to make the world worse also affects other Ford employees who were consulted about making the improvement. Everyone involved in the production of Pintos contributed to making the world worse. Those who knew about the problem, but worked on the Pinto anyway, violated their personal responsibility. Of course, each of us has many responsibilities that sometimes conflict. A plant worker also has responsibilities toward their family, and I am not insisting that workers with no say in the decision were morally obligated to quit their jobs. I am not even saying that every Ford employee who knew about the problem but did not speak out acted immorally, since we need not fight every moral battle. (See "When to Fight a Battle" above.) But, since those who were aware of the problem did have a personal responsibility not to engage in the making and selling of Pintos, each of them had to weigh this responsibility carefully against their other responsibilities, values, and duties. They each had a difficult moral choice to make.

However, there is always a trade-off between safety and cost. For this reason, there is sometimes a conflict between safety and consumer autonomy. It is not always obvious where to draw the line between offering customers a cheaper, less safe option and refusing to design or approve a product or process as not safe enough. Ford could have revealed the problem to consumers and made the $11 fix an option, arguing that this solution respects consumer autonomy.[3] After all, many automobile manufacturers today offer expensive safety features as an option. Would an ethical engineer be willing to send to market the Pinto without the $11 (or alternative) fix? (See Schlossberger 2013 for more about this issue.)

Example 5

Engineering and business decisions often effect the environment. Discussions of the effects of business operations on the environment tend to center on the rights individuals may have to a liveable environment, on the economics of externalities, on the social justice of apportioning negative effects, and on the intrinsic value of existing ecosystems. My point is that however these issues are decided, the individual engineer or executive has a duty to ensure that the operations she oversees do not make the world worse, and so she has a personal responsibility to mitigate or eliminate environmental harm caused by corporate operations. Just as in Case 16 (The Ford Pinto Case), this responsibility applies to everyone involved in the process that harms the environment, from the CEO to assembly-line workers, and must be weighed against other relevant ethical factors.

Example 6

There is some controversy about whether a television executive who fires an anchorperson because she is "too old" violates moral or, depending on the jurisdiction, legal rights. But he is certainly making the world worse by supporting a system based on ageism and/or sexism, and has a personal obligation, if not to take a stand against, at least not to help foster a system (of public perceptions and corporate behavior supporting those perceptions) in which age is seen as an unattractive liability. The same point applies to hiring decisions about receptionists, managers, and department heads.

Example 7

Sometimes, it requires some thought to see that one's decisions leave the world worse off than it was before. P.V. Pumphrey, for example, points out that banking decisions can affect communities in six important ways:

> Commercial loan policies can 'favor' certain industries, thus affecting competition and growth rates; mortgage and home improvement loans can be made to or withheld from specific residential areas, thus affecting stability and values in the area; new construction can be encouraged or discouraged by availability of financing, with long-term impact on an area's tax base; municipal financial stability can be encouraged or jeopardized by [a] bank's willingness or unwillingness to purchase notes and bonds; personal loan policies can favor or exclude certain population groups, promoting or restricting social mobility; and international loans can support oppressive foreign regimes; they can support the internal controls exercised over populations.
>
> *Pumphrey (1984, 208)*

Similarly, television advertisements as a group heavily influence the values of our society. A series of television advertisements for Time-Life books, which showed families and friends discussing books, exerts a positive influence, while advertisements that promote less healthy values exert negative influences. (See "Advertising" in Chapter 14.) Each of us has a responsibility to examine the less obvious consequences of our work decisions.

One final case helps answer an important objection. Some people have argued that, as the Shareholder Model discussed in Chapter 3 suggests, one's personal moral responsibilities stop when one enters the plant or office. The job of the executive or engineer, they suggest, is to make money for the company, and that job must take precedence over personal moral qualms. If they are right, then you should ignore the duty to leave the world no worse when making business decisions and engineering decisions that affect profitability. The next case illustrates the conflict between their view and what I am suggesting.

Case 17: He-Man Cigarettes

Jones is the president of a tobacco company. Smith, a bright young member of the advertising department, suggests to Jones that the company could raise its current return on investment of 12% to 13% by marketing "He-Man" cigarettes, which are extra high in tar and nicotine. Although at present no one would purchase such a product, Smith's research shows that a demand for He-Man cigarettes could be created by an ad campaign that plays on buyers' fears about their masculinity, using slogans such as "only wimps smoke low-tar cigarettes" and "real men smoke He-Man cigarettes." Jones has no reason to think that if they do not authorize marketing the product, some other tobacco company will market a similar product. Jones must decide whether to authorize the product and ad campaign.

Discussion: This case presents a clear conflict between maximizing profits and not making the world a worse place. If Jones does not authorize production, the company will (probably) earn less than it otherwise would. If Jones does authorize production, with the relevant advertising, they must admit, if they are honest, that the world will be worse as a result of their decisions: they are quite deliberately setting out to induce others to do things that would make the world a worse place, a place of greater misery and hardship.[4] Now, you might be tempted to say "no one forces the buyer to purchase He-Man cigarettes. It is his health and his money, and he must be the one to make the decision whether to smoke He-Man cigarettes or not. It is the buyer, not the manufacturer, who must take the responsibility." This may or may not be true: perhaps I do not owe it to you not to help you hurt yourself, and perhaps I do not even owe it to you not to try and induce you to hurt yourself (though some of the considerations in this book suggest otherwise).

But there is another issue here. By marketing He-Man cigarettes, the businessperson fails to meet one of their personal obligations as an individual, namely their duty to try not to make the world a worse place. This is not a duty they owe just to the particular consumer who would be hurt by smoking He-Man cigarettes. Rather, not trashing up the world is an entrance requirement to the moral community. If one wants to live anything more than the consumer life, one must take personal responsibility to see that one's life not be a blight. The executive who authorizes the marketing of He-Man cigarettes shirks this responsibility. You may bear the responsibility for destroying your own life with heroin, and it may be, in some sense, your choice. But my commitment to what I value requires me not to try to induce you to use the heroin (and, normally, frowns upon assisting you by providing the heroin). The executive may not adopt the morality of the drug pusher, washing their hands of responsibility for providing others with what they want. Even if drug pushers don't violate the rights of addicts, what they are doing is wrong. It is sometimes said that an executive, as an agent of the corporation's owners, has no right to substitute their own moral judgment for the corporation's explicit purpose, to make money. This is so, however, only if it is morally permissible to offer and accept employment that requires violating moral duties. But there are real limits on what one may licitly accept money to do. After all, it is no excuse for paid assassin that they are just doing what they are paid to do. It is a defensible principle of contract law that contracts calling for immorality are not enforceable since they are contrary to public policy, and it is a defensible principle of ethics that one generally ought not to keep a promise to do wrong. Thus, a contract asking engineers or executives to act immorally by violating their personal moral responsibilities as human beings has no moral standing.

WORLD RELIGIOUS TRADITIONS

In making ethical decisions, people have drawn upon their religious traditions in two different ways. They may make decisions on the basis of *faith in specific elements of religious doctrine*. For example, someone might refuse to do something because a religious leader says that it is wrong or because they interpret a sacred text in a certain way. Someone whose Church leaders oppose a new reproductive technology might, for that reason, make a *personal decision* not to work on that technology. Second, they may draw upon a *broad religious perspective and framework of values* that speaks to (can be explained in terms that make sense to) others of different religious beliefs (including atheists) when making *engineering decisions*. It is this second way religious traditions may inform ethical decision-making that this chapter addresses.

There are at least three reasons why, whatever personal force specific religious doctrines may have for an individual and may influence people in making personal decisions, they do not belong in a book on engineering ethics, while the broad values and perspectives of religious traditions can provide useful insights into engineering decision-making.

1. Engineering is collaborative in nature and engineering decisions frequently affect populations outside the engineering firm or group.

A safety decision in aircraft design can end or save the lives of passengers in remote nations. An engineering choice for a nuclear reactor can result in cancers a generation removed and half a world away from the engineers making the decision. Thus, it is important that engineering ethics be a public discussion, in the sense that ethically sound choices should be defensible to a wide range of people who might be affected. The sister of a passenger on a plane you helped design may not agree with your decisions but you should be able to give reasons supporting them that are logically defensible in the public arena. To non-Moslems/non-Christians, "Because the Qur'an/New Testament says so" is not a justification. "Because that is how we do things in Kenya" is not a good reason to someone outside of Kenya. Thus, engineering decisions differ from personal decisions, where individuals are free to draw upon specific religious tenets.

2. Engineering is, by nature, committed to clean, clear, rigorous decisions, made on the basis of evidence and justifiable interpersonally. No engineer would say "I have no need of structural and material science. I do not need to calculate. I have faith the building will stand. I feel it in my heart, and that is enough." Such thinking is contrary to the engineering way – it goes against the nature of the engineering profession. In the same way, engineering ethics should be based not just on faith and on what an individual engineer "knows in his heart," but on sound reasons that make sense to critically thinking individuals from a wide variety of backgrounds.

3. Reason and logic are interpersonal and provide a check on inclination. Faith alone does not. I may be inclined to think unethical anything personally repulsive to me and I may be inclined to think permissible anything I want to do. What our "gut" or "heart" tells us is influenced by a wide variety of extraneous things. Having to give reasons that make sense to others with different sensibilities and different faiths offers a genuine check on extraneous influences. Faith does not; there is no check on what faith tells you. People have different faiths and, unless they appeal to nondenominational reason and logic, all they can do when their faiths diverge is refuse to talk or resort to force. When Mark Hambrick killed his daughter in 2018 because he believed God told him to do it, he acted on faith. When, during the First Crusades, the followers of Peter the Hermit and others slaughtered countless Jews during the Rhineland Massacres,

they acted on faith. The torturers of the Spanish Inquisition, Shah Ismail I, and numerous Sikhs and Moslems after the British left India in 1947 all killed in the name of faith.

Of course, religious faith has also led individuals to perform acts of great moral heroism. The question here is not whether faith, in general, is a good or a bad thing. The point is just that individuals entering a discussion of engineering ethics should participate using general, nondenominational reasons that would appeal to others of different faiths, reasons that, it is hoped, might justify or support their own faith-based beliefs. There are reasons to believe that it is wrong, other things being equal, to kill other people, quite apart from the religious authority of the Ten Commandments, Atharvaveda 10.1.29 ("It is definitely a great sin to kill innocents"), and Buddhism's Five Precepts. Engineering ethics discussions should avoid invoking specific religious doctrines and faiths.

However, religious perspectives and values that also speak to people of diverse religious and cultural backgrounds can have a constructive place in engineering ethics, precisely because they can be interpersonally evaluated without depending on a particular faith that others may not share. We can explain why knowledge should be valued, as the Jewish religious tradition counsels, to those who do not believe that the Torah is the word of God. We can explain why long-established cultural traditions deserve respect, as Confucius teaches, to those who do not acknowledge the Tianming (Mandate of Heaven). Confucius' reminder of the value of resilience and perseverance, that the green reed that bends in the wind is stronger than the oak, speaks to people across the globe. (Indeed, a version of the same adage is a Sukuma proverb.) We can explain the value of forgiveness and charity to those who doubt Jesus existed. Shinto's affirmation of the love of nature can be embraced by atheists and theists alike. Thus, engineering ethics may draw upon broad religious perspectives that may also speak to atheists and those of different faiths.

Religions are complex and sophisticated systems of belief. Most of the world's major religions are further divided into branches or sects that disagree upon significant points of doctrine or interpretation. The discussion in this chapter merely highlights some general ideas from diverse religious traditions that may resonate with others outside those traditions. Engineers are encouraged to read more about these traditions to gain a more detailed and nuanced picture of them than these brief, general remarks could hope to provide.

Judaism traditionally values learning and education, and that respect for knowledge may inform an engineer's understanding of ethical issues. The original basis of that value might have been the religious duty to study the Torah (the five books of Moses) and, later, the Tanach (Rabbinic teachings), but the broad value of knowledge and learning that is an important part of the Jewish tradition transcends particular religious beliefs and speaks to many who are not Jewish. As in many religions, the Golden Rule is a central idea in Judaism, encapsulating many Jewish values, including charity (Tzedaka), compassion, and kindness. (See "The Golden Rule" below.) Lawfulness and respect for law is another central concept in Judaism that non-Jews can apply to engineering.

Christianity shows its Jewish roots by affirming the Golden Rule and emphasizing love and forgiveness and abjuring hatred. The Ten Commandments, common to Judaism and Christianity, emphasize, among other things, respect for life, cherishing family, speaking truth, avoiding envy, theft, fraud, and fidelity. All of these ideas can be embraced by engineers who do not accept the divinity of Jesus and other articles of Christian faith.

Much of what was said about Judaism and Christianity applies as well to Islam, chronologically the third of the three major Abrahamic religions. The Kitab al-Kafi has the Prophet commanding the Golden Rule. Giving to charity is one of the Five Pillars of Islam. Key elements of Islamic ethics that speak to those of different faiths include *khayr* (goodness), *haqq* (truth), *birr* (righteousness and good character), and two terms indicating aspects of justice, fairness, and balance, *qist* and *adl*. Mohammed (2013) claims that "According to Islam, whatever leads to welfare of the individual or society is morally good and whatever is injurious is morally bad." The Qur'an 4:36 tells us to treat well and show compassion and kindness to our family, neighbors near and far, strangers, orphans, those whom we encounter in our travels, and those within our power.[5]

Similarly, in Hinduism and Buddhism, Ahimsa (non-violence toward all living things) might have its origins in the doctrine of re-incarnation, but it, too, speaks to many who do not believe in re-incarnation. Engineers' perspective on ethical issues may be, in part, shaped by respect for life in all its forms: it might, for instance, help an engineer decide to forego work on pesticides in favor of non-lethal methods of crop protection, such as repellants. Among the cross-culturally recognized virtues articulated in the Bhagavad Gita (Mahabharata Book six, Bishma-Parva 40) are perseverance, charity, compassion, forgiveness, and freedom from anger, truculence, envy, and vanity. Buddhism's Five Precepts enjoin refraining from (1) killing humans and animals; (2) stealing (theft, fraud, forgery); (3) sexual misconduct (adultery, promiscuity); (4) falsehood (lying, gossip); and (5) intoxication (alcohol, drugs), an extension of which might include infatuation with a person, organization, or idea. These Five Precepts underlie the Kammapatha, that is, the ten courses of action or wholesome ways of being. The ten good courses of action in Buddhism consist in refraining from the ten bad courses of action, namely: (1) taking life, (2) stealing, (3) sexual conduct causing unhappiness for others, (4) lying and deception, (5) slandering and promoting conflict and disharmony, (6) harsh speech, (7) empty speech (gossip, idle chatter), (8) greed and envy, (9) wishing others harm (malice), and (10) wrong view (misconceptions and mistaken beliefs).

The southern African ethical concept of *ubuntu* is a complex notion that emerged from religion but is now frequently viewed as religion-independent. Ubuntu means "humanness," in the sense of human excellence, somewhat akin to "being a Mensch" in Yiddish: ubuntu encapsulates the ideal of self-realization as a form of human excellence expressed in good social relationships. "Our deepest moral obligation is to become more fully human. And this means entering more and more deeply into community with others" (Shutte 2001). Like Buddhist teachings, ubuntu emphasizes the connectedness of all people and all things.

People from varied cultures can feel the moral force of the Chinese notion of *Junzi*, a term originally meaning the son of a prince but which, in the *Analects*, suggests a person of ethical nobility or an exemplary person. Linked to Junzi are the five constant virtues (wu chang): *ren* (caring for others, benevolence); *yi* (righteousness); *li* (propriety), which includes not only rituals but standards of courteous and respectful treatment; *zhi* (wisdom, knowing the right thing to do); and *xin* (fidelity, trustworthiness, honesty). These virtues help one follow the *Dao*, the Way or Path. For Laozi, following the Dao is "living in a simple and honest manner, being true to oneself, and avoiding the distractions of ordinary living" (Singer 2021). Xunzi understands morality "as a way of harmonizing the desires of individuals so that destructive conflict is replaced by productive harmony" (Wong 2021). The five virtues all apply to engineering. For example, subordinating speed and cost of production to safety is an example of ren. Accurately reporting adverse safety test results is an example of xin.

Mino Pimatisiwin, which means "the good life" in Cree, includes a link to the land and to nature. According to Hart (2012), it emphasizes harmony and balance (wholeness), commitment to the good of the community, and "the showing of honour, esteem, deference, and courtesy to all." An oft-quoted apothegm of unknown purported Native American origin affirms, "We are earth people on a spiritual journey to the stars. Our quest, our earth walk is to look within, to know who we are, to see that we are connected to all things, that there is no separation, only in the mind." The concepts of connection to nature, striving for balance and harmony, dedication to the communal good, and respecting individuals have wide applications in engineering practice.

Below is a partial list of various lessons we can all learn from diverse religious traditions, including the various forms of Judaism, Christianity, Islam, Buddhism, Hinduism, Jainism, Confucianism, Taoism, Baha'i, Shinto, Sikhism, and Indigenous religions. (Most of these insights are found in multiple religions.)

- The value of generosity
- Respect and reverence for nature
- The importance of family
- Respect for long-established culture
- The value of community
- The value of knowledge and learning
- The value of forgiveness and charity (e.g., zakat and tzedakah)
- The value of personal responsibility
- The value of social responsibility
- The importance of love and kindness
- The importance of honesty
- The importance of sincerity
- The importance of flexibility and perseverance
- Respect for life and living things
- Appreciation for the non-material elements of human life
- Living in harmony with all things and all people
- Self-fulfillment through benefitting others

Finally, here are a few quotations that may sometimes prove helpful in thinking through an ethical choice:

- "African ethics is, thus, a character-based ethics that maintains that the quality of the individual's character is most fundamental in our moral life" (Gyekye 2010).

- "Sub-Saharan conceptions of the good life characteristically understand self-realization in terms of communal or harmonious relationships" (Metz 2019).
- "The Taoist ideal is for a person to take action by changing themselves, and thus becoming an example of the good life to others" (BBC 2009).
- Ibn Rushd (Averroes) argued that "the Qur'an enjoined the use of reflection and reason and that the study of philosophy complemented traditionalist approaches to Islam. He asserted that philosophy and Islam had common goals, but arrived at them differently" (Nanji 1991).
- "A Muslim is advised, among others, to be friendly, forgiving, compassionate, generous and helpful, and also to be inclined towards chivalry and gallantry, to give rather than to take, to sacrifice rather than to grab, to make way for others' need, to say kind motivating words, to be humane to one and all" (Hashi 2011).
- "There are two fundamental principles that every Hindu applies to determination of right and wrong in questions of conduct or conscience. The first is ahimsa, noninjury. The second is nearness to God leading to moksha, spiritual liberation" (Anon. 1997).
- "Even in one single leaf on a tree, or in one blade of grass, the awesome Deity presents itself" (Shinto saying).
- "World peace is not only possible but inevitable" (Universal House of Justice 1985).

PROMOTING GOOD CONSEQUENCES[6]

One important question to ask when deciding whether to do something is "what will happen if I do that?" In other words, we can do a cost-benefit analysis of the different options open to us. It is important to remember to consider all the costs and benefits, long-range as well as short-range, psychological and moral as well as physical, indirect as well as direct, and less obvious as well as more obvious. For example, while harsh penalties for non-performance might improve productivity in the short run, such a policy may have very bad indirect, long-term psychological consequences, such as souring the atmosphere and diminishing team spirit. Moreover, while sometimes we should look at the consequences of particular acts, sometimes we should look rather at the consequences of policies.[7] For example, while in one particular case it might be easiest to transfer a particular employee without telling them in what respects their work is unsatisfactory, it is a bad policy: the best policy is to be direct with employees, to let each employee know exactly where they stand. Knowing when to look at an act in isolation, and when to think in terms of general policy, is one of the hardest and most important things to learn in thinking ethically. (See the discussion of consequentialism and utilitarianism in Appendix II for more on promoting good consequences.)

Several factors count toward thinking in terms of policy. There is a reason to prefer enforcing a general policy when it is important for people to be able to form secure expectations about this sort of thing, when treating each case individually would lead to unfairness, or when the thing in question is important to the institutional character of the professional or corporate setting. Other factors count toward thinking in terms of particular acts. There is a reason to treat cases individually when cases are substantially different from one another, or when the harm or unfairness in a particular case outweighs the benefits of a policy. These factors have to be weighed when making particular decisions.

For example, the best approach to dealing with a struggling employee varies considerably from employee to employee because people are very different. An approach that would be successful with one person would be disastrous with another. Thus, there is some reason to treat struggling employees on a case-by-case basis. Since, however, it is unlikely that failing to inform a particular subordinate, Jaspers, of his shortcomings will make him a better engineer, this factor does not give us a strong reason for skirting the general policy of letting engineers know how they stand and in exactly which ways their work is unsatisfactory. Moreover, there are strong reasons for thinking that, in the case of Jaspers, we should follow the general policy rather than treat Jaspers' case individually. Employees need to be able to have secure expectations about how dissatisfaction with their work will be handled and secrecy about such matters undermines the community atmosphere of the corporation. (See "When to Break the Rules" in Chapter 6 for further guidance.) Thus, one should follow the policy and tell Jaspers frankly in which respects his work is unsatisfactory.

It is important to understand, I suggest, that although promoting good consequences is an important aim of ethical decision-making, it is not the only one. Most people recognize that some things are wrong to do, however good the consequences, and there are some things one must do, however bad the consequences (see Appendix II). The conflict between promoting good consequences and other moral factors is one of the major sources of ethical dilemmas. Two factors that set limits on promoting good consequences deserve mention. It is important to live in a moral environment, to be part of a moral community. In addition, you should live in a way that proclaims your values and ideals. (See "Proclamative Principle" below.)

It must be emphasized, however, that many philosophers are utilitarians/consequentialists of one sort or another. Some ethicists argue that promoting good consequences is the only ethical guideline. There is a vast literature debating all this. For example, there are several useful articles in the *Stanford Encyclopedia of Philosophy* (Zalta 2022).

Illustrating Example

Example 8

Jones and Smith are rival candidates for a promotion. Smith needs the promotion more and would be better in the job. Although the overall consequences of telling lies about Jones so Smith will get the job might be beneficial, it would be wrong to do this. Deceitfully shafting one person to help another person even more is inconsistent with a moral environment and undermines the moral community. Being part of a moral community requires an environment of trust in which deceitfully shafting people, even to help someone else, is not tolerated. Moreover, if you tell lies about Jones, your conduct, however good its hidden motives, does not proclaim your commitment to honesty and fair dealing.

THE GOLDEN RULE

It is sometimes said that the Golden Rule is all you need to know about ethics. The Golden Rule can be found in many of the world's major religions. Rabbi Hillel the Elder famously said, when challenged to summarize the entire Torah (Five Books of Moses) while standing on one foot, that the entirety of the Torah was the Golden Rule: all the rest is commentary. Udana Barga 5-18 says, "Whatever is disagreeable to yourself, do not do unto others." The Analects 15.24 reads, "'Is there one teaching that can serve as a guide for one's entire life?' The Master answered, 'Is it not shu, sympathetic understanding? Do not impose upon others what you yourself do not desire.'"

In fact, I suggest, the Golden Rule is not really a rule at all, or at least not a viable one. Rather, it is an important technique of moral thinking, a guiding idea.

It is easy to give counterexamples to the rule "always treat others as you would wish to be treated." First, people differ, and don't always want to be treated in the same way. For example, when I have a toothache, I would want someone to pull out my tooth. So, the Golden Rule seems to say I should go around pulling out everyone's tooth, since that is what I would want them to do to me. That is clearly wrong. Of course, as we saw in Chapter 2, one response to a counterexample is to modify the principle. Perhaps the Golden Rule should read "do unto others what you would have them do unto you if you were in their circumstances." That revision avoids the toothache counterexample but raises others. Either the other person's desires are part of their "circumstances" or they are not. Suppose their desires do not count as part of their circumstances. Then, if I am a masochist who would want others to hurt me, the Golden Rule tells me to walk up to strangers and hurt them (since this is what I would want done to me). To avoid this result, the other person's circumstances must include their desires (so part of their circumstances would be a desire not to be hurt). But now the Golden Rule obligates me to have sex with anyone who desires me, since, if I had their desires, that is what I would want done to me. Second, it is sometimes wrong to treat people the way they want to be treated. For example, consider a student who fails a course, because he never came to class and had no understanding of the material. He wants to be given a grade of A anyway. The Golden Rule requires that the professor give the student an undeserved A because the professor, like most people, would rather get an undeserved A than an F. Yet clearly it is wrong for the professor to give the student an undeserved A. Other versions of the Golden Rule have parallel problems. For example, one version runs "don't do unto others what you would not have done to you." Now consider a physician who wouldn't want to be told of a terminal illness. Since he wouldn't want to be told, the Golden Rule seems to insist that he not inform a patient who does want to be told the truth about her condition. And this seems clearly wrong.[8]

Perhaps we can rescue the Golden Rule by moving to a more general description of what I would want done to me. Perhaps what the physician wants is to have his wishes respected, so he should respect the wishes of the patient: the golden rule directs the physician to tell the patient who desires to be told.[9] However, this strategy of escaping to a more general description of the relevant desire is problematic. What is the right level of generality? Suppose, for example, that I would want to be given a strawberry-flavored candy (generality level 1). Then, the Golden Rule says that I should give my friend a strawberry candy, even though they prefer blueberry. Perhaps, however, the relevant desire is a more general description of my desire, for example, the desire to be given one's preferred flavor (level 2). Then, the Golden Rule yields the right result. But why not choose an even more general description, such as "I would want to be given something" (level 3)? If so, the Golden Rule would tell me simply to give my friend something or other – perhaps a ring or a box on the forehead. Choosing a level of generality makes all the difference, and there seems no principled way to choose between levels 1, 2, and 3. One must already know what is the right thing to do in order to determine that level 2 is more appropriate than levels 1 and 3. But then the Golden Rule is of no use. I must already know that I should give my friend the blueberry candy. It is only because I already know this that I can figure out what the Golden Rule tells me to do.

In short, three basic problems with the Golden Rule can be summarized simply. First, it is not always right to give people what they want (or what you would want if you were in their place). Second, people often want conflicting things but the golden rule is supposed to apply to everyone. For example, if I have to decide between keeping a more senior employee or a more efficient employee, the Golden Rule doesn't help me. I would want the decision to be made on seniority were I in the shoes of the more senior employee, and I would want the decision to be made on merit were I in the shoes of the more efficient employee. Finally, there is always some way of describing an action so that I would want it done to me, and some way of describing the same action so that I wouldn't want it done to me.

Nonetheless, the Golden Rule can play an important role in moral reasoning. I suggest we view the Golden Rule not as a rule, but as telling us three important things. First, *to understand the meaning of our actions, we must think about how they affect others.* What is it like to be on the other end of what I'm about to do? How would I feel if somebody did that to me? Thinking about what it would be like to be in the other person's shoes is an important part of understanding the moral character of what you're doing. Second, *we must take very seriously the way we affect others.* That is, the results of our "thought experiment" are important. Third, *we generally shouldn't ask of people things we aren't prepared to give.* This is not a strict rule: it may be fine in a particular case for me to ask for a particular thing of you that I wouldn't or couldn't give you because our circumstances differ. But it is a strong guiding idea. An employer should not, generally, ask loyalty from an employee unless the employer is similarly willing to be loyal to the employee.

Case 18: Giving Credit When Due

Ndiaye works very hard on a project. You are Ndiaye's superior. You put the finishing touches on the project and pass it along to your supervisor, M. M assumes that most of the work is your own and praises you for the great job you did.

Discussion: You might be tempted to allow M to continue thinking that the credit belongs to you. But the Golden Rule tells you two things. First, it says, place yourself, in your imagination, in Ndiaye's shoes: how would you feel if someone did that to you? You realize that you would be upset about not getting the credit you're due. Second, the Golden Rule tells you to take seriously the injustice to Ndiaye you are creating. So, the Golden Rule gives you a strong reason to tell M that Ndiaye did most of the work.

UNIVERSALITY

One important technique of ethical thought, closely associated with the writing of philosopher Immanuel Kant,[10] is to ask "what if everyone did that?" Morality, after all, plays no favorites: no one can justifiably say "none of you may lie, but I'm special, so I may lie whenever I feel like it." No one is a morally privileged character. This insight, that morality doesn't play favorites, is at the heart of the principle of universality (or "universalizability," as it is called in the philosophical literature). The principle of universality says that similar cases should be treated the same way, and so, generally, I should not do something if it is wrong for everyone else to do it. For example, suppose six farms are threatened by flooding unless a retaining wall is built. The sixth farmer refuses to help build the wall, knowing that the other five have no choice but to build the wall by themselves. Here, the sixth farmer violates the principle of universality. If everyone refuses to do his share, calamity results. The sixth farmer recognizes the need for the wall and wants the other farmers to do their share. He intends to get a free ride: he can afford to shirk his responsibilities because others do their duty. Clearly, shirking one's duty in this way is unethical. It is wrong to take advantage of the fact that others (and not you) are doing the right thing.

Again, if I'm not morally special, and if I steal, then I must be prepared to say "everyone should feel free to steal" – otherwise, I am insisting that I should play by different rules than everyone else. But the very idea of stealing only makes sense within a system of property. After all, if we all understand that anyone can take anything, then no one owns anything, so taking anything isn't stealing. (I am not stealing air when I breathe.) So, it is incoherent to say "everyone should feel free to steal." (That is, it is only possible to steal when people generally recognize property rights and feel they should refrain from stealing.) Moreover, we cannot really function together if we do not recognize some forms of property (or at least special usage rights). It seems irrational (the argument goes) to spend months building a house if anyone can kick you out in the rain at any moment. Thus, the principle of universality suggests that it is (generally) wrong to steal.[11] (See also the examples below about lying, paying taxes, and keeping promises.)

Applied wrongly, this technique leads to error. For example, while we need some people to become physicians, it would be disastrous for our society if everyone in it became a physician – no crops would be planted, no roads or houses would be built, no toasters or medicine would be produced, and no professional opera or sports would enliven leisure hours. This doesn't make it wrong to become physician. So, it is useful to keep in mind two factors that limit the principle of universality.

1. Duties such as firefighting and cleaning the streets may be allocated: I may do my part by paying taxes to support a professional fire department, or by doing some other form of volunteer work (you fight fires, I'll help the sick). This is why it is not wrong to become a physician. As long as we each do our part, social tasks, such as crop planting, road building, healing the sick, and singing opera, are allocated to different individuals. (Of course, when no one is doing an allocated task properly, we must take responsibility, individually or collectively.) The key thing is a question of fairness: don't be a parasite or a free-rider on others doing the right thing.

2. Minor obligations must be subordinated to more pressing ones. Remember that principles are only sources that must be weighed against other, conflicting sources. While the principle of universality urges us to keep our promises (see example 11), it doesn't tell us that we should never break a promise.[12] There are times when a promise cannot or should not be kept because keeping the promise

would violate some other value or duty. While you ought to keep your promises, it is also important to save a life if you can. Indeed, it is more important to save a drowning person than it is to keep your promise to me to be on time for my dinner party. So, you ought to stop to rescue the drowning person even if that makes you late to my party. However, the importance of promise-keeping means that (1) promises should not be made lightly: try to avoid making promises you might not be able to keep. (2) Only a strong duty or value can outweigh the duty to keep a promise. If keeping a promise means doing something immoral (e.g., keeping a promise to kill someone), or means allowing someone to die or come to serious harm, then there is a strong reason to break the promise. However, the more formal and solemn the promise, and the more harmful the consequences of breaking the promise, the stronger reason one would need to break it. An offhand promise to be on time to my dinner party is less pressing than a legal contract to safeguard a child. (3) When a promise cannot be kept, the promisee is due something. Suppose I promise to go with Smith to the opera, and later find I can't keep my promise. I let Smith down. So, I should make every feasible attempt to make it up to Smith. Perhaps we can go to the ballet. Perhaps I can go with her to the next opera. At the very least, Smith is due an apology and an explanation. (See Case 23 in Chapter 6.)

Example 9: Hiring Away

In engineering, there is a kind of "gentleperson's agreement" not to raid other firms: although there is no law against it, and no legally binding contractual duty, there is an implicit understanding among engineering firms that they will not "hire away" employees from each other. A Human Resources officer who violates this agreement, while relying on it to prevent other firms from inducing away her own employees, is violating the principle of universality: she is free-loading, enjoying the benefits of the rule without doing her part to pay the price. She is, in effect, saying, "the rules apply only to other people, not to me. I'm special. You people must obey the rule, and pay the price of not taking engineers away from other firms, but not me." Morality does not play favorites this way: no one is a morally privileged character. (See also Case 52 in Chapter 9.)

Example 10: Lying

Lying is not just saying something false. Suppose you watch a movie in which the lead actor says, "I was born on Planet Ictoo." This isn't a lie, even if the actor knows quite well he's from Earth, not Planet Ictoo. He's just performing his lines in the script. But if I, a New Jersey US native, say

"I was born in Minsk" in a job interview, this is a lie. I'm lying because we usually have a social understanding that we will tell each other the truth: the social rules of job interviewing are based on a mutual understanding that you're supposed to tell the truth. When we go to a movie, however, we all understand that the actors in the film will be reading a script, not telling the truth about themselves. In other words, you can only lie in a situation governed by the social rule of truth-telling. Lying is abusing a social rule, trying to deceive someone by taking advantage of the social rule that we're supposed to tell the truth. You can lie to someone about where you were last night only because we have a social rule that such questions are to be answered truthfully. We couldn't function as a society without this rule (imagine what it would be like if any question you asked of a fellow worker, a physician, a spouse, or a stranger were just as likely to be answered falsely). Moreover, we can have a social agreement to tell the truth only if most people tell the truth most of the time. So, unless most people tell the truth most of the time, lying is impossible. Thus, the liar is a kind of parasite or free-loader. He gets something for himself, the chance to deceive you, by shirking his part in a common duty, telling the truth. The liar gets his advantage only if other people generally forego the advantages of lying for truth-telling.

From the principle of universality, thus, it follows that engineers must show great concern for truth and honesty. Only in extraordinary circumstances is it ethical to lie.

Example 11: Promising

We could not get on as we do without promising: promises are central to marriage, and most business transactions, including employment ("if you work in my lab, I will pay you so much at the end of the month"). A promise allows me to get something from you. You mow my lawn now because I promise to mow your lawn next week. You wouldn't mow my lawn if you expect I will not keep my promise. So, if people in our society generally don't keep their promises, my saying "I promise to mow your lawn next week if you mow my lawn today" means nothing, and I won't have my lawn mown. Promising works only if most people keep their promises. The promise-breaker is thus a parasite – he is able to get something from you for nothing by exploiting the fact that most people keep their promises. It follows from the principle of universality, then, that engineers must place a high value on keeping their promises.

Example 12: Paying Taxes

No one who is employed, or has a business, can get on as she does in our society without the benefits that tax dollars bring. The tax-dodger gets the benefit of taxes (schools, roads, etc.) for nothing because she shirks a duty that most people don't. Thus, engineers must not engage in dubious transactions for the purpose of avoiding taxes.

Example 13

You promise a co-worker that you will keep confidential what he is about to tell you. He tells you he is going to falsify safety figures for a nuclear power plant. The harm of keeping the promise is, potentially, widespread and deep, and keeping silent about it is immoral. The duty to prevent an unsafe plant from being constructed is more pressing than the duty to keep your promise to your co-worker. (Perhaps you should have specified limits when you promised confidentiality. Be careful with your promises.)

Example 14

Although engineers are responsible for the safety of the projects in which they are engaged, no one can oversee and guarantee the safety of every aspect of a project. Thus, allocation of safety aspects to different teams or individuals is permissible. However, if an engineer becomes aware that some safety aspect is not being properly addressed by the team or person in her firm responsible for that aspect, she may not ignore it by saying "that's not my area."

PROCLAMATIVE PRINCIPLE (DUTY TO SET A MORAL PRECEDENT)

The proclamative principle suggests that the way we live should proclaim our values and standards, what we stand for as a person. (It was proposed in Schlossberger 1989 as an alternative way to capture some of the insight behind universality.) This means that we should aspire to ensure that how we act and live is a worthy example. In particular, we should try to live our lives so that they are fit to be part of the moral precedents that people in our society should draw upon in leading their lives. Precedents are highly regarded actions, decisions, persons, systems, or institutions after which we want to model ourselves. For example, an engineer who has to make a tough moral decision might ask himself "how would X handle this," where X is someone they respect and admire. Similarly, a supervisor might better understand how to balance patience and firmness by looking at the way Y, an outstanding supervisor, treats her employees. Precedents and principles work together: examples help make clear what principles mean and how to apply them, while principles help us see what is morally significant about past decisions. (See also "Case Studies" below.)

In saying I should try to be a good moral precedent, I don't mean that everyone should copy me. For example, there are many good kinds of friendships, not just one, and hence many, very different friendships are valid precedents. There are different kinds of good marriages. My marriage may not be perfect, but I should want it to be good enough that people thinking about what kind of marriage they want might, if they knew the whole story, legitimately recognize it as a good kind of marriage from which they might learn. They might prefer a different kind of marriage, but my marriage should present a valid kind of option among those available from which to choose. It is a valid option from which a positive takeaway message can be drawn. When making a decision affecting my marriage, I should ask myself if acting in that way, taken in context, diminishes the worthiness of my marriage to serve as a valid precedent. If it would, that would be a factor telling against doing it. (Remember that in making moral decisions, we are always balancing relevant factors.) To the extent that I fall short of this, I should try to correct it and do better in the future. But also remember that while we should strive for excellence, good is good enough.

The same applies to engineering decisions. Engineers should strive to ensure that the way they function as an engineer is worthy to serve, to those who know the whole story, as one model of a good way to be an engineer. (For more about the proclamative principle, see Schlossberger 2016.)

Example 15

Looking over a safety study report, Araya notices that someone accidentally entered the number 13 instead of 18. Since 13 is well within the margin of safety (and 18 is even better), Araya thinks that there is no harm in leaving the matter alone. However, is this really the way a good engineer should handle things? After all, accuracy is a key value of engineering. Thinking about the proclamative principle, Araya corrects the figure in the report.

CASE STUDIES

The case study method is often used in business and law schools. Not surprisingly, it is often used in ethical thinking: looking at other cases often gives us some guidance in making our own decisions. Three kinds of examples are particularly useful: precedents, that is, *paradigm examples* of good/bad moral decisions we can try to emulate/avoid; *hard cases* that raise tough questions and get us to re-examine our assumptions; and *pure cases* (perhaps imaginary) that serve as good subjects for the "ethics lab." (All three sorts of cases appear in this volume.)

Precedents, as discussed above in "Proclamative Principle (Duty to Set a Moral Precedent)," are good examples, of persons, actions, institutions, or relationships, that we can try to emulate. Christians often ask themselves "what would Jesus do?" A beautiful example of a moral precedent is what Gandhi did in Calcutta in 1947. When India gained independence, most observers anticipated that there would be widespread violence between Moslems and Sikhs. Calcutta was expected to be a site of particularly violent conflict. Gandhiji went to Calcutta and began a hunger strike until violence ceased. Because of his great stature, leaders of both communities came to Gandhi and said, "Eat. We will put down our arms." Gandhi said, "that is not enough—you must also put down the violence in your hearts." Because of Gandhi's selfless actions, placing his life on the line, Calcutta wound up experiencing much less bloodshed than almost anyone anticipated. Few of us have the moral courage or stature of a Gandhi. But the story of Gandhi's moral heroism in Calcutta can inspire us, in ordinary life, to do more or go further, and, in a very difficult situation, to make the moral choice, to become, in some small measure, a moral hero.

Hard cases get us thinking about the issues and keep us from becoming too rigid and inflexible. For example, suppose your aunt is very near death, but is of clear and sound mind. Her son has just been killed in an automobile accident. She asks you "how's my son?" Should you tell her the truth, and make her last hours miserable, or should you lie to her? Thinking about a case like this helps us see the tensions between honesty and promoting good consequences, and may cast light on how we should treat other situations where being honest would produce bad consequences.

Pure cases play the same role in ethics that lab experiments play in science. In real life, it is hard to isolate all the relevant factors. Thus, it is sometimes useful to create an artificially controlled environment. The same is true in ethics. For example, the "hard case" above is not a pure case, since lying might also produce bad consequences – the dying aunt might be hurt that her son is not visiting her, or she may discover the truth and be hurt that you lied to her. So, it is sometimes useful to invent a pure case of a conflict between honesty and avoiding bad consequences. Again, in real life, few competent employees wholly lack personal integrity, and few employees with much personal integrity are highly incompetent. If we want to know whether competence or personal integrity is more important in a given situation, it is useful to imagine a highly competent scoundrel and an inept saint, and ask which you would rather have working for you in that situation. This helps tell you, for that situation, which of the two is more important. The counterexample of Herbert and the doomsday bomb in Chapter 2 is another example of an illuminating artificially pure case.

RESPECT FOR PERSONS

The philosopher Immanuel Kant tells us that we should always treat people as ends, never merely as means. For example, when I put money in a candy machine, I am using the machine merely as a means of getting candy. If the machine jams, I might kick the machine: I don't worry about the machine's feelings. When I give 50 cents to a store clerk for a candy bar, I am also treating the clerk as a means of getting candy. But the clerk is not merely a means: I don't kick the clerk if he is too slow in giving me my change. This is because I recognize that the clerk is also a person: he has feelings that count and a sense of personal worth and dignity that I must respect.

Respect for persons has three related aspects. First, you shouldn't use people. "Don't exploit or manipulate," one of the rules discussed in "Treating Others Fairly and Well," is just an application of the idea that one shouldn't use people. There are many other applications for engineers in their professional and personal lives. It is unethical, for example, to let a subordinate think she will get a promotion if she completes a task well, when that is not the case. True, her misunderstanding does serve to motivate her, and in the short run you will increase productivity by not correcting her mistaken impression. But it is unethical to use her this way, and, in the long run, it is a bad business practice. It is also unethical to pursue a social relationship with someone you don't like because that person can advance your career. When you do that, you are exploiting the person because your offer of friendship is a fraud. In some ways, this is worse than lying about a termite problem to a prospective buyer of your house. In both cases, you're obtaining something by fraud that you would not get if you were honest. In the termite case, you are taking advantage of another person as a buyer. When you pretend to feel friendship in order to get ahead, you are taking advantage of another person as a person. The second aspect is that you should treat others with dignity and respect. It is easy to make others feel small and to do little things that undermine their sense of dignity and worth. While there are, necessarily, things that distinguish an executive from a secretary, try to avoid unnecessary things that say to the secretary "you're just a secretary." While in some cases it may serve a useful purpose to have separate entrances for executives and subordinates, having separate entrances is often just a way of making subordinates feel that they don't count as much. On the positive side, there are many subtle ways in which one can communicate respect to a subordinate. For example, saying "please do x," or "could you do x?" is better than giving an order. Similarly, calling out and demeaning a subordinate in front of their peers for making a minor error fails to respect their dignity as a person. The third aspect is that you should respect people as autonomous decision-makers. (See "Rights" and "Autonomy" below.)[13]

RIGHTS

One crucial rule that applies not only to engineering but to every facet of ethical conduct is "respect the rights of others." Our society is founded on the principle that people have rights. People are entitled to free speech, for example, and so we have to allow people to speak freely even if their speaking causes us much inconvenience or does much harm. There are three basic kinds of rights: permission rights, liberties, and entitlement rights. If society is obligated to provide everyone with a minimal education, that is an entitlement right: people are entitled to receive an education. The right to free speech, by contrast, is a permission right. It does not mean that everyone is entitled to free air time or newspaper space but that society may not restrict, interfere with, or prevent people from trying to get their views across. People must be given an education (entitlement) and must be left free to do what they can to get their views aired (permission). Liberties are permission rights that call for support in exercising them. For example, I have a permission right to touch the tree in front of my house, but society has no duty to support me in doing so. By contrast, the right to vote, in the US at least, is not just a permission right but a liberty. For example, assistance is provided to help disabled voters get to the polls (Schlossberger 2008).

In addition, there is a difference between legal rights and moral rights. Legal rights are the entitlements and permissions the law gives people. Moral rights are justified moral expectations constraining others' conduct. Legal rights change as the law changes. For example, suppose the Parliament of the Democratic Republic of Georgia passes a law (signed by the President) requiring employees to give two weeks paid vacation to all full-time employees. As a result, all full-time employees in Georgia have a legal (entitlement) right to two weeks paid vacation. This is a legal right that employees in Azerbaijan do not have. In contrast, moral rights are not a matter of law but of ethics. Although infidelity is not a crime in many jurisdictions, spouses in a traditional marriage (as opposed to an "open" marriage) have a moral right to expect their spouses to be faithful. An unfaithful spouse in such a marriage fails to fulfill legitimate moral expectations. (Of course, many moral rights are also legal rights, such as the right not to be murdered.) Engineers must be scrupulous about observing legal rights and sensitive to moral rights. Moral rights can generally be overridden by more pressing moral concerns. Telling a lie to save the planet from a crazed or evil scientist is, most people would agree, the right thing to do.

There are two kinds of moral rights: general moral rights and special moral rights. Special moral rights stem from particular relationships, such as marriage. My spouse and my children have special moral claims on my time and loyalty that other people do not have. Fiduciary relationships, such as doctor-patient and attorney-client, create other special moral rights (that are often also legal rights): an attorney owes special moral and legal duties to her client that she does not owe to other people. General moral rights are those moral claims people normally have on each other. For example, most people think that we have a general right to the truth. That is, while other people normally do not have to tell us everything we want to know, they should (normally) not tell us a lie. Normally, people who lie to us fail to uphold a legitimate moral expectation.

General moral rights come from different sources and vary in strength and importance. (A right can be generated by more than one source.) Natural rights (sometimes called "human rights") are claims all individuals have on their society. Natural rights are controversial. Jeremy Bentham famously called them "nonsense on stilts." I argued (in Schlossberger 2008) that natural rights are claims to adequate provision for those things, such as the ability to make rational decisions and the ability to feel happiness and satisfaction, that make it wrong to use violence against people without a strong justification. (By contrast, we may rip out a dandelion from our lawn simply because we feel like it.) Since, for example, a reasonable amount of free discussion is a precondition for making rational decisions, I argued, there is a natural right to adequate provision for free speech. It is important to notice that, on this view, the right to free speech doesn't mean that you may say anything at all. Rather, it means that each person must be given adequate opportunity to debate, reason, voice her views, and so forth. The right to education doesn't mean that you must be taught anything you wish to learn. The right to education is really two rights: an entitlement right that each person be given enough education to be able to function as a rational citizen, and a permission right to adequate opportunity for self-development. Other accounts of natural rights make more absolute claims. Many view natural rights as lines in the sand a government may not cross. The U.N.'s Universal Declaration of Human Rights (UN 1948, Article 24) insisted that people have a natural right to a paid vacation.

Duties of team loyalty, I suggest (Schlossberger 2008), also generate moral rights. Members of a sports team, such as the Brazil National Football Team or the Los Angeles Dodgers, owe each other duties as fellow teammates engaged in a common enterprise to which they are jointly dedicated. For example, other things being equal, teammates should help each other win and must not undermine their teammates' efforts to contribute to the team's victory.[14] Those of us who are dedicated to morality, to the common enterprise of making this a better (or at least a good) world, also form a kind of team – team morality. Members of team morality must not (other things being equal) undermine teammates' efforts to make this a better world and to live a good life. Since lying generally impedes making good and autonomous decisions, we have a general moral right not to be lied to (which can be overridden in certain circumstances).

Similarly, some general moral rights stem from the basic needs people have in order to participate as full citizens and moral beings in their society. Ignorant people cannot participate rationally in the public life of the community, and so education is a basic moral right. People cannot participate in the public life of the community if they are not free to argue and express their views, so free speech is a basic moral right. Similarly, people cannot function as moral agents if they do not have reasonable opportunities to get involved in shaping their own destinies, nor can they function as distinct individuals if they have no privacy.

The right to privacy, discussed in Chapter 13, the rights of future generations, discussed in Chapter 10, and intellectual property rights, discussed in Chapter 7, are other examples of rights that affect engineering practice.

It is important for engineers and engineering organizations to operate in a way that respects these rights, so central to our way of life. (For more about the nature of rights, see Schlossberger 2008.)

Case 19: Freedom of Speech

Jackson is an engineer employed by Worldwide Universal Chemical, a large company. Worldwide Universal manufactures DN3, a chemical with several industrial uses. Were the Defense Department to resume production of materials for chemical warfare, a significant amount of DN3 would be used in the manufacture of chemical weapons. Writing as an individual (not mentioning Worldwide Universal Chemical), Jackson publishes a strong letter in a national magazine denouncing chemical warfare as immoral and urging citizens to tell their representatives to oppose the bill authorizing the production of chemical weapons.

Discussion: In writing this letter, Jackson was properly exercising her legal and moral right to advocate her moral views. The letter did not involve Worldwide Universal Chemical. The company has a moral obligation not to cast a "chilling effect" on this right by penalizing Jackson in any way, direct or indirect, for writing the letter, even though the letter urges a course that would eliminate a financial benefit to the company. For example, Jackson's superiors must make an effort not to view the letter as evidence that Jackson lacks team spirit. Moreover, it is important that Jackson not be given the impression that her letter will count against her.

Several factors are important here. Jackson's letter did not breach the duty of confidentiality by revealing trade secrets or sensitive company information. In addition, Jackson's letter did not attack the company or criticize its actions and did not present its author as a representative or employee of Worldwide Universal Chemical. Thus, writing the letter constitutes neither disloyalty to the company nor avoiding proper channels. DN3 has other industrial uses, and Jackson did not suggest that DN3 be withdrawn or outlawed. By contrast, if Jackson's claim were that DN3 posed a danger to public safety and should not be produced, Jackson would have a duty to address the problem within corporate channels, if feasible, before writing a letter to the magazine. (See "Whistleblowing" in Chapter 7.)

AUTONOMY

Autonomy is, in the most basic sense, the ability to make our own decisions about our life. The three elements of autonomy are the *ability* to select, the *independence* of our choice, and the *efficacy* of our choices. People have the ability to choose to stay standing in a way that trees do not, but, if I am sick and weak, I may not be able to stay standing even if I choose to do so (my choice lacks efficacy). If I am told that I will be shot if I don't remain standing, my choice to remain standing was not independent. Considerable controversy surrounds the conditions of independence needed for a choice to be autonomous. Manipulation and coercion reduce the independence of a choice, but it is not clear what counts as coercion and manipulation. Threatening to shoot you if you sing violates your autonomy, but threatening not to clap if you sing does not. I manipulate you and violate your autonomy if I get you to do what I want by telling you lies but not by giving you sound and logical reasons. Ethicists also differ about the importance of autonomy: some argue that we may never violate others' autonomy, while others treat autonomy as a lesser consideration. Conly (2012), for example, argues that autonomy is not the primary ethical factor because people do not reason well about their own goals. Not all kinds of choices count equally. Mere-choices, such as whether to eat broccoli or peas, are less important than proclamative choices, that is, choices that proclaim a central value or affirm what we stand for as a person, such as choice of religion or choosing to die rather than tell a lie. High-impact choices, which significantly affect our lives (such as whether to undergo a possibly lethal surgery that may restore one's vision), are also more important than mere-choices. (See Schlossberger 2013.) Engineers need to consider the autonomy of individuals (such as co-workers), the autonomy of users/consumers (such as passengers in the Ford Pinto), and the autonomy of society (e.g., when building a nuclear plant). For more about the nature of autonomy, see Schlossberger 2020).

Autonomy is always a value, and, in some cases, a right as well. Luegenbiehl (2004) argues that in some parts of Asia, autonomy is valued less than flexibility, fitting in, and team identity, in part because, he claims, those regions have no history of capitalist greed or notions of professionalism. However, even though cultures differ about the role of community, autonomy is one of the most important values of a free society.

A central premise of modern societies is that, whatever other responsibilities people might have, people should be responsible for themselves. Unlike sheep, people are moral agents and have a responsibility as well as a right to play a role in shaping their own futures. We are a community of people, not of ants. Ants have particular roles to play (gathering food, defending the colony, etc.), and they just do their roles without question. One worker ant is very much like another. People are not like that. People have individual needs and abilities, and each person has her own ideas about what is good, about how one ought to live, and about what is important. Each person must participate in her own way in the common life of the community.

What this means for engineering ethics is that engineers must treat people as of special value (see "Respect for Persons"), and respect other people's desire to make decisions for themselves. Treating people as of special value means that the ethical engineer does not treat people as a commodity, interchangeable and replaceable. The loss of a life should not be treated as just another dollar figure, as in the Pinto case. Respecting other people's desire to make decisions for themselves means that the ethical engineer does not treat adults like small children unable to make intelligent decisions about their own lives. This is why, in Case 1, you should try to avoid taking steps that might have a major impact on Jones' career without consulting Jones. Jones should, if feasible, have some say in what happens to her. This is why a community should have some say about the risks imposed on it by an engineering project, such as a nuclear power plant. This is why voluntarily taken risks are preferable to involuntary risks. (See "Safety" in Chapter 4.)

However, although autonomy is an important value, it is not an absolute value. Other things count too. A company does not have to scrap its plans to build a plant because one resident doesn't like it or doesn't think the risk is worth the benefit. And it is not necessarily ethical to market a cancer-causing lipstick just because buyers know the risk and are willing to take it. (See Case 4 in Chapter 4.) Life is full of compromises and autonomy is one of the things that must sometimes be compromised. Autonomy is a value that must be weighed against other values. However, when Jones' autonomy is so severely compromised that she is denied a say in the most crucial and important decisions about her life, her rights have been violated and it is rarely ethically acceptable to violate other people's rights. (See "Rights" above.)

Again, it should be noted that some philosophers, contrary to what I suggest, consider autonomy a side-constraint on action, meaning that actions are permissible only if they do not violate autonomy. In their view, autonomy trumps all other moral considerations (see Nozick 1974).

To summarize, two points must be kept in mind. First, the ethical engineer should always try to respect autonomy, though autonomy must be balanced against other values and rules. So, in deciding between different options, it counts against an option that the option violates someone's autonomy, though other factors may count more strongly in favor of the option. Second, if an option would very severely violate someone's autonomy, this might violate a right, in which case choosing that option would violate a specific rule to respect the rights of others.

Example 16

You are designing a toy truck for children made from a cheap plastic that can easily break, posing the risk that a small piece may break off in a child's mouth, potentially causing choking. The current design with the cheaper plastic does not violate any safety regulations. Research indicates a viable market for the toy. Substituting a more ductile plastic, which would bend rather than break when bitten, would increase the cost by 70%. Should the engineer use only the safer plastic (increasing the cost to consumers), use only the cheaper plastic (reducing consumer cost), or produce both models, giving consumers a choice? What effect might the first option have on financially challenged families? If the cheaper plastic is used and, as a result, a child dies, is the engineer who specified the cheaper plastic responsible for the child's death?

Example 17

"Before a nuclear power plant is built near a residential area, engineers should help the public understand the potential benefits (such as reducing carbon footprint) and risks, both long and short term (such as accidents and the problem of storing nuclear waste), of nuclear power, and participate in (and listen to) the public debate. If, after a thorough and informed public debate, the residents who most bear the brunt of the risks are opposed to the construction of the plant, the fact that the construction of the plant violates their autonomy is a major factor in deciding whether it is unethical for engineers to work on the plant's construction" (Schlossberger 2020, 562).

NOTES

1. We influence people's choices in a wide variety of ways, such as giving facts and logical arguments, showing emotions, pointing out consequences, strongly suggesting, choosing what to mention, and much more. Not all of these count as

manipulation. For example, threatening someone who makes a choice we don't like generally counts as manipulation while evidencing minor displeasure generally doesn't. Where on this continuum is the line?

2. A few remarks about the limits of this principle are in order. It is not meant to resolve all moral dilemmas, for several reasons. First, what must be assessed are the on-balance, long-term likely effects of courses of action, rather than the effects of individual actions. In some sense, a dentist leaves the world worse by drilling a hole in her patient's tooth, though the patient will be better off, ultimately, as a result of the whole course of treatment. Since there is always some ambiguity about what counts as a "course of action," we can't apply the principle too rigidly. Second, not leaving the world worse than one found it is neither a sufficient nor a necessary condition for right action. We might find ourselves in a "no-win" situation, in which any decision we might make will leave the world somewhat worse than we found it. Moreover, in some cases, other moral considerations are more pressing. For example, a course of action that would make the world slightly worse might be required in order to avoid violating someone's rights. Again, rule-utilitarian considerations might force us to choose a course whose result will leave the world worse than we found it. And, of course, some small "worsenings" are permissible in the course of a life generally devoted to improving the world. To take the extreme case, a saint is entitled to some small pleasures that have tiny detrimental effects on the world. Third, there will be real disagreement about what counts as being "worse." In part, this is because agents must make difficult on-balance, long-term assessments of the probable effects of their actions, and in part because there is disagreement about the values in terms of which such assessments are made. In some cases, it is reasonably clear that a given course of action would make the world worse. In others, there is substantial disagreement, and each individual has no choice but to act according to her own best judgment, informed by relevant facts and arguments. Some possible actions fall into the "gray area" about which we have no reasonable grounds for strong conviction. Whether acting within this gray area is permissible depends on what is at stake – how bad is the risk, and how great the potential benefit? My point is that the duty to try to ensure that one's participation in the world does not leave it a worse place is something individuals must consider and address. The Toyota executive must ask himself "is the way of life made possible by the automobile worth 1.35 million worldwide deaths a year?" This is a difficult, perhaps an unanswerable, question, but one that cannot, in good faith, be avoided.

3. As Richard T. DeGeorge points out, "if I choose to take a risk to save $6.65, it is my risk and my $6.65. But if Ford saves the $6.65 and I take the risk, then I clearly lose" (DeGeorge 1990, 284). Here, it cannot reasonably be argued that consumers "vote with their dollars," since they were unaware of their options. In general, says DeGeorge, "the assumption that American drivers are more interested in styling than safety is a decision that has been made for them, not by them" (p. 287).

4. Jones may be tempted to deny that her ad campaign would make the world worse. One can always find *some* benefit to any course of action, and it is tempting to say "but think of the extra jobs the stimulus to the economy may produce," or "perhaps those who smoke high-tar cigarettes will smoke fewer of them." But such remarks ring hollow. Unless Jones is a fool, she cannot honestly believe that her campaign will not result in more cases of emphysema and lung cancer, or that a few possible extra jobs balance out the resulting terrible suffering and numerous deaths. No one who looks at the situation clearly and honestly can deny that, overall, in authorizing

the product and ad campaign, Jones will make the world a place of greater suffering. The fact remains that no one currently desires to smoke extra high tar and nicotine cigarettes, and no one will smoke them unless Jones sets out to induce them to do so by playing on their insecurities. If her ad campaign succeeds, many people will not live as long as they otherwise would, and the additional cases of emphysema and lung cancer caused by smoking He-Man cigarettes will lead to great suffering that would otherwise have been avoided. Thus, Jones must admit that she is quite deliberately setting out to bring about a worse world.

5. The phrase is translated as "that which your right hand possesses," sometimes interpreted as "slaves you own."

6. Some philosophers have argued that promoting good consequences is all that matters in ethics. Act utilitarians suggest that one should always perform the action that has the best overall consequences for everyone concerned. Rule utilitarians suggest that one should always follow the set of rules that would produce the best overall consequences if those rules were generally followed. (See Appendix II.) Unlike strict act and rule utilitarians, I treat the promotion of good consequences as one moral factor that must be weighed against others.

7. This corresponds to the difference between act and rule utilitarianism.

8. A parallel case to the professor and the undeserving student can be given for the negative version of the Golden Rule. I do not want to be locked away (incapacitated), and I would not want to be incapacitated even if I had committed multiple assaults. Yet, it is not wrong to protect innocent persons by incapacitating a frequent assaulter, just because I would not want to be incapacitated if I were an assaulter.

9. Of course, there may be other reasons than the Golden Rule for informing the patient, such as patient autonomy.

10. While Kant treated the principle as a test that every action must pass in order to be considered ethical, I treat the principle as one source of ethical decision-making that must be weighed against others.

11. Of course, the philosophical issues about property and stealing (as well as the interpretation of Kant) are much more complex than this simplified discussion suggests. See, for example, Waldron 2020 and Schlossberger 2008.

12. Kant himself thought that it did. Suppose, however, that we choose to follow the maxim "break a promise only once every five years." This maxim would not undermine the practice of promising, since most people, in fact, break promises much more often than this.

13. Kant thought that treating people as ends and universalizability were (along with two other formulations) functionally equivalent versions of the categorical imperative. How and whether this is true is the subject of much debate among Kant scholars.

14. There are obvious limitations on this principle. For example, teammates need not help another member of the team to cheat. Helping one teammate may hurt another. More pressing moral concerns may be more important than the team's winning.

REFERENCES

Anon (1997) "Swami, Bill Clinton Has a Question," *Hinduism Today* https://www.hindu-ismtoday.com/magazine/june-1997/1997-06-swami-bill-clinton-has-a-question/

Baura, G. (2006) *Engineering Ethics: An Industrial Perspective* (Amsterdam: Elsevier).

BBC (2009) "Taoist Ethics," https://www.bbc.co.uk/religion/religions/taoism/taoethics/ethics_1.shtml

Conly, Sarah (2012) *Against Autonomy: Justifying Coercive Paternalism* (Cambridge, MA: Cambridge University Press).

DeGeorge, Richard (1990) "Ethical Responsibilities of Engineers in Large Organizations," in David Appelbaum and Sarah Verone Lawton, eds., *Ethics and the Professions* (Prentice Hall).

Dowie, Mark (1977) "Pinto Madness," *Mother Jones* Magazine) Sept./Oct., 18–32.

Echold, J. S. (1972) "Ford Motor Company Internal Memo. Fatalities Associated with Crash-Induced Fuel Leakage and Fires," reprinted in D. Birsch and J. Fielder, eds. (1994) *The Ford Pinto Case: A Study in Applied Ethics, Business, and Technology* (Albany, NY: SUNY Press), 165–173.

Gyekye, Kwame (2010) "African Ethics," *Stanford Encyclopedia of Philosophy* https://plato.stanford.edu/entries/african-ethics/

Hart, Michael Anthony (2012) "The Mino-Pimatisiwin Approach and Sustainable Well-Being," https://www.eswbrg.org/uploads/1/2/8/9/12899389/hartkeynote.pdf

Hashi, Abdurezak A. (2011) "Islamic Ethics: An Outline of Its Principles and Scope," *Revelation and Science* 1:3, 122–130 http://irep.iium.edu.my/18564/1/Islamic_Ethics.pdf

Luegenbiehl, Heinz C. (2004) "Ethical Engineering and Autonomy in a Cross-Cultural Context," *Techne: Research in Philosophy and Technology* 8:1, 57–58.

Metz, Thaddeus (2019) "The African Ethics of Ubuntu," *1000 Word Philosophy* https://1000wordphilosophy.com/2019/09/08/the-african-ethic-of-ubuntu/

Mohammed, J.A. (2013) "The Ethical System in Islam – Implications for Business Practices," in Christoph Luetge, ed., *Handbook of the Philosophical Foundations of Business Ethics* (Dordrecht: Springer) https://link.springer.com/referenceworkentry/10.1007%2F978-94-007-1494-6_3

Moss-Racusin, Corinne A., Dovidio, John F., Brescoll, Mark J. Graham and Handelsman, Jo (2012) "Faculty's Subtle Gender Biases Favor Male Students," *Proceedings of the National Academy of Sciences* 109:41, 16474–16479 DOI: 10.1073/pnas.1211286109

Nanji, Azim (1991) "Islamic Ethics," in Peter Singer, ed., *A Companion to Ethics* (Blackwell), 106–118.

Nozick, Robert (1974) *Anarchy, State, and Utopia* (Basic Books).

Pumphrey, P.V. (1984) "Chase Manhattan Bank, N.A.: A Case Narrative in Company Values," in Hoffman and Moore, eds., *Business Ethics: Readings and Cases in Corporate Morality* (McGraw-Hill).

Schlossberger, Eugene (1989). "Is Morality Universalizable?" Colloquium Paper, Central Division American Philosophical Association Meeting, Chicago, IL.

Schlossberger, Eugene (2008) *A Holistic Approach to Rights: Affirmative Action, Reproductive Rights, Censorship and Future Generations* (Lantham, MD: University Press of America).

Schlossberger, Eugene (2013) "The Right to an Unsafe Car? Consumer Choice and Three Types of Autonomy," *Journal of Applied Ethics and Philosophy* 5, 1–9.

Schlossberger, Eugene (2016) "Engineering Codes of Ethics and the Duty to Set a Moral Precedent," *Science and Engineering Ethics* 22:5; 1334–1344. Published online 2015 https://link.springer.com/article/10.1007%2Fs11948-015-9708-3.

Schlossberger, Eugene (2020) "Autonomy in Engineering," in Diane P. Michelfelder and Neelke Doorn, eds., *The Routledge Handbook of the Philosophy of Engineering* (Taylor and Francis), 558–568.

Shutte, Augustine (2001) *Ubuntu: An Ethic for the New South Africa* (Cluster Publications).

Singer, Peter (2021) "Ethics," *Encyclopedia Britannica* https://www.britannica.com/topic/ethics-philosophy/China

UN (1948) "Universal Declaration of Human Rights," https://www.un.org/en/about-us/universal-declaration-of-human-rights

Universal House of Justice (1985) "The Promise of World Peace," https://www.bahai.org/beliefs/universal-peace/

Waldron, Jeremy (2020) "Property and Ownership," *The Stanford Encyclopedia of Philosophy* https://plato.stanford.edu/archives/sum2020/entries/property/

Wong, David (2021) "Chinese Ethics," *The Stanford Encyclopedia of Philosophy* https://plato.stanford.edu/entries/ethics-chinese/

Zalta, Edward N., principal ed., (2022) *Stanford Encyclopedia of Philosophy* (Metaphysics Research Lab, Stanford University) https://plato.stanford.edu/

6

Additional Ethical Sources (Part 2)

PERSONAL VALUES, MORAL BEAUTY, AND THE GOOD LIFE

Since ethics is, at heart, about what is worthwhile and good in life, one basic ethical source is a vision of the good life, a sense of what kind of life is worth living and striving for. There are many different kinds of good human lives, ranging from those celebrated as moral heroes (Gandhi, Mother Teresa, Dr. Martin Luther King, Raoul Wallenberg, the Dalai Lama, Malala Yousafzai, etc.) to farmers with a love for and kinship with the earth; from scholars and humanists like Martin Buber to Dostoevsky's character Prince Myshkin in *The Idiot*; from the dedicated and inspired artist to the good parent and spouse living an ordinary life. A good human life does not have to be perfect but each has something unique, so that the world would be a poorer place, to a degree great or small, without any one of them.

Who is to say what is a good life? One guide is social convergence: good human lives are those that enduring societies, interacting in the right way (holding thoughtful discussions, becoming friends, reading/viewing each other's literature or films, and so forth), would come to recognize as good, that is, come to see as morally beautiful in their own way. For example, life is hard and relatively short. Most people in enduring societies (as opposed to hermits or short-lived cults) need to feel connected to something bigger than themselves, something that endures after they die (otherwise, their lives are just radar blips on the screen on time). Social convergence, thus, will tend to recognize as good-only lives imbued with a sense of connection. (For more about social convergence, see Schlossberger 2021.)

Having a vision of a good life is a key not only to ethical decision-making but to living well. People often do foolish things because they don't think about what really matters. No one really thinks that the most important thing in life, the thing worth risking everything else for, is not letting

DOI: 10.1201/9781003242574-8

anyone get ahead of you.[1] Yet, we often see people on highways risking their lives to prevent other cars from getting ahead of them.[2]

Thinking about what really matters in life may change how you act, what you feel, and how you make ethical decisions. If you come to see that participating in a family is one of the really important things in life, you will spend more time with your family, and you will regard going to a parent-teacher conference not as an annoying distraction but as part of what really matters to you. If you come to see personal integrity as more important to a good life than consumer goods, you will be unwilling to lie to a prospective buyer of your used car about the fact that it tends to stall on cold mornings, because being a person of integrity is more important than the new couch you would buy with the extra money.

At the core of a vision of the good life is a set of values, of those things we think are important for their own sake. Someone who values honesty not only wants to be honest but also appreciates honesty in others. Thus, she will help and encourage others to be honest. For example, circumstances permitting, she would want to provide an incentive for employees with a drug problem to come forward, perhaps by establishing a program in which those employees who admit to a drug problem are treated at company expense, while their job is kept open for them (perhaps with appropriate monitoring) upon successful completion of the program. Similarly, a dedicated engineer and a great tennis player may place a high value on excellence for its own sake. They may care less about the fame or money or success they derive from their work than they do about excellence as such. Therefore, the tennis player takes delight in the skill of an opponent or rival. (Professional baseball players often express the sentiment that they would rather play against the best team, rather than against the worst.) The engineer wants to improve not only their own work but the achievement level of the profession as a whole. Again, if I value the search for knowledge, I will not fudge the data in a scientific paper for presentation, even if presenting the tainted paper would further my career.

Closely connected to the idea of a good human life is the notion of moral beauty. Beauty, in general, is a big part of what makes life worth living. Beauty makes life human and it makes life bearable. It is unpleasant to have to sit for an hour in an ugly room that smells bad. It is a treat to sit for an hour in a room of beauty, of sweet fragrance, lovely textures, beguiling sounds and pleasures for the eye. Moral beauty is the most compelling beauty of all, for it remakes the world entirely. The world of the happy person, said the philosopher Ludwig Wittgenstein, is a different world from that of the unhappy person, even though it contains the same things. Being happy is much more than a feeling. When you're happy, everything you see and hear and touch is bright and inviting and alive. In the same way, in a world suffused with moral beauty, everything is different. Getting a well-deserved raise doesn't just mean more cookies in the cookie jar. It's an affirmation of justice and the worth of your efforts – it means something much more than a few extra dollars in your paycheck. Without moral beauty, you can want another person the way you want a cookie. My love for my wife is so much more than that. In part I care about her flourishing because I genuinely value what is good and worthy in her and her life, because I see her as a good person whose thoughts and feelings are lovely things, and thus demand my affection. The special and lovely sense of deep closeness we feel would be impossible without a common deep commitment to a shared vision of a good, worthwhile, beautiful, illustrious life. The rich, deeply moving, intensely satisfying love I feel for her, the love that goes so far toward making my life worth getting up for in the morning, would be impossible without a sense of moral beauty that pervades my entire life and is at the center of my way of perceiving, thinking, and feeling. In a world without moral beauty, all that would be lost. It would not be the same marriage. A world without moral beauty is a world of meaningless, random events. A world of moral beauty is full of small delights, from the sunshine dancing on a blade of grass signaling hope to a decomposing leaf giving itself back to the earth, one small contribution to the song of the universe, the great cosmic struggle toward fulfillment.

Moral beauty also helps explain why we should be ethical. Sometimes all we have to hold on to is our commitment to moral beauty, to living the best way we can, whether or not anyone wants it or notices or cares. It is so often easier to forget our integrity. Doing what one believes in doesn't always feel good. It may not win us friends or fortune. Why then be moral? One answer is moral beauty: when we act well, live our lives ethically, we bring moral beauty to the world. That is why doing the right thing is its own reward. By doing right, we gain, literally, the entire world, the only world worth living in, the world of moral beauty.

This is why you can't be moral on Tuesday but not on Wednesday when it's inconvenient. Moral beauty is not something we can turn on and off like a faucet. You gain the realm of moral beauty only by being a certain kind of person, a person who sees the world in moral terms. You can't turn that off because it is who you are inside. So, either you make a lifelong commitment to morality or you don't, and making the commitment gives you, overall and in general, a better life.

VIRTUES

Another way of approaching ethical decisions is asking "what kind of person do I want to be?" A good person, for example, is courageous rather than rash or cowardly: they show a healthy respect for danger but face it when necessary. They take reasonable risks for reasonable purposes. When deciding whether to fight a particular battle, in addition to the factors in "When to Fight a Battle" (Chapter 5) one may ask what a courageous person would do – would it be rash to fight this battle or cowardly to run from it? Moreover, by striving to be a courageous person generally, one will come to have a better sense of when to fight and how to persevere. The same is true of the other virtues, such as generosity (being neither too stingy nor profligate and extravagant but giving and spending what is appropriate); self-regard (being neither selfish nor a doormat); and wisdom.

Virtues, as Aristotle said, are achieved by practice. We begin by imitating what we see as children but as we live, we learn from our experience, reflect thoughtfully and make changes, then continue to adjust and strengthen as we live, imitating, observing, reflecting, and adjusting. We all make mistakes. When we do, we should reflect on it, make amends, make adjustments in ourselves and our work and our relationships, resolve to do better, and move on. If we do it right, it gets easier with time, and we get it right the first time more often. (See further the discussion of virtue theory in Appendix II.)

THE PRECAUTIONARY PRINCIPLE

The precautionary principle suggests that when there is uncertainty about a significant risk, steps should be taken to protect against possible harm. There are two major and oft-cited formulations. The Rio Declaration Principle 15 in 1992 states that, "Where there are threats of serious or irreversible damage, lack of full scientific certainty shall not be used as a reason for postponing cost-effective measures to prevent environmental degradation" (Rio Declaration on Environment and Development 1992). The Wingspread Consensus Statement in 1998 proclaims that:

> When an activity raises threats of harm to human health or the environment, precautionary measures should be taken even if some cause and effect relationships are not fully established scientifically. In this context the proponent of an activity, rather than the public, should bear the burden of proof. The process of applying the Precautionary Principle must be open, informed and democratic and must include potentially affected parties. It must also involve an examination of the full range of alternatives, including no action.

Wingspread Conference 1998

Even in the absence of certainty, precautionary steps should be taken to address a potentially serious risk.

Case 20: Banqiao Dam Failure, Henan, China

When the clay dam over the Huai River was built in 1952, hydrologist Chen Xing warned that over-construction of dams in the region could raise water tables to dangerous levels. Xing also recommended 12 sluice gates but only five were built. As a result of his criticisms, Xing was removed from the project (MacLeod 2019). Nonetheless, 87,000 reservoirs were built in China between 1950 and 1970 (Fish 2013). Early warning systems and evacuation plans were lacking (Britannica, The Editors of Encyclopedia 2014). Cracks and leaks appeared shortly after construction but were repaired by Soviet engineers. In 1961, Chen Xing was brought back onboard, continued his criticisms, and was again removed. In 1975, Typhoon Nina dropped 60 inches of rain over Henan Province in three days. Communication lines went down, leaving workers unsure what to do. The record-breaking, intense rainfall, coupled with insufficient drainage sluices, substandard construction, and faulty maintenance, led to the failure of the dam, releasing a seven-mile-wide surge of water almost 20 feet high. The floodwater, in turn, caused the failure of 62 additional dams. "Nine days later there were still over a million people trapped by the waters. These people relied on air-drops of food and were unreachable by disaster relief groups. Epidemics and famine devastated the trapped survivors" (USDIBR 2015).

Total deaths from the flood plus disease and dehydration suffered by the stranded are estimated to be in the vicinity of 200,000.

Discussion Question: What might have been done differently had the Precautionary Principle been observed?

RESPONSIBILITY (DUAL-USE)

For what are we morally accountable? Put another way, what is my responsibility as an ethical engineer?

An ethical engineer takes responsibility for the projects in which he or she is involved. Indeed, as Neelke Doorn and Ibo van de Poel (2012, 1) remind us, "there is something special about the relation between responsibility and technology and engineering." How far does that responsibility go, especially when the project is used or managed by others in ways the engineer might not have intended? Do engineers and researchers, as Stephanie J. Bird (2014) says, "have a responsibility not only to oppose the misuse of their work but further, to attend to its foreseeable societal impacts"? That a technology can be used either for good or for ill (or may have unintended bad consequences) has been called the "dual-use dilemma" (Awais et al. 2009). (For example, if two people use hand-held industrial lasers to battle each other, then this "duel-use" is an example of dual-use.) Several types of scenarios raise issues:

- The particular project on which an engineer works may be legitimate in itself but may be part of a larger project that is morally questionable. Working on a more durable nozzle is legitimate but not if the nozzle is being developed as part of the design of a Nazi gas chamber.
- Technology developed by an engineer may be incorporated by remote teams, perhaps years later, into an objectionable project.
- Products of engineers' work may be put to unapproved uses by end users. Pharmaceuticals may be used off-label and glue meant as an adhesive may be sniffed to get high. Individual police officers in Chicago and Minnesota used police databases to look up personal information of girlfriends, neighbors, or family members. Google Maps has been used by terrorists and thieves to help locate their targets.

- Projects may have a variety of indirect effects not intended by the original engineers as a result of the actions of others. The management of hydroelectric dams, touted as reducing flooding, sometimes produces more severe floods. Damn managers seeking to maximize the efficiency of hydroelectric dams keep water levels high, which lessens space needed to receive flood surges. Managers trying to keep water levels high may wait too long before starting emergency release of water during heavy rainfalls. In addition, agriculture and industry clearing land above the dam accelerates siltation, which shortens the expected life of the reservoir and may lead to coastal erosion.

To what extent should engineers be concerned about harm caused by others' use of their technology? "Can, and should," ask Doorn and van de Poel (2012, 2), "designers design technological systems so that they enhance rather than limit the responsibility of the users of these systems?"

As with all difficult ethical questions, making decisions about this requires balancing several moral factors. At least eight ethical considerations are important in thinking about this question: (1) the nature of moral responsibility; (2) promoting good consequences (the utilitarian perspective); (3) safety and future generations; (4) leaving the world no worse than one found it; (5) the Weak Samaritan Principle; (6) respect for autonomy; (7) the doctrine of intervening wills (*novus actus interveniens*); and (8) four concluding questions.

These considerations, of course, are in addition to whatever other factors are relevant. For example, if the harm is an environmental one, then ethical considerations about the environment, such as partnership with nature, may be important. As always, engineers making particular ethical decisions that involve the possibility of dual-use should use the five-step process to balance all the relevant factors for their particular case.

The Nature of Moral Responsibility

Moral responsibility is a complex topic about which ethicists have debated for over two thousand years. While some philosophers would disagree, I suggest that the first fully developed example of what has come to be called "*attributionism*" may be particularly helpful here (Schlossberger 1986, 1992, 2021).

The view and the issues involved are complicated but, briefly put, the idea is that people are morally responsible for their worldviews. My worldview is the complex network of attitudes, values, feelings, ways of perceiving, and so forth, fleeting and long-term, specific and general, that, together, constitute the way I make sense of and respond to the world, making me the particular person I am. More specifically, I'm responsible for embodying the traits or features that reveal my worldview, that make me the moral agent I am. For example, if, out of laziness, I don't stand up on a crowded bus when a sick or disabled person experiencing obvious difficulty enters the bus, I might be responsible for being selfish or uncaring – not having to spend the effort of getting up mattered more to me than another's distress. If I didn't stand because I didn't notice, I might be responsible for not paying attention to my surroundings, which, depending on the circumstances, might not be morally bad. If I didn't stand because I sincerely thought it might embarrass the other person, I get some credit for trying to be considerate, and perhaps (depending on the situation) some blame for poor judgment. What I'm responsible for depends on the particular story about me. A single action can make me responsible for embodying several traits, some good and some bad (or neutral). Our "moral scorecard" can be complex and nuanced.

The most relevant feature here of these extremely brief remarks is that the engineer should ask, "what does how I respond to the possibility of dual-use in this case (e.g., ignoring the issue or building safeguards into my work to limit the potential harm) show about my values and attitudes? What does it show about the kind of person I am?" For example, does my not taking the time to think carefully about this show, given the circumstances that I do not really care about the environment? Is my response to the possibility of dual-use in this case a morally valid precedent for being an engineer? The kind of soul-searching appropriate here obviously varies for each case and each person.

Promoting Good Consequences (the Utilitarian Perspective)

Some ethicists insist that all of ethics amounts to striving to bring about the greatest good for the greatest number of individuals. While many disagree with this strong version of utilitarianism, most ethicists agree that promoting good consequences is an important ethical goal. (See "Promoting Good Consequences" in Chapter 5 and "Consequentialism" in Appendix II.) One thing that ethical engineers should keep in mind is the consequences of the projects in which they participate. That means considering and weighing the magnitude, likelihood, and proximity of the benefits and harms (see below).

Safety and Future Generations

As noted in Chapter 4, the severity of a risk depends on how likely the risk is, how widespread the risk is (how many people or how much of nature is affected), and how severe the risk is (a risk of death is more severe than a risk of a minor rash). The same three factors apply to benefits: how likely is the benefit, how widespread is the benefit, and how large is the benefit to each person affected. If a project is almost certain to greatly benefit a large number of people but poses a very remote risk of minor harm to a few, then, other things being equal, an engineer need not refrain from working on the project (especially if there is no way to lessen the risk of harm.) There is no easy calculus for balancing troubling cases. When the atomic bomb was first tested in 1945, the scientists involved felt it was extremely unlikely that the test blast would ignite Earth's atmosphere.[3] Was it irresponsible to take a remote risk of destroying virtually all life of earth? Engineers and scientists working at Los Alamos had to be aware of the dangers of a world with nuclear weapons and of the horrors suffered by civilians if the bomb were deployed (as it was), as well as the danger of the planet coming under Nazi rule if the Germans developed nuclear weapons first. Inevitability plays a role as well. Even had Oppenheimer and the others refused to work on nuclear weapons, it seems inevitable that, within the reasonably near future, someone, somewhere would have developed them. Whether or not engineers in a particular firm agree to build a dam in a developing nation, modernization (and resulting famine, environmental degradation, and cultural disruption) will occur. Engineers should ask themselves whether building the dam is the least bad way in which modernization will occur. As noted in Chapter 10, however, this is not an all-purpose excuse. One simply should not do some things that are straightforwardly evil, even if someone else will do them anyway. In 1973, when Attorney General Elliot Richardson resigned rather than follow President Nixon's order to fire the Watergate Special Prosecutor Archibald Cox, Richardson knew someone else would do it. (Robert Bork, later Ronald Reagan's unsuccessful nominee for the Supreme Court, fired Cox after William Ruckelshaus also resigned rather than carry out Nixon's order.) Richardson and Ruckelshaus did the right thing, even though their integrity in resigning did not prevent the firing. (The "Saturday Night Massacre" did, however, help hasten Nixon's resignation.)

The proximity of the risk – how close the risk is in space, time, and the causal chain – also plays a role. If the company for which an engineer is developing software with a legitimate use is planning to sell the software to local terrorists, the risk of harm from the software's other use is immediate in all three dimensions: the chain of events linking the development of the software with the deaths of innocents is not a long one and the deaths will occur both soon and nearby. For engineers working on burying nuclear waste in concrete vaults, the biggest risk is to future generations. Even if current vault technology improves to the point where engineers are certain the vaults will not leak during the next five hundred years, the stored material will still be dangerous in 50,000 years. No current storage technology is effective for anything close to that length of time. The risk is remote in time but severe and foreseeable. For engineers working on cooling systems for nuclear reactors, the storage risks are also more remote in the causal chain. It is not the cooling system as such that poses a risk to future generations but the waste material produced by the reactor using the cooling system. (Of course, risks posed by leaks, accidents, and meltdowns are more proximate in both cases and may pose dangers to current generations, which is a separate issue.) Because of this remoteness (and the uncertainty it brings), our obligations to future generations are moderately discounted: they are real but not quite as strong as a similar obligation to living persons. (See Schlossberger 2008.) The risk posed to future generations by storing long-lived nuclear waste in concrete vaults is not quite as compelling as the same risk for harm five years from now would be (which is why some prefer nuclear reactors to burning fossil fuels). But it is not negligible either.

Leaving the World No Worse Than One Found It

An engineer, like everyone else, has a duty to leave the world no world than she found it (see Chapter 5). Engineers working on a project that will leave the world significantly worse off need to take this seriously in their moral decision-making. Other things being equal, such projects are better refused, though, as always, this factor must be weighed against others. Of course, when a particular harm is inevitable, an ethical engineer might help bring about the least bad form of that harm. True, one is helping to make the world worse than one found it. But the world will be less bad than it would have been without one's efforts.

The Weak Samaritan Principle

Another factor engineers must consider when deciding how far their responsibility goes is the "Weak Samaritan Principle." The principle says, "it is a moral shortcoming (it is morally less than ideal) not to rescue from serious harm or avert serious harm to an innocent other when (a) it is within one's power to do so, (b) without significant, non-tangential cost or risk to oneself or to the innocent pursuits of others, and (c) without doing a wrong or violating a right" (Schlossberger 2015, 198). For example, suppose I come across someone who is drowning in a pool and crying out for help. There is no one else around. I can easily toss him the life preserver on the ground next to me, and, if I toss the life preserver, no one else is harmed or wronged and there is no risk to me. Refusing to throw the life preserving would be a moral shortcoming. This Weak Samaritan Principle applies to taking measures to mitigate the risk of harm posed by others' use of one's technology. Of course, the Weak Samaritan Principle is only one factor that applies. Engineers need to consider how likely, severe, and widespread the risk is. As the risk becomes more likely, severe, and widespread, what counts as a significant cost or risk changes. Engineers should tolerate more cost or risk to avert a very likely, severe, and widespread harm. The more foreseeable and serious the harm posed by others' use of one's technology, and the more one has control over and responsibility for that technology, the worse it is not to do something about it when one can. Engineer Arkan designs refrigerators. There is a remote risk that someone or other will kill a family member by hurling the family member against one of Arkan's refrigerators. Arkan can mitigate that risk by encasing the refrigerator in heavy padding. However, the risk, while severe (death), is remote and local (the harm is unlikely to happen and would involve a single individual). The heavy padding interferes both with the usefulness of the technology (making it more expensive and significantly larger, thus providing considerably less storage space for the same amount of wall space) and with the refrigerator's aesthetic desirability to consumers. A company that tried to market such a refrigerator would almost certainly put itself out of business. Yet, even were the padded refrigerator somehow commercially successful, the padding would probably not prevent a single death. However, the likelihood, severity, and range of harm from glue-sniffing (solvent abuse) is high, especially in Latin America. Adding Allyl isothiocyanate (found naturally in horseradish and wasabi), as Testor did in 1968, does not decrease the effectiveness or suitability of glue for its proper purpose but makes it unpleasant to sniff, thus mitigating the likelihood of harm from misuse of the glue.

Respect for Autonomy

The next two factors are based on the fact that people are responsible for their own lives. As discussed in Chapter 5, engineers should show respect for others' autonomy. Some people would argue that it is the glue-sniffer's decision whether or not to sniff glue, not the decision of the engineer deciding on the glue's formula. "Engineers who refuse to build [a] dam out of concern for the foreseeable consequences are, in effect, denying the people of Kenya the ability to decide for themselves (through their government) whether the benefits of the project are worth the foreseeable harms" (Schlossberger 1997, 322). Respecting the autonomy of those who may decide to misuse one's technology is an important factor countering some of the other factors discussed above. "On the other hand, even when we are not entitled to prevent other people from harming themselves, we should, in general, avoid helping them do it. While ripping the cigarette out of the hand of a 40-year-old neighbor recovering from lung cancer violates her autonomy, assisting and encouraging her smoking is morally unsavory. 'If she is going to kill herself,' one might say, 'she must do it without my help'" (Schlossberger 1997, 322). In addition, misuse of one's technology may affect children, where autonomy is much less of an issue. That consideration, in turn, raises the issue of who should make decisions about a child's welfare. For example, building a dam in Kenya affects children. If the dam results in outbreaks of schistosomiasis (by providing a fertile breeding ground for the snails that host schistosomiasis), children will suffer. Should the decision affecting the welfare of those children be made by the Kenyan parents of those children (a parent-centered view), the government of Kenya (a government-centered view), or with the "global village," including the engineers tasked with building the dam? Again, even if the decision rests properly with the people of Kenya, an ethical engineer might say "but not with my help."

In general, autonomy-related questions about the effects of an engineer's projects raise three questions discussed in Chapter 4: Do those who bear the risk receive the benefit? Do those who bear the risk do so voluntarily? Are those who bear the risk aware of the full extent of the risk?

The Doctrine of Intervening Wills (*Novus Actus Interveniens*)

In many cases, the individual misusing an engineer's technology is causing harm to third parties (not themselves). Respecting the autonomy of a terrorist does not mean respecting his decision to harm innocent victims. The responsibility of engineers in such cases is complicated by the doctrine of intervening wills (*novus actus interveniens*). The doctrine says that people are not responsible for an outcome when another person's will intervenes. For example, when Jeremiah throws a rock through a window after Abib told him something about the owner that made Jeremiah angry, the doctrine says that it is Jeremiah, not Abib, who is responsible for the broken window, even though Abib knew Jeremiah would break the window. Again, Watkins informs Clark that Clark's wife, Clarissa, is having an affair. Clark then murders Clarissa. The doctrine says Watkins is not responsible for Clarissa's death. Clark' choice to kill Clarissa came, in the causal chain of events, between Watkins's action and Clarissa's death. Similarly, the doctrine says that engineers working on the Kenya dam are not responsible for the outbreak of schistosomiasis because the people of Kenya freely chose to build the dam, that programmers who worked on Google Maps are not responsible for the deaths of innocents because terrorists freely chose to use the software to kill people, and that glue manufacturers are not responsible for the deaths of glue-sniffers.

As a universal principle, the doctrine is surely wrong. Someone who pays an assassin to kill his spouse can't evade responsibility for the murder because the assassin freely chose to carry out the contract. Both are responsible for the killing. Still, people are not responsible for everyone else's bad decisions. A witness who truthfully testifies in court that he saw the defendant shoot, in self-defense, a member of organized crime is not responsible for the death of the defendant when the criminal organization kills him in retaliation, even if the witness knows that is likely to be the result of his testimony. However, if Watkins gleefully goes out of his way to inform Clark of Clarissa's infidelity, even though Watkins is fully aware that Clark will kill Clarissa, and if there is no morally, legally, or prudentially compelling reason for Watkins to tell Clark, then it seems clear that Watkins acted wrongly. When Clarissa's sister blames both Watkins and Clark, she is not mistaken. Whether or not glue manufacturers are morally responsible for the deaths of glue-sniffers, the fact that another's will intervenes does not shield them from moral reproach if they can easily do something to diminish the risk but choose not to. Doing nothing would violate the duty to leave the worse no worse, the Weak Samaritan Principle, and so forth. However, recall the case of Arkan in "The Weak Samaritan Principle" above.)

When deciding whether one bears responsibility for doing act w, which, as the result of the choices of another person, led to a bad outcome, x, the four concluding questions below may prove helpful. In particular, one may ask:

- How foreseeable was x and how likely was it that x would occur?
- Was there something, z, I could have done to prevent x? How difficult or onerous would it have been for me to do z?
- Was there a strong and legitimate reason for me to do w despite the risk of x?
- How proximate in the causal chain is w to x?

Four Concluding Questions

While there is no clean line to be drawn, four factors may be articulated. (i) Is the activity that foreseeably leads to harm one that I am morally entitled to do? To avoid blame, I must have clean hands. After all, I am not morally entitled to ask someone to kill my aunt, while I am morally entitled to answer truthfully a question on the witness stand. (ii) Does the activity have an important point and purpose other than the foreseeable harm? There is a good reason to tell the truth on the witness stand, a reason quite independent of organized crime's desire for revenge. Similarly, the pursuit of knowledge constitutes a good reason to publish one's research, even if one foresees that other people may misuse and misrepresent the data. By contrast, there is no good reason to seek out and inform organized crime that Jones killed Smith. (iii) How severe, widespread, and likely is the foreseeable harm resulting from one's action? Publishing data that one knows will result in the end of life on earth is inexcusable, however scientifically valuable the data might be. (iv) How important is the activity that leads to the harm? How much will one suffer by abstaining from the activity? Refusing to tell the truth on the witness stand, for example, renders one liable to incarceration, a serious harm. By contrast, one loses nothing by keeping quiet instead of seeking out organized crime.

TWO PRINCIPLES OF INSTITUTIONAL RESPONSIBILITY

The engineer is only one person in an organization, and that organization is only one organization within a society. The individual engineer can't take responsibility for everything. She can't even take responsibility for all the consequences of her own actions, since, very often, we can't know or control all the consequences of what we do. Moreover, the engineer has to balance her personal responsibility against the willingness to be part of a team.

Addressing this balance are two principles of institutional responsibility. I have to put considerable trust in the larger institution of which I am a part. I cannot be a one-person watchdog committee. But my trust in the company should not be blind trust: I should keep an ear open, and I should not ignore signs of trouble. In short, I should be aware of the effects of my actions on the company as a whole, and I have some duty to keep an eye on how the company operates. We can summarize this ideal in two sentences.

The first principle of institutional responsibility is that I am responsible for seeing to it that my participation in a project, in my company, in the profession and in society generally supports and leads to an ethical outcome. For example, if I am asked to treat an employee unfairly, falsify data, etc., it is very clear that doing so would violate the first principle of institutional accountability: my conduct would be aiding and abetting unethical corporate behavior. The first principle applies to many cases in this volume, such as Case 16 in Chapter 5 (Ford Pinto) and Case 85 in Appendix III.

The second principle of institutional responsibility is that I am responsible for monitoring, to the best of my (often limited) ability, the ethical character of my company, profession and society, and taking whatever steps are warranted when my company, profession or society goes astray. One application of the second principle (the duty to monitor the company's activities) merits special notice. Engineers must be alert for what David Frew calls "synergism."[4] Frew interviewed employees of a heavily polluting company, from the President down, and realized that "although each interviewee recognized his organization's role as a polluter, none either was acting in a way [that] would be interpreted…as [directly] causing the pollution, nor was any individual in a position to end or…significantly change the process" (Frew 1990, 231). The ethical engineer will look out for such effects, give some thought to what could be done to change them, and bring these observations to the attention of appropriate superiors. The second principle also applies when my action or project, in itself, may not be clearly harmful. I may come to have knowledge or suspicions about other projects of the company, or about the big picture of which my project is but a small part. (Case 26 in Chapter 7 is an example.) The second principle of institutional responsibility does not require me, single-handedly, to take responsibility for correcting the wrong. It requires rather that I ask myself "what feasible and warranted steps are available to me?" The answer will depend on the particular case. Guidance in making that decision is given elsewhere in this volume.

INSTITUTIONAL DUTIES

Rules and duties sometimes conflict. When they do, engineers need to decide which rule has greater weight. One factor in weighing conflicting duties is that institutional duties have special importance.

Institutions such as engineering, journalism and medicine have, as noted in Chapter 3, special duties. Because medicine is an institution devoted to health, physicians have a special duty to promote health. So, a physician has a special responsibility to stop and give aid to the victim of an auto accident when he drives past a serious collision. Similarly, while "don't spread (unsupported) rumors" is a rule that applies to everyone, it is much worse for a journalist to spread an unsupported rumor than it is for me to do so. This is because journalism as an institution has a special commitment to accurate reporting, and so the journalist has a special obligation to report accurately. (This is one feature that separates actual journalists from the so-called "journalists.")

Engineers also have special institutional duties. They include the duties to protect public safety, to use technological know-how to further human welfare, to keep accurate records and perform adequate tests, and to work in partnership with nature. The rules that reflect these duties have special weight, and this gives them an edge when they conflict with other rules.

It is helpful to understand just what institutional duties are, and why they are of special importance.

By an "institution," I mean an important organization or clearly defined body that has a socially recognized role it is expected to perform, and is governed by socially recognized rules, procedures, and practices.

For example, in our society, the health care profession is an institution. Because health care professionals are licensed by the state, the health care profession is a clearly defined body of doctors, nurses, hospitals, nursing homes, etc. The health care profession has a clear, socially recognized role, namely, attending to health. The way that physicians and therapists attend to health is determined by certain well-understood roles, such as that of physician and nurse, and health professionals are expected to follow socially recognized practices. (So, faith healers and Aunt Tillie's starving your cold and feeding your fever don't count.)

Engineering is also an institution. It is a clearly defined body (it is usually clear who is an engineer and who is not).[5] Engineering has a socially recognized role: safely advancing the progress of the human community, in partnership with nature, through know-how used in a systematic practice of clear, clean, practical decision-making. And it does this by establishing well-understood roles and practices.

Institutions create special duties. Institutional duties are created when an institution cannot function without public trust in the institution's faithfully performing its task according to its practices and roles. For example, if the public could not trust audit accountants to use standard accounting procedures to give an accurate and exact picture of a company's financial situation, there would be no reason to hire audit accountants. The institution of accounting, one might say, makes a kind of public promise to do this, and functions as a profession only because people trust accountants to keep this promise. So, accountants have an institutional duty to keep or create accurate records in accordance with standard accounting procedure.

In the same way, engineering as a profession depends upon public trust. If company X is known to run shoddy, inaccurate, or falsified tests, no rationally run society will allow X to build a bridge, a nuclear facility, or a chemical plant. Similarly, no manufacturer will purchase a synthetic plastic from X unless the specs of that plastic are reliable. In general, if people could not rely on the implicit promise of engineering as a profession to perform its social role faithfully, the profession of engineering could not function as it does.

Therefore, engineers have special institutional duties to protect public safety, to use technological know-how creatively to further human welfare, to work in partnership with nature, and so forth. An engineer who does not go out of her way to protect the safety of the public, thus, violates a special trust, a trust she invites by becoming an engineer.[6] Everyone, of course, has a moral duty not to endanger other people. Everyone also has a stake in advancing human welfare. But engineers have a special institutional duty to safety and human welfare that goes well beyond the general moral duties non-engineers have: by joining the institution of engineering, engineers become obligated to give greater weight to safety and welfare, and go further in their pursuit of safety and welfare, than other people. Similarly, engineers have a special institutional duty to perform thorough and reliable tests, and keep accurate and precise records of those tests, to respect nature, and to obey legal requirements, in letter and in spirit.

These institutional duties put great pressure on the engineer not to falsify tests, even if that means losing one's job or harming the company. As always, there are exceptions: an engineer forced by the Nazis to develop efficient methods of exterminating Jews ought to sabotage that project if she can. But the exceptions are rare. Usually, because institutional duties have great weight, they must not be violated.

WHEN TO BREAK THE RULES

One important and difficult ethical question professionals often have to face is "when should I break the rules"? When should an engineer not "work through channels"? When should a judge overlook a legal technicality? When should a professor make an exception to the rules for a student? When should a physician deceive a patient for his/her own good (by, for example, prescribing a placebo)? When should an editor not print a newsworthy story?

There is no simple answer to these questions but there are considerations that help guide professionals in answering them.

We might begin by asking "why have rules at all?" There are three reasons for following rules.

Reason (1): The Need for Reasonable Expectations. Rules help us know what to expect, and it is often necessary for people to know what to expect, and to be able to count on their expectations. Attorneys can't prepare their cases in a reasonable way unless they know how the court will proceed, what it will admit as evidence, and so forth. So, unless the court follows clear rules of evidence and procedure, attorneys cannot adequately represent their clients. Rules, in short, make life anticipatable, so that we may form reasonable plans, engage in long-term projects, and so forth.

Reason (2): Fairness. Rules help insure that like cases are treated alike. The rules governing the earning of merit badges in the Boy Scouts prevent John, who can only tie a half-hitch knot, from getting a knot-tying badge denied to Jim, who can tie both half-hitch and slip knots.

Reason (3): The Need for Rule-Governed Practices. Much of our lives depends on "rule-governed" practices, that is, social activities defined by a set of well-understood rules.[7] Chess, courtroom trials, square-dancing, company audits, and political elections are rule-governed practices. So is baseball. I can swing a bat and throw a baseball just to "loosen up," or to practice, or because I like the feel of a bat in my hands. That isn't playing baseball. My friends and I are playing baseball when we follow the rules about when to bat, when a run is scored, when a batter is out, etc. If you don't follow those rules, you're just not playing baseball. Most professions are also rule-governed practices. Not everyone who makes people healthier is a physician: the role of the physician is spelled out by certain rules and norms, and only by following those rules and norms is one acting as a physician. There are many ways to teach but only those who follow certain norms and rules about classes, lectures, discussions, grades, and so forth are professors. (Gandhi, for example, was in many ways a powerful teacher. But he was not a professor.) The same is true of engineers, judges, attorneys, and journalists. So, one should generally follow the rules of engineering, judging, and so forth, because a) unless one follows those rules one is not being an engineer or a judge, and b) society benefits from having engineers and judges. That is, the profession as a whole is a useful one, and the profession exists only when its members follow the rules of the profession.[8]

Example 18

Late in the season, the Cubs lead the Dodgers by one run in the 8th inning. The Dodgers are in contention for the Division title, while the Cubs are in last place. The Cubs have a runner on third base. The Cubs' batter lifts a fly to shallow right field. The runner tags and races home, preceding by a fraction of a second the throw from right field. According to the rules of baseball, the runner is safe and the Cubs score a run. One can imagine the Dodgers' manager arguing that it would be more useful to call the runner out, because this would make the game and the pennant race closer, providing more pleasure for the fans, which, after all, is the point of baseball.

Discussion: We all agree that the umpire should be unmoved by this argument. We think so because baseball is a rule-governed activity: you can't have professional baseball if umpires bend the rules to make the game more exciting. The integrity of the rules is the backbone of the sport: it is what makes baseball a major league sport, and not something else. True, the fans might have more pleasure if this particular call didn't follow the rules. But for baseball to be a real sport, the rules must be followed, and baseball as a sport gives a lot of pleasure.

Example 19

Because of a technicality regarding search and seizure procedures, evidence proving beyond a doubt that Green murdered his landlord is ruled inadmissible, thus permitting Green to go free.

Discussion: We don't have a system of law if judges feel free to bend the rules whenever the rules would let a murderer go free. If judges are free to bend the rules, then we have replaced the rule of law with the rule of individuals. Now the rule of law is a good thing, a better thing than the rule of individuals. Citizens need to know what to expect from their courts. So, judges have a duty to obey the rules of law, even when following the rules in a particular case results in some unfairness. (It is a separate question, of course, whether the law should be changed.)

Now that we understand why rules are important, we can see when they should be broken. This is because the four reasons for having rules do not apply equally to all situations. Reason (1) is the need for expectations. Clearly, not all expectations are equally important: it is more important to know when it is illegal to buy stock (insider trading laws) than to know when your dancing partner will do-se-do. Reason (2) is fairness. Here again, not every kind of unfairness is equally important, and not every case of treating people differently is equally unfair. Reason (3) is the importance of rule-governed practices. Notice first that not all activities are equally rule-governed. Professional tennis depends on strict adherence to the rules. A friendly game of tennis between husband and wife does not. The couple is not playing tennis if they follow no rules at all but ignoring a few technicalities does not seem out of place, just as a legal contract is more rule-dependent than a casual promise to a friend. Second, not all rules are equally important to a practice. Third, some practices are more important and useful than others.

From these considerations, we can formulate four major questions to guide us in deciding whether to break a rule.

1. *To what extent is the practice rule-dependent?* Some practices are, by nature, "looser" than others. Professional sports need strict rules, while the rules of etiquette can be fairly loose without making it impossible to be polite. So, it is more important for World Cup football referees to adhere strictly to the rules of football than it is for a host to adhere to the rules of etiquette. One key factor is anticipatibility, the extent to which it is necessary for others to be able to anticipate precisely how the professional will act. For example, lawyers and therapists require their clients to divulge confidential information. Unless clients tell their attorneys and their therapists very sensitive and personal information, their attorneys and therapists cannot help them. The very nature of legal representation and therapy means that clients must rely on the confidentiality of what they say. So, it is crucial for clients to be able to anticipate when attorneys and therapists will reveal information confided to them during therapy or as part of preparing a defense. It is much less important for me as a guest to be able to anticipate precisely how my host will greet me when I walk in the door.

2. *How central to the practice is this rule?* There are several ways of assessing the centrality of a rule. First, to what extent would the practice be recognizable without this rule? Most practices have peripheral rules that can change without destroying the practice. In bridge, for example, whether "honors" scores 50 points or 100 points is peripheral – if the rule about this were changed tomorrow, we would still think of the game as bridge. But a game in which reneging is allowed would not be bridge at all: the essence of the game, its fundamental strategies and logic, would be lost. Similarly, a minor technicality of procedure is peripheral to the rule of law, while convicting someone without a hearing strikes at the very roots of lawfulness. In short, breaking some rules have less effect upon the integrity of the practice than others. Another relevant question is "*how important to the practice is standardizing this aspect?*" For example, accounting would not be accounting without standard bookkeeping practices, forms of ledger entry, etc. Following the rules about these matters is crucial to what makes accounting a profession. However, a professor who uses good but unusual grading methods and teaching styles is still recognizably a professor – standardization of these things is not crucial to teaching's being a profession. A third relevant question is "*what kind of rule is it?*" Of the first importance are the rules without which the

practice could not exist, such as rules of confidentiality for attorneys and therapists, rules of evidence for courts, and so forth. Of second importance are rules governing the tone, setting and public role of the profession, such as the rule that English Barristers wear wigs. Least important are rules for ease of communication and co-operation, which may usually be broken without harm when all parties involved are comfortable with and understand the situation. A final question in judging the centrality of a rule is "*how does keeping the rule in this case relate to the point of the rule?*" For example, the rules of confidentiality concerning audit accountants are meant to ensure that businesses can obtain accurate audits without having their legitimate business secrets revealed. The rules are not meant to protect fraud. So, an audit accountant may have to break confidentiality in order to reveal a fraud, as long as legitimate business secrets are protected. (By contrast, the rules of attorney-client confidentiality in the US law do cover some possible illegalities since the point of the rules is to enable clients to prepare a defense by discussing actions that might have been illegal.) Similarly, the rules of bidding are meant to promote fair commerce. If following the rule in a given case would undermine instead of promote fair commerce, this is a reason for breaking the rule (which must be weighed against the other reasons for not breaking the rule).

3. *How does the value of the point or aim of the practice measure against the harm caused by following the rules?* Not all practices are equally important. Baseball, whose purpose is to give pleasure, is less important than medicine, since health and life are more important than the sort of pleasure one gets from a baseball game. So, if following the rules of baseball places people's lives in jeopardy, it is better to give up playing baseball than follow the rules.

4. *How unfair would it be to others if this case were treated differently?* This question has two aspects. *Would there be any real unfairness if not all cases are treated alike in this regard?* For example, it is important that how well a defendant's case goes does not depend on which judge tries the case. It would be unfair for defendants who get a "lenient" judge to be treated better than defendants who get a "strict" judge. Thus, a judge must hesitate before she bends the rules, even in the cause of justice, if she feels that other judges might not act as she does. The other aspect concerns appearances. *How sensitive to the appearance of irregularity is the situation?* In many cases

the appearance of impropriety or favoritism is almost as bad as the actuality of it. This is especially true of positions of public trust, such as the judiciary. In general, *strict uniformity is more important in conflictful than in co-operative situations, in formal than in informal settings, and in public than in private settings.*

These questions articulate factors that must be balanced: the factors in favor of breaking the rule in this case must be weighed against the factors in favor of following the rule. For example, if breaking the rule would appear highly irregular in a very conflictful situation, this would outweigh a small actual unfairness that would result if the rule were followed. In general, rules should be followed unless the case for breaking them is very strong. *Give the benefit of the doubt to following the rules.* Also follow the rules if the reasons for breaking them are only slightly stronger than the reasons for following them.

Case 21: When to Take Bids

Is it ethical to buy a product without an investigation/bidding process, when the product is fairly good, and although there might be a better product (or equally good, cheaper product), the advantage gained by procuring the other product is probably a minor one, not worth the effort/cost of an investigation?

Discussion: The engineering profession cannot operate as it does without public trust: reliance in the strict probity and fairness of bidding and supplying procedures is a precondition for engineers to function as they do. Hence there is an institutional duty of strict propriety in these matters. The rules of propriety here are central to the practice of engineering: to buy materials casually, haphazardly, or on a friendship basis, rather than on the basis of a clean-cut, rigorous, rational decision process, would strike at the heart of engineering as a professional practice. It is important to the practice of engineering that bidding be reasonably standardized, in part because the interests of commerce (for both buyers and suppliers) is best served when suppliers can anticipate how buyers will go about making purchases. Moreover, the situation is conflictful (the firms involved in supplying and bidding on contracts are competitors), and so the appearance of impropriety is almost as damaging as actual impropriety. The legitimate gain in bending these rules, the avoidance of a brief bidding/

investigation process, is minor. Thus, there is a very strong case for strict adherence to rules of propriety: the engineer should institute a bidding/investigation process. In general, engineers should err in the direction of caution in giving or accepting anything that smacks of a favor or gift, in acquiring an interest in a supplier or potential contractor, etc. The rule of thumb is: if there's any doubt, don't do it.

However, it is clearly silly to institute a bidding procedure before buying a single 50 cent green pencil. Our four questions help us to see why this is so. Neither the possible saving to the buyer, nor the possible profit made by the seller, comes close to the cost and trouble of taking and submitting bids. There is no point to extending the rules of bidding to tiny purchases. Thus, keeping to the rules about bidding when buying one 50 cent pencil seems to conflict with the point of the rules, which is to facilitate fair commerce. Rules about tiny purchases are peripheral rather than central. And, because only a few cents are involved, there no real unfairness involved, and this is not a conflictful situation in the way that major purchasing is.

Case 22: "Adjusting" the Records

Making a minor "adjustment" to records would greatly simplify matters (for example, back-dating an order one day would significantly simplify calculating taxes).

Discussion: The same considerations apply to bookkeeping procedures. The legitimate use of such procedures depends upon reliance on their strict accuracy. Record-keeping as a practice is highly rule-governed, and demands strict adherence to rules – lax records serve little purpose. It is crucial to record-keeping as a practice that records be standardized. Back-dating an order, however harmless it may appear, casts in doubt the very practice of order-dating itself, and without public trust in the accuracy of such matters the profession cannot function as it does. (For example, taxing agencies could not trust firms to do their own record-keeping and would have to require things such as government-kept books, a practice that would be costly and invasive to the government, to engineers, and to their customers.) Thus, there is an institutional duty of strict adherence to these rules, and to avoiding any appearance of impropriety. Finally, the benefit of bending the rule, namely avoiding a complicated accounting procedure, is minor compared to the benefits conferred by the practice of strict record-keeping.

Case 23: Keeping a Promise

You are the head of the Civil Engineering Dept. of City N. You promise Washington, who wants to take a year's unpaid leave, that their job will be there for them when they return. Meanwhile, the budget is severely cut. The best way of adjusting to the reduced budget includes eliminating Washington's job.

Discussion: Drawing upon the discussion of Example 11 in Chapter 5, two conflicting arguments could be made. Argument (1) goes as follows:

Promises should be kept.
You promised Washington their job would be kept for them.

_____

Therefore, you should keep Washington's job.

Argument (2) centers on the duties of a public servant:

A civil servant should do what is in the best interests of the public.
It is in the best interests of the public to eliminate Washington's job.

_____

Therefore, you should eliminate Washington's job.

As in all cases in which moral factors pull in opposite directions, you must weigh, for this particular case, the importance of keeping your promise against the importance of best serving the public. If, for example, the hardship for Washington of losing their job would be severe but the gain in public service of cutting their job would be slight, then you should keep Washington's job. Another factor is the nature of the promise made to Washington. A casual promise is less binding than a solemn oath. To what extent can Washington reasonably rely upon the kind of promise you made to them?

CODES OF ETHICS

Most major engineering societies have published a code of ethics. (Links to a number of such codes of ethics from across the globe of may be found in Appendix I.) This section first characterizes the contents of common codes of ethics (and their rationale) and then discusses the role codes of ethics should play in engineer's ethical thinking. For a more detailed discussion of these points, see Schlossberger (2016). At the heart of engineering codes of ethics is the proclamative principle, which "directs engineers to strive to insure that their actions, thoughts, and relationships be fit to offer to their communities as part of the body of moral precedents for how to be an engineer" (Schlossberger 2016, 1334).

Contents of Codes of Ethics

Engineering codes of ethics consist of seven *core clauses*, versions of which are found in most of the major societies' codes of ethics:

- *The Paramountcy Clause* (public safety and welfare): Engineers should hold paramount the safety and welfare of the public.
- *The Environment/Sustainability Clause*: Engineers' work should protect the environment and aim toward and promote sustainable practices.
- *The Competency Clause*: Engineers should work only within the areas of their competence.
- *The Honesty/Integrity Clause*: Engineers should avoid bribery, deceitful or deceptive conduct, and conflict of interest.
- *The Collegiality Clause*: Engineers should support the professional development of fellow engineers.
- *The Faithful Agent Clause*: Engineers should act as faithful fiduciaries or trustees of the client or employer (private or public) who engages them.
- *The Dignity Clause.* Engineers should increase (or at least not diminish) the prestige of the engineering profession.

In addition, there are *code-specific clauses* found in the codes of only one or a few organization's codes. Two code-specific clauses of special interest are:

- *The Duty of Association Clause*: Engineers should not associate with disreputable individuals or organizations (ASME).
- *The Public Outreach Clause*: Engineers should participate in civic life and extend knowledge of and appreciation for engineering (NSPE).

The rationale for the core clauses should be clear, especially after reading Part Two. For example, the Paramountcy[9] and Environment/ Sustainability clauses are applications of core values of the engineering profession (safety, advancing human welfare, partnership with nature), while the Collegiality clause is an application of community. Less obvious is the rationale for the more controversial code-specific clauses.

As Schlossberger (2016) points out, the Association Clause might be deemed to apply to six different types of association. An engineer might *directly participate* in a morally unsavory project, *indirectly participate* (for example, by designing a nozzle that might have different uses but is being used in Nazi death chambers), working on an unobjectionable project with an organization whose unrelated project is objectionable, such as working on a biodegradable plastic for a company that also produces Ricin (*insulated professional relationship*), *publicly supporting* an objectionable organization, person, or activity either *as an engineer* or *as a private individual*, and *privately associating* with a morally objectionable person (such as Jeffrey Epstein).

Promoting good consequences and the proclamative principle provide a reason why engineers should think twice about some of these kinds of associations. One reason for not being close friends with a moral monster is that one's friendships should proclaim one's values. Is befriending an evil person a good moral precedent? (As always, this factor must be balanced against other factors, such as family loyalty.) Helping to make profitable a company that (unrelatedly) produces biological weapons does not promote good consequences – the consequences would be better if the company went out of business or at least had fewer resources to commit to biological weapon development. Moreover, some instances of direct participation may violate the Paramountcy Clause, while indirect participation may sometimes violate the responsibilities for dual-use. However, trying to regulate what engineers publicly support as private individuals might violate their free speech rights.

The Public Outreach clause is supported by the model of the engineer as social enabler and catalyst, which suggests engineers should help the public understand their needs and the ramifications of engineering decisions; by the coercive bargaining position created by licensing of engineers; and by the institutional duties of engineering as a profession.

The Role of Codes of Ethics

Michael Davis (2002) has argued that engineering ethics consists principally in an engineer's following the relevant codes of ethics. By contrast, I have argued in this volume (and elsewhere) that ethical decision-making requires balancing, in a particular case, various ethical factors, some (but not all) of which are articulated by relevant codes of ethics. Moreover, people often agree about very general ethical statements but disagree about the specifics. Most people agree it is wrong to kill a person without a strong justification but disagree about what counts as a strong enough justification (for example, about whether executing a murderer is morally permitted). Thus, codes of ethics must be very general to gain general acceptance in the field. As a result, however, they are often of little help in settling difficult moral issues. Is hurting fish a strong enough environmental reason for holding back human welfare by not building a hydroelectric dam that would provide power for sewage treatment plants providing large numbers of impoverished residents with clean water? Most engineering codes of ethics are not helpful on this, since they require both promoting/advancing human welfare and respect for the environment. Even taking "paramount" literally, so that human welfare trumps environmental concerns, does not help, as harming the environment generally also harms human beings in the long run. Often, different individuals are more affected by the available options. Whose welfare counts more? The codes do not say.

However, a professional society's code does add moral weight to factors mentioned in this volume, especially for engineers who are members of the society. First, codes of ethics play a rhetorical role: a society's code expresses and proclaims to members and to society what that society stands for (it is proclamative), and the engineer chose to become (and remain) a member of that society. In addition, engineering societies' codes, collectively, help form the public understanding of the profession upon which is based the public trust that creates institutional duties. That is, one factor in creating the public trust in engineers (which allows engineers to design bridges, skyscrapers, and power plants) is the set of codes of ethics proclaiming to the public the commitments engineers make as members of their profession.

In sum, while no code of ethics replaces the need for individual decision-making of the kind described in Chapter 2, engineers should give extra weight in their deliberations to the factors emphasized by the relevant codes of ethics. Codes of ethics can also be a useful starting point in thinking about what to do.

NOTES

1. To see this, ask yourself if anyone would really give up a full and prosperous life in which others are more prosperous for a miserable life in which everyone else is more miserable. Virtually, no one would. What this shows is that what really matters is how good *your* life is, not whether others get ahead of you.

2. The same goes for revenge – people who make great personal sacrifices or take large personal risks to get revenge are not thinking clearly about what is important in life. Virtually, no one really thinks that having a full and rich life is less important than making sure no one gets away with wronging or harming them. Similarly, some people are dedicated to owning a 14-carat diamond, large wardrobes, a status car, going to the "right" college, outplaying their friends at golf, or giving a "better" party than their neighbors. People who see such things as the most important things in life have not thought clearly and rationally about what matters in life. (See "The Consumer Life v. the Life of Values" in Chapter 1.)

3. Before the Trinity Test, Hans Bethe felt that it was "absolutely clear...that nothing like that would happen" but the director of the Chicago Laboratory, Arthur Compton, was not absolutely certain (Horgan 2015).

4. Frew (1990, 230) defines "synergism" as "that property by which...a system...takes on an identity [that] is essentially different from the aggregate of the parts of the system."

5. Interestingly, however, an NSF study of the 1960 Census showed that "73,000 technically qualified persons...had been missed because they had identified themselves with nontechnical positions, presumably managerial" (cited by Layton 1989, 487).

6. Cf. Flores (1982, 273): "By accepting membership into a profession or any other institution, one voluntarily takes on a set of role-defined duties above and beyond one's general moral obligations."

7. Cf. Sullivan (1988, 41): "a practice is an activity socially organized and defined by impersonal standards through which certain goods are realized."

8. This version of rule utilitarianism is suggested by Rawls' famous 1955 paper, "Two Concepts of Rules," by Mabbott's 1939 theory of punishment, and explicitly defended by Stephen Toulmin (1950).

9. I question the term "paramount," as it suggests "above everything else," while, as pointed out in various places in the text, safety must be balanced against other factors. "Give special weight to" would be, I suggest, more accurate than "hold paramount."

REFERENCES

Awais, Rashid, Weckert, John, and Lucas, Richard (2009) "Software Engineering Ethics in a Digital World," *Computer* 42:6, 34–41.

Bird, Stephanie J. (2014) "Socially Responsible Science Is More than 'Good Science'" *Journal of Microbiology and Biology Education* 15:2, 169–172 https://www.asmscience.org/content/journal/jmbe/10.1128/jmbe.v15i2.870#b16-jmbe-15-169

Britannica, The Editors of Encyclopedia (2014) "Typhoon Nina–Banqiao dam failure," *Encyclopedia Britannica* https://www.britannica.com/event/Typhoon-Nina-Banqiao-dam-failure

Davis, Michael (2002) *Profession, Code, and Ethics* (Ashgate).

Doorn, Neelke and van de Poel, Ibo (2012) "Editors' Overview: Moral Responsibility in Technology and Engineering," *Science and Engineering Ethics* 18, 1–11.

Fish, Eric (2013) "The Forgotten Legacy of the Banqiao Dam Collapse," *The Economic Observer* http://www.eeo.com.cn/ens/2013/0208/240078.shtml

Flores, Albert (1982) "The Philosophical Basis of Engineering Codes of Ethics," in James Schaub and Sheila Dickison, eds., *Engineering and Humanities* (John Wiley & Sons), 269–276.

Frew, David (1990) "Pollution: Can People Be Innocent While Their Systems Are Guilty," in David Appelbaum and Sarah Verone Lawton, eds., *Ethics and the Professions* (Prentice Hall), 229–232.

Horgan, John (2015) "Bethe, Teller, Trinity and the End of Earth," *Scientific American* https://blogs.scientificamerican.com/cross-check/bethe-teller-trinity-and-the-end-of-earth/

Mabbott, J. D. (1939) "Punishment," *Mind* 48:190, 152–167.

MacLeod, Fiona (2019) "Reflections on Banqiao," *The Chemical Engineer* https://www.thechemicalengineer.com/features/reflections-on-banqiao/#:~:text=On%208%20August%201975%2C%20about%20one%20hour%20after%20midnight%2C%20the,died%20of%20disease%20and%20starvation.

Rawls, John (1955) "Two Concepts of Rules," *Philosophical Review* 64:1, 3–32.

Rio Declaration on Environment and Development (1992) Principle 15, United Nations Doc. A/CONF. 151/26.

Schlossberger, Eugene (1986) "Why We Are Responsible for Our Emotions." *Mind* 95:377, 37–56.

Schlossberger, Eugene (1992) *Moral Responsibility and Persons* (Philadelphia, PA: Temple University Press).

Schlossberger, Eugene (1997) "The Responsibility of Engineers, Appropriate Technology, and Lesser Developed Nations," *Science and Engineering Ethics* 3:3, 317–326.

Schlossberger, Eugene (2008) *A Holistic Approach to Rights: Affirmative Action, Reproductive Rights, Censorship, and Future Generations* (Lantham, MD: University Press of America).

Schlossberger, Eugene (2015) "Bad Samaritans, Aftertastes, and the Problem of Evil," *Philosophia* 43:197–204, 198.

Schlossberger, Eugene (2016) "Engineering Codes of Ethics and the Duty to Set a Moral Precedent," *Science and Engineering Ethics* 22:5, 1334–1344. Published online 2015 https://link.springer.com/article/10.1007%2Fs11948-015-9708-3.

Schlossberger, Eugene (2021) *Responsibility beyond Our Fingertips: Collective Responsibility, Leaders, and Attributionism* (Lantham, MD: Lexington).

Sullivan, William (1988) "Calling or Career: The Tension of Modern Professional Life," in Albert Flores, ed., *Professional Ideals* (Wadsworth), 40–46.

Toulmin, Stephen (1950) *The Place of Reason in Ethics* (Cambridge University Press).

USDIBR [United States Dept. of the Interior Bureau of Reclamation] (2015) "Dam Failure and Flood Event Case History Compilation," https://www.usbr.gov/ssle/damsafety/documents/RCEM-CaseHistories2015.pdf

Wingspread Conference (1998) "Wingspread Statement on the Precautionary Principle," *Johnson Foundation* Racine, Wisconsin https://view.officeapps.live.com/op/view.aspx?src=https%3A%2F%2Fwww.who.int%2Fifcs%2Fdocuments%2Fforums%2Ffo rum5%2Fwingspread.doc&wdOrigin=BROWSELINK

Part III

Problems and Issues in Engineering

7

Honesty and Professionalism

Honesty and professionalism sometimes require that engineers make tough decisions. No one wants to blow the whistle on her own company, inform customers of the defects of the product she is trying to sell, turn down a promotion (when the job is beyond her competence), or censure her friends and colleagues. Good people are sometimes tempted to avoid taxes or red tape by taking a "shortcut": modifying the records slightly often saves much work and money. Should one "give in" and "cook the books"? The material in this chapter will help you to make these decisions.

WHISTLEBLOWING

Although many companies and firms recognize that being unsafe or unethical generally does not pay off in the long run, it is unfortunately true that some companies or firms will try to cover up or ignore serious problems, such as unsafe products or structures, violations of environmental law (such as illegal dumping of toxic waste), falsified test results, or discriminatory hiring and promotion. The concerned engineer's efforts to get the company to correct the problem may prove futile. If the problem is serious enough, engineers may feel a need to "blow the whistle" on their firm or employer. Whistleblowing is disclosing information, outside of ordinary reporting mechanisms,[1] to a client, the press, the public, or an appropriate agency, organization, or responsible individual, indicating a serious issue of ethics, safety, or law, to which the whistleblower seeks to draw attention either in order to redress the problem or because the information is being withheld from someone who has a right to know. However, loyalty to their firm or company, confidentiality, and the fear of

DOI: 10.1201/9781003242574-10

reprisal all make engineers understandably reluctant to blow the whistle. The decision to blow the whistle is never an easy one. Whatever decision you make will leave you feeling uncomfortable, because whatever you do will involve betraying a loyalty. Whistleblowing always involves a conflict between loyalty to the organization and loyalty to society, and sometimes involves a conflict between moral duties such as honesty and fairness. Thus, every attempt should be made to address the problem within the company, or in a way that does not harm the organization. In some rare cases, this is not possible. So, the ethical engineer needs some guidance about when whistleblowing is permissible, and when whistleblowing is morally required. Here are some factors and remarks that may help if you are ever in this situation.

Duska (1989, 321) argues that whistleblowing is required when there is a clear harm to society; "it is the 'proximity' to the whistleblower that puts him in the position to report his company in the first place;" there is some chance of succeeding; and no one else is more able to blow the whistle and more proximate. de George (1982) suggests that whistleblowing is justifiable if it concerns a serious harm to the public and all avenues within the company have been exhausted. Whistleblowing is required, says de George, if, in addition, the employee has documentation of the problem and the employee has good reason to believe that whistleblowing will bring about the necessary changes to safeguard the public.[2]

Five factors count in favor of whistleblowing. The duty to promote good consequences means that one should be prepared to take major steps to prevent catastrophic harm. The duty to leave the world no worse suggests that one has an obligation not to participate in activities that will leave the world significantly worse than one found it. The institutional duties of engineering, as well as the values of engineering as a profession, require putting a high premium on public safety and acting to ensure that public safety is not compromised. The principles of accountability require that you take feasible steps to ensure that your company does no wrong. The points mentioned in "When to Fight a Battle" suggest that if you are in a special position to know about or correct the harm, or if you are closely connected to the institution doing the wrong, then you have a special responsibility to do something about it. (Of course, many additional factors may be relevant in a given case.)

However, *three factors count against whistleblowing.* Whistleblowing generally conflicts with loyalty toward the employer and with maintaining a community atmosphere of teamwork and mutual interest within the company. Treating others fairly means that you may not publicly accuse or take other harmful action against another without hearing the other persons' (or the company's) side and weighing the evidence very carefully. Whistleblowing often requires a breach of confidentiality. (For example, it is often impossible to document a danger without releasing company records, test results, or trade secrets.)

These factors must be weighed and a compromise solution reached. For example, exhausting all in-house avenues before going public would take some account of loyalty to the firm and the importance of community, and would accommodate the need to hear the other person's side, all without severely compromising the five factors in favor of whistleblowing (except in the case of a pressing emergency). Thus, to the extent feasible, *an employee should not blow the whistle without exhausting all in-house remedies*: the employee must go the extra mile to deal with the problem within the company in a supportive and constructive way. However, the company cannot operate as a community if it is devious and/or immoral, and thus if the company spurns all of an engineers' many attempts to get it to address a well-documented, serious or unethical danger it poses to society, the demands of loyalty and community become less relevant. That is, if the community atmosphere of the company cannot be restored until the harm is redressed, and if whistleblowing is the only way to redress it, then loyalty and community do not weigh very heavily against whistleblowing.

Violating confidentiality is particularly troubling since the engineer came into possession of the information only because she was trusted to keep it confidential. Thus, in publicizing the information, she is violating a trust and breaking an implicit promise. (See the discussion of promising in "Universality," Chapter 5.) There is a conflict, in other words, between keeping a promise and participating in making the world worse. The way to resolve this conflict is to ask "is the firm or company entitled to ask for such a promise on my part?" It is wrong to ask someone to promise to help you steal, and an unethical promise of that sort has little or no moral standing. (That is why, in contract law, contracts contrary to public policy are not enforceable.) Similarly, it is wrong to ask you to help cover up theft, embezzlement, or fraud. So, while the duty of confidentiality is very strong, it does not include covering up a crime. In general, releasing confidential information without your employer's consent is permissible only when every feasible attempt has been made to obtain the employer's consent and it is absolutely necessary to document extreme and severe danger to the public, serious violations of the law, or grossly immoral conduct.

Finally, the whistleblower must realize that the personal consequences of whistleblowing are often severe. The whistleblower may well lose their job. Even Whistleblower Protection Laws leave plenty of loopholes: the company can "reorganize" a department, and eliminate the whistleblower's job in the process. It is not easy to prove that the reorganization was a form of retaliation and not a legitimate business decision. Even if the whistleblower keeps their job, they may suffer harassment on the job, such as unwelcome transfers or working conditions ("welcome to your new office in the boiler room," or "we need you in our Antarctica office"). When a company transfers an engineer to a remote location, it is hard to prove that the transfer was a form of retaliation rather than a legitimate business judgment. Moreover, future employers are often wary of hiring someone who blew the whistle. Fortunately, as more companies and firms actively seek ethically minded engineers, "going public" may not mean "the kiss of death." After all, suppose I am the head of a large R&D department. I am an ethical person and would not tolerate falsified test results. Unfortunately, I cannot personally oversee the integrity of every aspect of every project. Would I rather hire an engineer who will give in to project leader N's demand to falsify results, or an engineer who would speak up and come to me with the problem? So, provided you can show me that you made every attempt to work within the company before blowing the whistle, and that the problem involved was well-documented, real, and serious, I would want you working for me. (Remember that, because I am ethical, I know that you would not need to go public if you came to me with a well-documented, serious problem.)

Nonetheless, life is not easy for the whistleblower. Apart from being fired or harassed, she can expect to be publicly attacked, to have her motives, her personal life, and her professional competence questioned in the media. Several prominent whistleblowers died in suspicious circumstances, including Babita Deokaran, a key witness in a state corruption case, who was gunned down on August 23, 2021. The potential whistleblower should thus ask herself whether her motives and her competence can withstand this kind of scrutiny, whether she and her family have the ability to weather the storm, and whether her documentation will stand up to a determined attack. Public reaction to whistleblowers varies. Some whistleblowers, such as Jorge Enrique Pizano in 2015,[3] Ashish Panday in 2001, Jeffrey Wigand in 1996, and Karen Silkwood in 1974, have been celebrated as heroes, while some, such as Julian Assange in 2010 and Edward

Snowden in 2013, remain controversial, with many supporters and detractors (See Touchton et al. 2020).

The whistleblower, therefore, must consider the issues very carefully before going public. Peterson (2020, 106) offers "a first piece of advice to any potential whistle-blower is to book a meeting with a lawyer." Which questions he must ask herself depends on the situation. Suppose an engineer is actively involved in an unsafe project. Then he must ask himself whether he can live with being an accomplice. After all, if I am helping to build an unsafe nuclear power plant, and I do not speak out, I must recognize that, in the event of a serious accident, I am personally responsible for the death and suffering that ensues. Suppose an engineer is tempted simply to quit instead of going public. Then she must ask herself whether this is something she can walk away from (she must weigh the factors mentioned in "When to Fight a Battle"). After all, the problem remains even if she leaves the company or firm. Consider an engineer who is in a special position to know about problems with an unsafe nuclear reactor. The danger is likely, severe, and widespread; she is in a special position to know about the problem; and her speaking out is likely to lead to change. Can she live with herself after a lethal accident if she didn't speak out but simply changed jobs? So, she must ask herself "*is there a neutral option*," "how bad is it" (morally and in terms of how likely, severe, and widespread the risk is), and "*how directly am I involved*?" That is, she must ask "am I involved in the infraction," "am I closely tied to the infracting organization," "*am I in a special position to know about or correct the problem*," and "would doing nothing compromise doing my job faithfully?" Suppose she is tempted to pass on the information anonymously. Then she must ask herself whether an anonymous "leak" would be effective in solving the problem, and whether the leak would really remain anonymous. The answer to both questions is often "no." Sometimes leaking the problem without providing documentation or testimony would not result in the problem's being corrected. If you do provide the documentation, it may not be hard for the company to figure out who leaked it. Sometimes, thus, leaking the information anonymously is simply not a viable solution. Finally, she needs to ask herself how vulnerable she is: "*can I afford to go public?*" The answers to these questions are, of course, connected. If the problem is bad enough, you are directly involved, your speaking up would correct the problem, and no one else will speak up, then you have to be prepared to bear the cost.

If the engineer does decide to do something about the problem, she must make every feasible attempt to work within the company, and she must *document everything* (both the problem itself and her attempts to resolve the problem within the company). Leaving a paper trail is often the key to successful whistleblowing.

The engineer in a potential whistleblowing situation must keep in mind the duty to set a moral precedent. (See "The Proclamative Principle" in Chapter 5.) Engineers in such a situation should ask themselves: "what kind of example do I think should be set here? About which overall course of action do I want to proclaim 'this is how to be an engineer' and, more broadly, 'this is how to be a good person'?"

Finally, there are three ways in which whistleblowing situations can be avoided. First, *choose your employer judiciously*. Don't work for unethical firms or companies. Second, *deal with potential problem situations early*. If you notice a tendency to cut corners about minor matters, deal with it before a major catastrophe looms. If you think there may be safety or construction irregularities in a project, check it out and deal with it before deadlines approach and correcting the problem would cost the company dearly. The earlier you deal with a problem, the easier it is to resolve it, both for you and for your company. Third, before any problems arise, suggest that your firm or company establish troubleshooting mechanisms for dealing with such problems. *Prevention is generally the best way of dealing with whistleblowing.* Lynch and Kline (2000) suggest shaping engineering practice to avoid heroic moments when engineers must confront management. As Westrum (1991, 315) puts it, "whistleblowing is essentially a last-ditch measure. It should be used when everything else has failed. It cannot solve the problems of most organizations…." Westrum quotes Samuel Florman's remark (1989, 19) that "a system that relies on heroism is neither stable nor efficient." Instead, Westrum recommends that companies establish troubleshooting mechanisms such as ombudspersons. (See also Appendix IV.)

Case 24: Cadmium and Outflow Sampling

Engineer Alomar of Z consultants is asked to sample the outflow from Y company's plant into a river used for drinking water. The sampling site was picked by the Department of Environmental Quality. The task of sampling is given to Alomar, who discovers that the sampling site is actually upstream of the main point of discharge. Although samples taken at the selected site are uniformly under the limit, samples taken from below the main point of discharge indicate an average of 0.020 milligrams per liter of cadmium, well above the legal limit. Alomar notifies both the Head of Z consultants and the President of Y company. Both order Alomar to remain silent about what they have discovered.

Discussion: Recall that cadmium, which accumulates in the kidneys, is linked to kidney disease and high blood pressure, and remains in the body a long time (it has a half-life of 10–30 years). The factors discussed under "Safety" in Chapter 4 suggest that Alomar should speak up. The risk involved is serious (kidney disease), widespread (it applies to everyone who drinks the water), and fairly likely. The danger to the public is not voluntarily undertaken, nor is the public aware of the risk. The individuals taking the risk are not, in general, those receiving the benefit. Alomar would not want their family to undergo the risk of drinking cadmium-laden water for these benefits, and, if Alomar ignores the problem, they are not being a good trustee of the public welfare. Moreover, the community can take steps to protect itself if the danger from cadmium pollution is known, which makes publicizing the risk all the more important. In addition, dumping dangerous doses of cadmium into a river hardly constitutes acting in partnership with nature, and if Alomar does not speak up, they are participating in making the world a worse place than they found it. Alomar has already done everything feasible to work within the company. If Alomar remains silent, they become an accomplice. Drawing on the factors mentioned in "When to Fight a Battle," we find that Alomar is directly involved, that remaining silent does interfere with doing their job properly, and that the infraction is serious. This is not a problem Alomar may walk away from, since they are deeply involved, have documentable evidence of a serious harm, and are in a special position to know about and do something about the danger. Of course, speaking up would conflict with loyalty to the client. However, the point of the duty of loyalty is to protect the legitimate interests of a client, and the client's interest in getting away with causing kidney disease is not a legitimate interest that deserves protection. Moreover, unlike some of the more controversial cases of whistleblowing, Alomar's speaking out does not require them to break confidentiality since the site of the discharge is not a trade secret. In any case, not speaking up would constitute fraud, and promises that require fraud are not binding, since the point of promising is not undermined by refusing to keep promises that constitute committing fraud (see "When to Break the Rules" in Chapter 6). Thus, Alomar should document everything carefully, and then inform DEQ of the problem, in a way that does minimal harm to Z consultants. If DEQ does not change the sampling site, Alomar must consider taking the problem to the governor or going public. Taking those steps would set a good precedent for how to be an engineer.[4]

Case 25: Going Easy on Safety Assessments

A company, in order to remain competitive, instructs an engineer in its employ to "go easy" on a safety assessment of a procedure or product. The engineer believes that the product or procedure may jeopardize workers' safety but also believes that no cost-effective way of resolving the problem exists. How should the matter be handled?

Case 26: Installing a Rapid Transit System (B.A.R.T.)

Your company is responsible for installing an expensive automated subway system. You are aware that (a) the automatic train control is unsafely designed, (b) contractors building the system are inadequately monitored, (c) there are continual problems with the software designed to run the system, and (d) plans for training operators and performing safety tests before public use of the system are inadequate. You write a series of memos about these problems to your employers, including your immediate supervisors and two higher levels of management. No action is taken. [Historical note: This case shows some points of similarity to the Bay Area Rapid Transit (B.A.R.T.) case. In the B.A.R.T. case, the engineers who wrote to the Board of Directors of the city's mass transit system were eventually fired. Problems did plague the system when it became operational. The computers were prone to temporarily losing sight of trains. A crystal oscillator malfunctioned, and on October 2, 1972 a train hit a sandpile (Westrum 1991, 313–314).]

Consider these possibilities:

1. Although you work for the company, you are not specifically assigned to monitor the safety of this project.

2. You decide to go outside of normal channels and contact the board of directors of the city's mass transit system. They wish you to provide them with signed documentation of the problem, to be released to the press. Alternatively, the board does not dispute your conclusions but for political reasons decides to take no action.
3. You are told by your supervisor to keep your concerns to yourself.

Case 27: Challenger Disaster

On January 28, 1986, the Challenger Shuttle exploded, killing six astronauts and a school teacher. The contract to design the rocket boosters using a segmented booster was awarded to Morton Thiokol (MT), although it was rated fourth in "design, development, and verification" (Rogers Commission 1986), behind Aerojet General (who proposed to make the booster from a single tube). Morton Thiokol had strong backing by the Church of Latter-Day Saints as well as Utah state officials. Dr. James C. Fletcher, head of NASA in 1973, was a member of the Church of Latter-Day Saints as well as a member of Pro-Utah, a Utah state lobbying organization. Although questions remain about this decision, it should be noted that "a congressional investigation has found no evidence that Dr. James C. Fletcher, the administrator of the National Aeronautics and Space Administration, or other space officials violated conflict-of-interest regulations in 1973 when they chose Morton Thiokol Inc. to produce the solid-fuel booster rockets for the space shuttle" (South Florida Sun-Sentinel 1987.) Authority to oversee the project was transferred from Lewis Research Center to Marshall Space Flight Center. Between 1970 and 1985, quality control personnel at Marshall were reduced from 615 to 88. Roger Boisjoly was the engineer responsible for the design of the booster segment joints. Early experiments showed that low temperature affected the seals. In 1985, Boisjoly wrote a memorandum to MT management, insisting that this problem with the seals could result in "catastrophe." On January 27, a teleconference was held between members of NASA (Marshall Space Flight Center and Kennedy Space Center) and Morton Thiokol (including Boisjoly and Robert Lund). Boisjoly urged that, due to the low temperature, the launch be delayed. At first, MT management agreed with Boisjoly. Lund said the Challenger should not be launched at temperatures below 53 degrees. However, NASA officials asked MT management to reconsider. Pressure from management led Robert Lund, Vice President of engineering at MT, to agree to the launch (Rogers 1986; Westrum 1991, 253–255). In later remarks, Roger Boisjoly (1987, 11) said that he would not sign off on anything he would not want his wife and children to use.

Discussion: Engineers must give high priority to the safety and welfare of the public (including the astronauts). The principles of accountability suggest Boisjoly and Lund should contribute to an ethical outcome. That suggests speaking up. Although the cost to Roger Boisjoly of speaking up is significant, the consequences of whistleblowing that stops the launch (saving seven lives) are better than the consequences of keeping silent. The golden rule, applied to the astronauts, suggests that Roger should speak up to try and stop the launch (were the positions reversed, Roger would want the engineer to speak up and try to save his life). Roger Boisjoly is in a special position to know about the danger. Roger has exhausted all remedies within Morton Thiokol. The harm is severe (death) and likely. Roger can document the danger (although see below). If Roger blows the whistle, he will be ostracized by his colleagues and harassed at work but this is a much smaller price to pay than death (what the astronauts may suffer). Roger is directly involved in giving a go/no go decision (he is part of the process). There is no feasible neutral option. It would appear, for these reasons, that this is a battle Roger has to fight by refusing to agree to give the flight a go and/or blowing the whistle if his refusal is overridden. When Lund was told (by an MT Vice President) to take off his engineering hat and put on his management hat, perhaps he should have tossed his hat in the ring (so to speak).

What, on the eve of the launch, can Roger do if his refusal is overridden? Roger Boisjoly came to my campus some years ago (he is now deceased), and I knew him because he had commented in print on an article of mine, so we had a chance to talk. Roger said he agonized for years about what he might have done. Roger pointed out that the data was complicated enough that he could not explain it beyond any doubt to non-scientists in a short phone call at a late hour. So, had he gone to the press, he said, they would have checked with his superiors, who would (correctly) have pointed out that before every launch someone had concerns. The press, he felt, would probably have said nothing before the launch. It had been my view that Roger should have called the chief of the astronauts (the crew itself is sequestered that close to a launch), who had veto power over the flight. Roger told me that while this would have been a good idea, it was not practical – he had no access. He mentioned as evidence that he had been trying for months to get Buzz Aldrin to sign a picture of the two of them and was never put through. He felt that, since he was blocked for months during normal business hours, he would have been blocked at this late hour when demanding immediate access.

This case demonstrates several other interesting things. NASA defended itself by saying that new launch criteria should not be added at the last moment and that there was no definitive evidence the 0-rings would fail, since no tests had been conducted at that temperature. It is worth pointing out, however, that at Congressional Hearings about the Challenger, Nobel Prize winning physicist Richard Feynman put a bit of the ring material in a glass of ice water and hit it with a hammer. The dent did not pop out. It took just five minutes, a bit of common sense, and a glass of ice water to demonstrate clearly that launching was problematic. Sometimes, bureaucracy should yield to logic. Additionally, it is worth noting that individuals not directly in the line of decision-making can influence decisions. William R. Lucas, Director of Marshall Space Flight Center, was not in the direct launch-decision chain. Lucas was not subject to any disciplinary action due to the Challenger. However, *The L.A. Times* reported that "under Lucas' leadership, [New York Representative James H.] Scheuer said, there was 'pervasive arrogance and smugness' at the center that made engineers deaf to complaints from contractors and astronauts" (Times Wire Services 1986). In addition, the *Times* stated, "critics charge he set up an autocratic empire in which negative information about the center's work was strongly discouraged" (Times Wire Services 1986). "Lucas was notorious for reprimanding—or, more accurately, verbally tearing apart—subordinates who made mistakes, in public meetings (R. Boisjoly, personal communication, July 17, 1997)" (Adams and Balfour 1998, 131). Lucas purportedly was adamant that Marshall should not be the source of any delay (Adams and Balfour 1998, 132). He was purportedly known for saying "don't tell me it can't be done– tell me how you are going to do it." Even though he was not involved in the decision chain, Boisjoly and others have said that his managerial style and attitude influenced subordinate NASA officials to pressure Morton Thiokol to agree to the launch. Lucas, in a 2020 Netflix documentary (*The Final Flight*), said that he does not believe he did anything wrong.

Case 28: The DC-10 Cargo Door

The DC-10's cargo hold and passenger cabin are both pressurized at high altitudes. If a midair collision, explosion or sabotage causes a hole in the fuselage, resulting in either compartment being depressurized, or if the cargo door should open in midflight, the cabin floor separating them will buckle, as it is not designed to withstand the force of the pressure differential. (Minimizing weight is a key design goal.) Since hydraulic control lines run through the cabin floor, serious buckling results in the loss of control of the airplane. The problem was noted during the design of the DC-10, resulting in what Daniel Applegate, director of product engineering for Convair, later called "Band-Aid" remedies that did not solve the problem. On June 12, 1972, the cargo door problem nearly resulted in a crash with 67 passengers aboard. This prompted Applegate to write a memorandum to his superiors predicting that a certain number of DC-10 cargo doors will come open, resulting in the loss of the airplane. In particular, Applegate noted that a design flaw makes it possible for the crew to think the cargo door was securely latched when the lock pins were not fully engaged. According to court documents, Convair decided to suppress the memo rather than pass it on to McDonnell Douglas (Witkin 1975). Convair reached an agreement with the FAA: no FAA action would be taken, on the basis of Convair's promise that the problem would be rectified. Convair did make some alternations, none of them adequate. It should be noted that McDonnell Douglas was experiencing financial difficulty and felt that delay in producing the DC-10 might jeopardize the existence of the company. Convair and McDonnell Douglas were already disputing how costs of design changes were to be apportioned between the companies, and Convair did not wish to risk footing the bill for expensive changes (Unger 1982). What, if anything, should Applegate do?

[Historical footnote: In 1974, the cargo door of a DC-10 opened in midflight. All 346 people on board were killed. *After the crash, the DC-10 cargo door was completely re-designed.* No subsequent cargo door blowout occurred (Beresnivicius 2019).]

Case 29: Cost Overruns (C-5A Galaxy)

A. Ernest Fitzgerald, an Air Force cost analyst, found over a billion dollars' worth of cost overruns on Lockheed's C-5A Galaxy transport.

He informed William Proxmire's Senate Committee of the problem. Although Fitzgerald had previously been given awards for saving the government money, he was investigated. After the investigation of Fitzgerald unearthed no improprieties, he was sent to work on bowling alleys in Thailand. Eventually his position was eliminated (President Nixon was recorded on the Oval Office tapes saying "I said get rid of that son of a bitch"), and Fitzgerald found himself blacklisted. He remained unemployed for four years, and it took 13 years until he was "fully reinstated" (Smith 2019; Westrum 1991).

Case 30: Surry Nuclear Power Plant Facility

Mechanical engineer Carl Houston inspected the welding operation for a nuclear power plant under construction in Surry, Virginia. He discovered improper welding techniques and underqualification of welders. Houston reported his findings to the manager and to his firm. Nothing was done to modify welding practices. Houston eventually notified Virginia Electric and Power Corporation, the reactor manufacturer, the AEC, the governor of Virginia, the Virginia Dept. of Labor, and two US Senators. Eventually, "some repairs were made to correct deficiencies, and increased surveillance of welding practices was instituted at this and other nuclear facilities" (AEC 1973). The license of the Surry plant was restricted and the inspection rate of the plant was increased (Unger 1982).

COMPETENCE

Engineers have an institutional duty of competence since the public places its trust in the competence of engineers. Competence is also a key component of professionalism generally, and a value of the engineering profession (excellence and clean decision-making are impossible without competence). Moreover, incompetent engineering endangers public safety, leads to bad consequences, and harms the firm or company.

Accordingly, engineers have a two-fold duty. *Engineers must make every effort to be as competent as possible and they should never undertake a task or responsibility beyond their competence.* The first duty means that engineering is a life-long learning experience. Engineers should strive to increase their knowledge and skills. This may include reading trade and professional books and journals, attending professional meetings, workshops, and seminars, taking additional courses or other training, and making the most of work opportunities. Ask questions, try to understand all the decisions pertinent to your project, speak to colleagues about what they are doing, try new things.

Another consequence of the duty to be as competent as possible is that you must make every effort to be at your best. Avoid coming to work tired. Don't let personal troubles interfere with your work. Take care of your health, both mental and physical. Exercise, relaxation, proper diet, a good attitude, and plenty of mental stimulation are all essential to functioning at your best.

The second duty means that engineers must know and be candid about their own limitations. You should never be afraid to say "I don't know," and never be too proud to look up information you're not certain about. When you are asked to do tasks beyond your abilities, you should speak up. When you need assistance, ask for it.

Case 31: Competence

You are a senior chemical engineer. The project you were supervising has just been completed. The engineer in charge of mechanical safety for a plant dies suddenly. Because the company is experiencing a temporary financial squeeze, it does not want to hire a new supervisor or promote anyone. You are therefore asked to assume the vacated position temporarily, until the financial picture improves or another senior engineer becomes available. Although you had some courses dealing with mechanical safety in college, your work over the last 15 years has been in a different area, and you do not feel competent to assume the position. How should you handle the matter?

KEEPING ACCURATE RECORDS AND OBEYING THE LAW

Three key rules apply to record-keeping and following regulations. First, *engineers must not falsify records nor break the law*. Second, *engineers should be faithful to the spirit as well as the letter of the law* and should not seek loopholes to circumvent either the law or sound record-keeping principles.

Third, *engineers must be scrupulous in avoiding the appearance of impropriety. Never say or write anything you would not wish to explain in court.*

Engineers must not falsify records because records are useful only if others can depend on their strict accuracy. Even what seem like "minor" irregularities can prove fatal (see Case 33). Back-dating an order, however harmless it may appear, casts in doubt the very practice of order-dating itself, and without public trust in the accuracy of such matters engineering cannot function as it does. (For example, the Internal Revenue Service could not trust firms to do their own record-keeping, and would have to require such things as government-kept books, a practice that would be costly to the government, to engineers, and to their customers.) Thus, there is an institutional duty of strict adherence to these rules, and to avoiding any appearance of impropriety. The considerations mentioned in "When to Break the Rules" also imply that records, especially records that have a legal use or status, may not be altered or falsified.

The case for obeying the law is even stronger. *Engineers have a special duty to obey the law.* "Engineers have an institutional duty to perform their engineering duties in accordance with law: only in the most extreme cases may engineers violate the law" (Schlossberger 1995, 164). The law represents the moral values of the community, as well the community's decisions about matters that affect it. When an engineer breaks the law, she is denying the public the ability to decide about its own future, thus violating the rights of citizens. This is especially problematic for engineers since it is only because of public trust that engineers are permitted to design and implement projects that might affect public safety. "That permission is an expression of trust that the engineer will faithfully represent society's judgment (through its legislature and regulatory agencies) about safety, record-keeping, and so forth" (Schlossberger 1995, 165). The law also constitutes the ground rules by which business is to be carried out. The considerations mentioned in "When to Break the Rules" indicate that, since business is competitive and public, those rules must be respected. Moreover, no company can survive unless people generally obey the law. No corporation could operate without a framework of laws to protect its property, without courts to uphold its contracts, or without the roads, sewers, financial system and so forth established by the community through its government (see "Engineering and Business" in Chapter 3). Thus, when a company breaks the law, it violates the principle of universality. (That is, a law-breaking company freeloads upon others' obeying the law.) It cheats the community without whose services the company could not survive.

It is equally important to *keep to the spirit of laws and regulations*. Because laws and record-keeping serve a valid purpose, it is generally wrong to defeat that purpose by using "loopholes" to circumvent the spirit of the law or create dubious or misleading records. In addition, using loopholes often violates the principle of universality. For example, the government passes pollution control laws and regulations in order to protect the health and environment of citizens. If all companies use a legal loophole to pollute beyond tolerable limits, the government will simply re-write the laws and regulations. If company A pollutes by exploiting the loophole, while companies B and C do not, then A is freeloading on B and C's compliance with the spirit of the law. Similarly, using misleading record-keeping techniques freeloads on those companies that keep honest records, for if all companies used "creative record-keeping," no one would trust any company's records.

Finally, *the appearance of irregularity can cause harm*, even when the letter and spirit of laws and regulations have not been violated. In particular, common sense dictates that you should be careful about what you say or write. Engineers should be aware that there is no such thing as a "private" comment or memo. Everything you say or write could conceivably wind up in court. Furthermore, every remark you make at a party, a church or social function, or while standing in line, reflects on your organization, firm, or company: people who know you work for company P will judge company P in the light of what you say. *Think of yourself as an "ambassador" for your organization, because in an important sense you represent the organization in everything you do or say.*

You should be particularly careful in anything touching on "sensitive" matters, such as anti-trust, discrimination, and regulatory laws. Give extra thought to every remark or memo that might be construed as expressing a bias concerning race, gender, national origin, sexual orientation, or religion. Avoid using terms or phrases that might suggest either power to control or collusion in controlling market factors. Avoid using terms or phrases that might suggest less than strict and voluntary compliance with laws and regulations.

Case 32: Tianjin Warehouse Explosion

Tianjin warehouse explosion, Tianjin, China. The Tianjin warehouse run by Ruihai International Logistics contained large quantities of illegally and unsafely stored hazardous materials. "Investigators found that Tianjin Ruihai International Logistics, the operator of the warehouse, illegally stored hazardous materials and that its 'safety management procedures were inept'" (Tremblay 2016). In August of 2015, Nitrocellulose in the warehouse became too dry and caught fire. The flames reached 800 tons of illegally stored ammonium nitrate, a flammable fertilizer. Firefighters were unaware of the illegally stored materials, including 700 additional tons of toxins such as sodium nitrate. Efforts to extinguish the fire, as well as the fire itself, resulted in two explosions. For example, calcium carbide reacts with water to create acetylene, which is highly explosive. The explosions, which together were the equivalent of 24 tons of TNT, damaged over 300 buildings and 12,000 cars, up to a kilometer away. One hundred seventy-three people were killed. The Number 2 Intermediate Court of Tianjin imposed a (suspended) death sentence on the Chairman of Ruihai International Logistics for paying bribes from 2013 to the time of the explosion to store the illegal materials (BBC 2016). Proper storage, effective and honest government oversight, and establishment of emergency response plans including worker training might have prevented the tragedy.

Case 33: GM Ignition Switch

GM Ignition Switch. In February and March of 2014, GM recalled 2.15 million Chevy Cobalts and HHRs, Pontiac G5s and Solstices, and Saturn Ions and Skys from model years 2003 to 2011, because of a potentially dangerous problem with the ignition switch: due to an improperly designed detent plunger, a jarring event or the weight of a heavy key chain could inadvertently switch the car to accessory mode (or even off) while driving, turning off power brakes and power steering (though manual brakes and steering remained operational) and creating a risk that airbags would not deploy in a collision. "Before the ignition switch went into production in 2002, some GM engineers knew that the switch could inadvertently rotate out of position. From approximately the spring of 2012, certain GM personnel knew that the low-torque switch presented a safety issue because it could cause airbag non-deployment associated with death and serious injury" (Raynal 2022). In 2006, a new ignition switch with a longer detent and greater torque was approved for 2008 models. However, the part number was not changed. The failure to change the part number made it difficult to determine which switches were faulty and which were not. As a result, some vehicles were repaired or serviced with the faulty part instead of the new part. The issue led GM to examine and ultimately recall the 2008–2011 model year vehicles listed above. At least 124 deaths and 274 injuries are alleged to have resulted from the faulty ignition switches (Associated Press 2021). GM acknowledges that 15 deaths have occurred in crashes involving 2003–2007 model year vehicles in which switch rotation occurred and may have caused or contributed to airbag non-deployment. In addition, during 2014, GM undertook "an extensive analysis across its platforms to assess the risk of unintended key rotation, and conduct[ed] full vehicle testing under extreme conditions" (Raynal 2022). As a result, GM also ultimately issued several additional key-rotation recalls that covered millions of vehicles worldwide (Lopez 2021).

SALES, HONESTY, AND DISCLOSURE OF PRODUCT'S LIABILITIES

Honesty and truth-telling are core values of the ethical life, and as such as normally govern engineering practice. One reason for being honest comes, as we have seen in Chapter 5, from the principle of universality: we all need to rely on a general social commitment to telling the truth, so someone who lies says, in effect, "all of you must tell the truth, but I will gain a special advantage by being the one who doesn't," and hence violates our understanding that morality doesn't play favorites. Duties of team loyalty, discussed in "Rights" (Chapter 5), provide additional reasons for telling the truth.

Case 34: Volkswagen Emissions

Volkswagen Emissions. Between 2009 and 2015, 11 million Volkswagen vehicles worldwide were outfitted with a defeat device. The device monitored steering wheel position, speed, and air pressure to detect when the vehicle was being tested. It then switched modes, running the engine at lesser power, in order to improve emission test results. When not tested, the device switched out of test mode and the engine ran at normal levels, emitting from 15 to 40 times the allowable level of nitrous oxide. Volkswagen said that the software was originally developed to lessen engine noise. One VW engineer who pled guilty was sentenced in 2017 to 40 months in prison for his role in the emissions scandal. Hanno Jelden, who was in charge of developing the software, blamed the problem during his 2021 trial on Volkswagen's corporate culture, which, he claimed, emphasized solving problems quickly rather than analyzing the root problem. The issue has cost Volkswagen over 30 billion Euros, with several suits still outstanding. (See EPA 2021; Jacobs and Kalbers 2019; Reuters 2021; Hotten 2015.)

Because many engineers become involved in the sale of their company's product, it is worth mentioning some special problems that sales engineers face.

Consulting v. Adversarial Sales

There are two possible sales relationships. Buyer and seller can be adversaries, or the seller can act as a consultant to the buyer. In adversarial sales, buyer and seller view themselves as opponents. Each is out for his own advantage at the expense of the other. They place no trust in each other and take no responsibility for meeting the other's needs. They are not interested in fairness or honesty, and will take any advantage, however unfair or dishonorable, that the law permits. In consulting sales, the seller acts as a consultant to the buyer, trying to figure out how the seller's company can best meet the buyer's needs. He invites the trust of the buyer, with the understanding that he will not abuse that trust. In short, the sales engineer can either present himself as an adversary, proclaiming "let the buyer beware" (caveat emptor), or present himself as a trustworthy advisor.

While adversarial sales may be appropriate to some contexts, they are generally not appropriate for engineering sales. First, the engineer is a professional, who offers his special expertise as a professional. Engineering sales usually involve fairly technical matters, and cannot proceed if the buyer does not place considerable trust in the seller. In these respects, engineering sales differ from selling a used car. When I purchase a used car, the salesperson does not present herself as an expert, and I am in a position to have the car thoroughly checked out by my own mechanic before buying it. I am in as good a position to assess the car as is the salesperson, and in a better position to gauge my own needs than she is. So, I do not generally need a used car salesperson to act as a consultant. In engineering sales, however, the buyer is usually not in a good position to evaluate the product on their own, and must rely upon the salesperson's representations to a significant extent. Moreover, the buyer is not necessarily well situated to determine which product best suits their needs. Thus, most buyers need the engineering salesperson to act as a consultant. And so, if a buyer has a reasonable choice between purchasing from an adversarial and a consulting salesperson, the buyer will almost always purchase from the consulting salesperson. Customers would rather buy from a trustworthy advisor than from a "snake oil" salesperson. Buyers prefer companies they can trust, and an engineer who misrepresents a job or product, or sells a buyer a job or product that does not suit the buyer's needs, does his company long-term harm. From the ethical standpoint, consulting sales promote the best long-term consequences since everyone benefits: consulting sales best suit the customer's needs and produce more sales for the salesperson's company. In addition, clean, clear decision-making is a value of the engineering profession. Clean, clear decision-making is not possible without the relevant facts, and so for the salesperson to hide relevant facts would be for him to attempt to undermine a value of the engineering profession. So, consulting sales promote this value, while adversarial sales undermine it. Finally, it is unlikely that adversarial sales will advance human welfare. Human progress is not best served by customers using inappropriate products out of ignorance. Thus, ethics and business sense dictate that the engineering salesperson establishes a consulting sales relationship rather than an adversarial one.

Anyone who presents himself as a consulting salesperson, inviting the buyer to rely upon him, has a moral obligation to live up to that promise. The value of honesty, the principle of universality and the rules for treating others fairly all forbid the salesperson from getting the sales by posing as a consulting salesperson, and then abusing that trust by steering the customer wrong. Indeed, someone who represents himself as a consulting salesperson has an institutional duty to do his best to meet the customer's needs, since he is able to function as a consulting salesperson only because customers rely upon his faithfully performing the duties of a consulting salesperson.

We can summarize these points with the following specific rule: *be a (faithful) consulting rather than an adversarial salesperson.*

Case 35: Sales Honesty

While discussing her product with a customer, a sales engineer realizes that the customer's needs are best served by a competitor's product. Although her product does have one or two advantages, it has features (level of accuracy, etc.) that the customer does not need for his application. It is thus significantly more costly than the competitor's product. How should the sales engineer handle the situation?

Discussion: Most writers feel that the company's long-term interest is best served by honesty. "If the sales engineer sells a customer a product [that] is not as satisfactory for a particular application as a competitor's product would be, the customer may very well discover the fact and thereafter be suspicious of the company [that] sold him the product, or even stop all business relations with it" (Alger et al. 1965,126).

The rule that one should be a faithful consulting salesperson requires that the sales engineer be honest. However, since the ultimate decision is that of the customer's, the sales engineer should not be shy about urging whatever legitimate advantages her product can offer, even if, in her judgment, the competitor's product is, overall, better suited to the customer's needs. In short, the sales engineer should be frank about the drawbacks of her product relative to a competitor's, given the buyer's needs but should emphasize whatever (legitimate) advantages her product can offer and leave the ultimate decision to the customer.

NOTES

1. Normally, the term whistleblowing is used to denote disclosing information outside the company or organization but it is also possible to blow the whistle on one part of the organization to a responsible individual within the organization.
2. See also Nader (1974) for a list of eight questions the whistleblower must ask.
3. Three days after Jorge Pizano's death from an apparent heart attack, his son died from drinking from a water bottle on his father's desk that had been laced with cyanide.
4. Not everyone would agree with me, of course. Milton Friedman's view, discussed in "Engineering and Business" (Chapter 3), suggests that the job of executives and their agents is to make a profit for their shareholders. On this view, it is up to the government to attend to the public welfare, not the engineers and executives, since the extra costs required to prevent pollution are tantamount to a tax on the shareholders of Y company, and only a democratically elected government may impose a tax. Thus, Friedman suggests, engineers and executives should do everything legally permissible to increase profits. It seems to be a consequence of Friedman's

view that it is up to DEQ to pick a suitable sampling site, and if DEQ chooses poorly, it is not the company's problem. As we saw, however, there are many difficulties with Friedman's view.

REFERENCES

Adams, Guy B. and Balfour, Danny L. (1998) *Unmasking Administrative Evil* (Thousand Oaks, London, and New Delhi: Sage).

AEC (Atomic Energy Commission) (1973) "Letter to Chairman, House Committee on the Judiciary," May 23, 1973 https://www.nrc.gov/docs/ML2008/ML20086N873.pdf

Alger, Philip, Christensen, N.A. and Olmsted, Sterling P. (1965) *Ethical Problems in Engineering* (John Wiley and Sons).

Associated Press (2021) "General Motors Settles with California for 5.75M over Ignition Switches," Detroit Free Press https://www.freep.com/story/money/cars/general-motors/2021/02/12/gm-ignition-switches-settlement-california/4469385001/

BBC (2016) "Tianjin Chemical Blast: China Jails 49 for Disaster," https://www.bbc.com/news/world-asia-china-37927158

Beresnivicius, Rytis (2019) "From Cargo Door Blowouts to One of the Most Reliable Aircraft—the DC-10," *Aerotime* https://www.aerotime.aero/articles/23156-dc10-history-cargo-doors-reliable-aircraft

Boisjoly, Roger (1987) "Ethical Decisions–Morton Thiokol and the Space Shuttle Challenger," *ASME Proceedings* (87-WA/TS-4), 1–13.

de George, Richard T. (1982) *Business Ethics* (MacMillan).

Duska, Ronald (1989) "Whistleblowing II," in Peter Windt et al., eds., *Ethical Issues in the Professions* (Prentice Hall), 317–322.

EPA (2021) "Learn about Volkswagen Violations," https://www.epa.gov/vw/learn-about-volkswagen-violations

Florman, Samuel C. (1989) "Beyond Whistleblowing," *Technology Review* 92:5.

Hotten, Russell (2015) "Volkswagen: The Scandal Explained," *BBC* https://www.bbc.com/news/business-34324772

Jacobs, Daniel and Kalbers, Lawrence P. (2019) "The Volkswagen Diesel Emissions Scandal and Accountability," *CPA Journal* 89:7, 16–21 https://www.cpajournal.com/2019/07/22/9187/

Lopez, Jonathan (2021) "GM Agrees to a $5.75M Ignition Switch Settlement with California," *GM Authority* https://gmauthority.com/blog/2021/02/gm-agrees-to-a-5-75m-ignition-switch-settlement-with-california/

Lynch, William T. and Kline, Ronald (2000) "Engineering Practice and Engineering Ethics," *Science, Technology, & Human Values*, 25:2, 195–225.

Nader, Ralph (1974) "An Anatomy of Whistle Blowing," in Ralph Nader, et al., eds., *Whistle Blowing* (Penguin).

Peterson, Martin (2020) *Ethics for Engineers* (Oxford University Press).

Raynal, Maria [Director of Corporate News Relations, GM] (2022) Personal correspondence 6/15/2022.

Reuters (2021) "WV Culture to Blame for Silence over Emissions Scandal, Ex-Manager Says in Trial," https://www.reuters.com/business/autos-transportation/vw-culture-blame-silence-over-emissions-scandal-ex-manager-says-trial-2021-09-23/

Rogers, William et al. (1986) "Report of the Presidential Commission on the Space Shuttle Challenger Accident," *NASA* https://history.nasa.gov/rogersrep/v1ch6.htm

Schlossberger, Eugene (1995) "Technology and Civil Disobedience: Why Engineers Have a Special Duty to Obey the Law," *Science and Engineering Ethics* 1, 163–168.

Smith, Harrison (2019) "A. Ernest Fitzgerald, Pentagon Whistleblower Fired by Nixon, Dies at 92," *Washington Post* https://www.washingtonpost.com/local/obituaries/a-ernest-fitzgerald-pentagon-whistleblower-fired-by-nixon-dies-at-92/2019/02/07/2f3277f4-2afe-11e9-984d-9b8fba003e81_story.html

South Florida Sun-Sentinel (1987) "NASA Administrator Cleared of Conflict of Interest," Nov. 4, 1987 https://www.sun-sentinel.com/news/fl-xpm-1987-11-05-8702030834-story.html

Times Wire Services (1986) "Embattled Director of NASA Rocket Center to Step Down: His Retirement Had Been Regarded as Inevitable Since Shuttle Disaster," *Los Angeles Times* https://www.latimes.com/archives/la-xpm-1986-06-04-mn-8845-story.html

Touchton, Michael, Klofstad, Casey, West, Jonathan and Uscinski, Joseph (2020) "Whistleblowing or Leasing? Public Opinion toward Assange, Manning, and Snowden," *Research and Politics*, 7:1, 1–9.

Tremblay, Jean-Francois (2016) "Chinese Investigators Identify Cause of Tianjin Explosion," *Chemical and Engineering News* https://cen.acs.org/articles/94/web/2016/02/Chinese-Investigators-Identify-Cause-Tianjin.html

Unger, Stephen (1982) *Controlling Technology: Ethics and the Responsible Engineer* (NY: Holt Rinehart and Winston).

Westrum, Ron (1991) *Technologies and Society* (Belmont, CA: Wadsworth).

Witkin, Richard (1975) "Engineer's Warning on DC-10 Reportedly Never Sent," *New York Times* https://www.nytimes.com/1975/03/12/archives/engineers-warning-on-dc10-reportedly-never-sent.html

8

Good Faith

The next set of issues deal with acting in good faith, both in fact and in appearance. Engineers often find themselves in situations where others must rely upon their integrity. In bidding, keeping information confidential, respecting patents and copyrights, and in potential conflict of interest situations, the engineer is expected to refrain from compromising the trust placed in them; their personal motivations must take second place to the integrity of the process.

CONFLICT OF INTEREST

Conflicts of interest are situations in which other interests place a strain upon the loyalty of an engineer to do the very best she can for her organization (or, in the case of a consulting engineer, for her client). Conflicts of interest are, to a certain extent, inevitable. "Conflicts of interest present some of the thorniest ethical issues for engineers because the practice of engineering, by its very nature, involves relations with parties often with conflicting interests" (NSPE 2019). The ethical engineer may experience a conflict between the interests of their organizations, firm, or company and moral imperatives such as protecting public safety or respecting nature. If a neighbor or cousin of a procurement engineer works for a supplier, there is a motivation to favor (or avoid) that supplier and this motivation may affect the judgment of the procurement engineer. There are thus three ethical aspects of conflict of interest. First, to the extent feasible, *avoid situations that might bias, or appear to others to bias, one's judgment on the organization's behalf.* Second, when this is not possible, *strict impartiality, in appearance and in fact, is called for.* Third, "an engineer has an

DOI: 10.1201/9781003242574-11

obligation to disclose all known or potential conflicts of interest to employers or clients by promptly informing them of any business association of interest or other circumstance that could influence, or appear to influence, the engineer's judgment or the quality of services" (NSPE 2019).

Some conflict of interest situations are obvious, while others are less obvious. "A conflict of interest exists in a situation where an independent observer might reasonably conclude that the professional actions of a person are or may be unduly influenced by other interests. This refers to a financial or non-financial interest which may be a perceived, potential or actual conflict of interest" (Australian Code 2018). At least five sorts of conflict of interest situations are common.

1. Conflicts of interest may arise when *engineers engage in activities that compete with the company*. This would include owning stock in competing firms or companies and marketing a product that competes with the company's product. Thus, an engineer who works for a company with extensive oil or mining interests should not speculate in oil and mineral rights.

2. Conflicts may also arise *when engineers have relationships with suppliers that may compromise their judgment*. Thus, engineers should avoid accepting gifts with more than nominal value, accepting extravagant entertainment or special discounts not available to other members of the company, accepting cash payments or loans, moonlighting for suppliers, non-company business transactions with suppliers, and owning stock in supplier companies. Most large companies and most governmental agencies have policies about allowable gifts. The US government policy in 2007 (still in place in 2020) allowed "gifts valued at $20 or less per 'source' per occasion, although the total value of such gifts must not exceed $50 in a calendar year from a single source" (Office of Government Ethics 2007). The German Government forbids all employees from accepting gifts and other benefits except when the responsible authority gives explicit approval (Bundesministerium des Innern 2014).[1]

    To help you in thinking about this, Shaw and Barry (1989, 305–306) list seven questions you should ask yourself about a gift. What is the value of the gift? What is the purpose of the gift? What are the circumstances under which the gift was given or received? What is the position and sensitivity to influence of the person receiving the gift? What is the accepted business practice in the area? What is the company's policy? What is the law?

3. Potential conflicts result when *engineers conduct business with their company*, such as selling real estate to the company, leasing equipment to the company, and selling engineering, contracting, managerial, or financial services to the company.

4. Conflict of interest results when *engineers use for personal gain their position in the organization, company facilities, or knowledge of the firm's or company's affairs*. Examples include speculating on real estate using inside knowledge of company plans and using company equipment or supplies for personal uses.[2]

5. Conflict of interest situations can arise when *engineers have personal interests affecting organizational decisions*. Controlling company employment relations with relatives, for example, includes hiring relatives and determining the salary, raises, promotions, or bonuses of close family members or others with whom the individual has close personal relationships. Conflicts can also arise when safety or ethical conduct adversely affects an individual's advancement or remuneration, for example, if delays caused by an engineer's refusing to approve a design that failed safety tests would cost the engineer a bonus.

Frederick (1983) suggests that firms and companies have a clear statement of company policy that is detailed about clear-cut cases and establish a mechanism (such as a committee) for dealing with borderline cases, which employees must consult in new or questionable cases. In addition, new policies must be made known to employees and there should be penalties for violating the policy.

Engineers Canada (2014) suggests that four options are open to engineers facing a conflict of interest situation:

1. "Proceed with the work.... It is wise to document this decision and the information that was considered in arriving at this conclusion."

2. "Proceed with the work and erect any necessary confidentiality screens.... this course of action is only suitable for professional/professional conflicts. It is not possible to effectively create confidentiality screens in personal situations (e.g.[,] when a spouse is evaluating their partner's bid)."

3. "Proceed with the work after having informed the client(s) (both new and existing, if applicable) and obtained consent....If agreement cannot be found, engineers have no option but to withdraw their services."

4. "Do not proceed with the work."

CONFIDENTIALITY AND TRADE SECRETS

Discretion is needed to protect eight overlapping kinds of information: (a) the *privacy of individuals* (for example, do not reveal the age or health of an employee); (b) the *integrity of decision processes* (for instance, do not reveal the contents of a confidential letter of recommendation for an employee); (c) *client information*, including information whose secrecy is desired by your company (do not reveal lists of clients and suppliers) and information whose secrecy is desired by the client (do not disclose client's trade secrets or business plans); (d) *business plans* (for example, do not leak insider information that might affect the stock market or reveal the location of a projected plant that might affect real estate prices, or information that might compromise your company's chances when submitting sealed bids), and four kinds of professional information; (e) *patented information*; (f) *trade secrets* (do not reveal unpatented secret processes, formulae, etc.); (g) *tricks of the trade*; and (h) *general knowledge*.

Laws pertaining to intellectual property vary across the globe. The World Intellectual Property Organization (WIPO), an agency of the United Nations, with headquarters in Geneva and external offices in Algiers, Rio de Janeiro, Beijing, Singapore, and elsewhere, helps coordinate intellectual property law and disputes among its 193 member states. Member nations include Afghanistan, Bangladesh, Chile, Ethiopia, Italy, Jamaica, Japan, Jordan, Madagascar, Mexico, Nepal, New Zealand, Uganda, and Vietnam.

Issues about the first four categories of sensitive information apply not only to engineers but to professionals generally. Privacy of individuals is discussed in Chapter 13. Respecting client privacy and protecting business plans are relatively straightforward. In thinking about protecting the confidentiality of the decision process, remember that the point of the practice of confidentiality here is to enable people to speak candidly without fear of reprisal or embarrassment. However, while candor is necessary for intelligent decision-making, evil, slander, and unfairness are not. Confidentiality here is meant to protect honest decision-making, not evil-doing. For example, engineers generally do not need to keep confidential the commission of a crime – in some cases, doing so may itself be a crime. Needless to say, engineers must respect all legal duties regarding confidentiality.

Dealing with professional information is perhaps the trickiest issue and presents some issues specific to engineering, especially when engineers leave one employer for another (see Schlossberger 2003). General knowledge is available publicly and not restricted by law. It is generally held to be the property of the engineer, even if the knowledge was gained during the engineer's employment, since the employer has no special claim on it, just as a university or teaching hospital has no claim on the general knowledge it taught students or residents. Tricks of the trade are things not publicly available but learned from one's own experience as an engineer (or shared by a colleague or mentor). For similar reasons, engineers are generally free to use tricks of the trade they acquired. Trade secrets, by contrast, are proprietary information, not publicly available, about specific processes or products, ingredients, etc. It is, in general, unethical for engineers to reveal trade secrets of a previous employer or use those secrets in a new employer's product or process. In the US, trade secrets are protected by the Defend Trade Secrets Act (DTSA) of 2016. Unlike tricks of the trade and general knowledge, the employer does have a special claim on trade secrets they worked hard to develop on their own. In practice, however, the distinction is not an easy one to draw. As Martin and Schinzinger (1983, 177) put it, "an engineer's knowledge-base generates an intuitive sense of what designs will or will not work, and trade secrets form part of this knowledge base." However scrupulous an employee is, she cannot help drawing upon the fruits of her past experience. For example, no one would expect an engineer to repeat the effort of trying out a design she already knows, from past experience at another company, will fail. Thus, engineers have to draw upon past employers' trade secrets, at least to the extent of knowing what not to try. Alger et al. (1965, 117) mention several relevant factors: "1) the extent to which the information is known outside the business; 2) the extent to which it is known by employees and others involved in business; 3) the extent of measures taken to guard the secrecy of the information; 4) the value of the information to holders and to competitors; 5) the amount of effort or money expended in developing the information; 6) the ease or difficulty with which the information could be properly acquired or duplicated by others." These factors are not always easy to assess.

There are limits on the duty to protect trade secrets. "Trade secrets that are themselves improper or immoral are not covered by the duty to respect trade secrets. That P Company's secret process illegally infringes a patent is a trade secret, but not one that engineers have a special duty to protect: the duty to respect that trade secret is voided" (Schlossberger 2003). In addition, an engineer deciding to blow the whistle to protect public safety may have to reveal proprietary information. (Of course, the engineer should try to reveal as little proprietary information as feasible, restrict access as much as they can, and, when possible, inform the employer beforehand.)

This suggests several points. Engineers should strive to avoid changing companies, when feasible. "To the extent that fewer professionals move from one employer to another, less knowledge would be communicated from one employer to another" (Png and Samila 2013). Make an effort to work things out with the present employer rather than changing jobs. Engineers should not accept employment if they believe that the motive for the offer is their knowledge of trade secrets. Engineers should not explicitly reveal trade secrets and must be very skittish about drawing upon special knowledge paid for by their previous employer. However, employment is a kind of partnership, and the general knowledge, skill, and experience an employee gains in her job are part of her remuneration. Since innovation is a key engineering value, engineers should seek new approaches to problems they have worked on under a previous employer, rather than seeking to duplicate or make minor variations to secret processes. Taking a thoroughly new approach minimizes the danger of using the trade secrets of a past employer and maximizes the possibility of advancing human knowledge and welfare. Honesty and communication are also key values and can often avoid or resolve problems.

Employers can also take steps to minimize problems. Baram (1989) suggests several measures, including:

> prohibition of consulting and other 'moonlighting,' dissemination of trade secrets on a strict 'need to know' basis to designated employees,…prohibitions on the copying of trade secret data…restrict research and other operational areas to access for designated 'badge' employees only and divide up operations to prevent the accumulation of extensive knowledge by any individual…distribute unmarked materials–particularly chemicals–to employees. [These policies] must be exercised with a sophisticated regard for employee motivation, however, because the cumulative effect may result in a police state atmosphere that inhibits creativity and repels prospective employees. (p. 308)

(It should also be emphasized that some of Baram's suggestions, such as distributing unmarked chemicals, might compromise safety and undermine a sense of community.) Baram also suggests that employers provide incentives for departing employees to opt not to work for a competitor. For example, such employees might be eligible for an annual consulting fee if they do not work for a competitor. Finally, Baram recommends debriefing of departing employees.

As always, *engineers should be aware of and follow relevant laws* for their jurisdiction.

Case 36: Trade Secrets

Migenes' old employer, D Drugs, developed an adaptation of standard equipment that makes it more efficient at holding drugs at a standard temperature during their manufacture. The adaptation is a trade secret. Migenes goes to work for a petrochemical company, M, not in competition with D Drugs. Migenes realizes that a similar adaptation might be made to a different machine to make a very efficient temperature controller in the production of a synthetic rubber. (See also Kohn and Hughson 1980.)

Discussion: This case walks the line between general experience and trade secrets – although the process is similar, it is made to a different machine for a different purpose and requires some imagination and insight on the part of Migenes. Here, the importance of honesty and communication become evident. Migenes should contact D Drugs and ask permission to make the change, after informing M that she will do so.

PATENTS AND COPYRIGHTS

Patented information is information whose application or employment, while publicly available, is legally restricted. It is certainly unethical to violate patent and copyright laws. Engineers are generally required to keep to the letter of patent and copyright laws. However, by making a few minor modifications to the patented or copyrighted item, it is sometimes legal to duplicate what is basically a patented or copyrighted product or process. Is it unethical to make a minor modification in a patented process or product (or copyrighted program) so as to avoid copyright infringement?

The patent and copyright laws are a compromise between two competing aims. On the one hand, creators and innovators deserve to enjoy the fruits of their labor. To deprive them of the benefits of their innovation would be unjust (a kind of theft) and it would be socially harmful: without patents, the incentive to innovate would be lessened, resulting in fewer innovations, and innovators would have to rely on secrecy, thus undermining the communication and co-operation necessary for progress. Thus, justice (treating others fairly and well), human welfare, and promoting good consequences require that ethical engineers do not infringe on patents and copyrights. On the other hand, society has an interest in the development and improvement of innovations, and in the widespread use of beneficial technologies. (That is why patents are granted for a fixed term, rather than forever.) In other words, ethics tries to balance being fair to the innovator and spreading and improving technology to advance human welfare.

This suggests four factors that engineers may weigh.

- *Has the innovator had a chance to benefit from the innovation?* Some patents yield profits immediately, while others take a long time to pay back investment costs. It is less troubling to market a modified patented product if the innovator has had a reasonable chance to profit from the innovation.
- *Is the modification a real improvement, or is it just there to get around the patent laws?* It is less troubling to market a modified product that truly advances human welfare. If the modification makes the product safer, more precise, more useful, more widely available (either by being cheaper or by fitting current procedures or machinery), or less environmentally harmful, marketing the modified product is more justifiable.
- *How significant a change is the modification?* Changing a few unimportant lines in a program is more troubling than making major alterations in machinery.
- *Is there any doubt about the legality of marketing the modification?* Engineers should not engage in legally questionable practices.

BIDDING

The engineering profession cannot operate as it does without public recognition of trust: reliance in the strict probity and fairness of bidding and supplying procedures is a precondition for engineers to function as they do. Hence, there is an institutional duty of strict propriety in these matters. The rules of propriety here are central to the practice of engineering. Moreover, the situation is conflictful (the firms involved in supplying and bidding on contracts are competitors). Finally, trust is an essential factor, and so the appearance of impropriety is almost as damaging as actual impropriety. The legitimate gain in bending these rules is minor. Thus, there is a very strong case for strict adherence to rules of propriety. (Recall "When to Break the Rules" in Chapter 6.) In general, *engineers should err in the direction of caution* in giving or accepting anything that smacks of a favor or gift, in acquiring an interest in a supplier or potential contractor, etc. The rule of thumb is: *if there's any doubt, don't do it.*

Engineers can be involved in bidding as either bid takers or bid submitters. Ethical issues can arise at either end of the bidding process.

1. *Conflict of interest* is perhaps the most common ethical issue pertaining to bidding. In addition to the issues discussed above, engineers must avoid offering or taking kickbacks (e.g., "pay to play" schemes) and bribes.

2. Putting out an RFP (request for proposals) and then *appropriating an idea from a proposal whose firm was not chosen* presents ethical questions. If you "use an RFP to shop for ideas and then have another agency deploy them, you're stealing. It's an unquestionable breach of ethics" (XD Agency 2018). A related concern is when a supplier designs and develops a product or process specifically for your company and you later decide to publicize it in an RFP. Is this unfair to the original supplier?

3. *Bid shopping* "occurs when a general contractor discloses the bid price of one subcontractor (or suppliers) to its competitors in an attempt to obtain a lower bid than the one on which the general contractor based its bid" (Gregory and Travers 2010). Gregory and Travers suggest that bid shopping has the following ill effects: "defeating the purpose of the competitive bid system; promoting lower-quality work; incentivizing corner-cutting; increasing claims and change orders; delaying project completion; and generally worsening the business environment."

4. A bidder's *use of insider information* can be problematic. "If it's insider information which he knows because he knows the area better than his competitors, it's considered 'smart' bidding. But if it's insider information he has because he received it in an unethical manner (called privileged information), then this practice is frowned upon and considered unethical" (Quality Mechanicals Inc. 2020).

5. It is unethical and generally illegal for bidders to collude in *price fixing or bid rigging* "Price fixing is an agreement among competitors to raise, fix, or otherwise maintain the price at which their goods or services are sold" (U.S. Dept. of Justice 2021). Bid rigging occurs when bidders reach agreement in advance of the bidding who will submit the winning bid. Forms of bid rigging include establishing a rotation for who will win, having firms agree to submit overly high bids, so the chosen firm will win ("complimentary bidding"), or agreeing that one or more firms will not submit or will withdraw a bid (bid suppression) (*Ibid.*).

Case 37: When to Cease Bidding (Bidding 2)

After many years of experience, you come to believe that Y is the best supplier. May you cease taking competitive bids, and is it ethical to take bids when the decision has already been made? How should the situation be handled? What are the governing considerations? (See also Case 21 in Chapter 6.)

Case 38: The Expense-Paid Trip

You are genuinely interested in the equipment of firm Y. Y offers to pay expenses for you to visit a plant in which Y's equipment is used or the plant at which the equipment is produced. Is it ethical to accept? Note: Alger et al. (1965, 136) says about a similar case that "if there is no other possible way for the engineer to get the necessary information…he should make the trip, with the understanding that, in accepting this invitation, he is incurring no obligation to the company." Do you agree? Are there other steps you might take (e.g., informing your superiors)?

Case 39: The Personal Discount

You are a purchasing agent responsible for ordering trucks and other automotive equipment. For several years, you have purchased reliable and inexpensive trucks from a particular dealer. This dealer also sells RVs. When you go to the dealer to purchase an RV for your own use, the dealer offers to sell you the RV at a price well below what it generally sells for.

Discussion: Clearly, the dealer is offering you a personal discount with the intention of influencing your judgment. He is not offering the discount to the firm (which would be a legitimate sales incentive), but to you personally. His purpose is not to persuade you to buy more vehicles for yourself but to ensure that you continue to place orders with him for the firm. A substantial amount of money is involved and the discount is not available to the general public. You cannot deny that the dealer is offering you a bribe. True, the dealer did not explicitly say, "I expect you to continue to purchase your company's vehicles from me" but his intention in offering the discount is clear. Thus, accepting the discount is unethical. After all, if you accept the discount, you make an unspoken promise: you create an obligation to the dealer that violates your duty as a purchasing agent for your company. Suppose, for example, that next month another supplier offers a better deal to your company. If you do not switch suppliers, you betray your company. If you do switch companies, you break a tacit promise to the dealer (at the very least, you treat him unfairly and take advantage of him). Accepting the discount creates psychological pressure on you to continue buying from this dealer: it motivates you to regard his offer in the best possible light, to look for flaws in other suppliers, and so forth. Even were you to say explicitly to the dealer when accepting the discount that your judgment on behalf of your company will not be influenced, you will find it difficult in future to be completely unbiased. In any case, accepting the discount creates the appearance of impropriety, and in bidding the appearance of impropriety should be avoided if at all possible. The Golden Rule helps resolve this case: how would you feel if you were bidding to supply the Z company, and discovered that the purchasing agent for the Z company had accepted a large personal discount from one of your competitors? (However, I must, in honesty, note that a fair number of my students over the years have remained adamant that you should take the discount because no explicit quid pro quo was mentioned.)

Case 40: Discussing One Vendor with Another

You are taking sales proposals for a particular piece of equipment. Is it ethical for you to discuss the features of one vendor's equipment with another vendor?

Discussion: "While no supplier can hope to keep a special feature or design for his exclusive use for long, he is entitled to full protection on any special design or feature he offers on any specific purchase" (Alger et al. 1965, 115). However, on products offered in the open market, protection of special features must come from copyrights and buyers are entitled to hear how one vendor would answer the claims of another. In general, then, if the feature is publicly available information, it is permissible, but if the feature is a trade secret or specifically designed for this proposal, it is not. This problem is to be contrasted with the next.

Case 41: Using an Unsuccessful Bidder's Idea

You are taking proposals for a complex project. Is it ethical for you to take one of the ideas in bidder N's proposal, and adopt it as your own by revising the specifications and then asking the other bidders for revised bids?

Discussion: To the extent that the idea taken from bidder N's proposal is innovative, or results from special expertise, taking the idea without paying for it is stealing. The ethical thing to do, generally, is to evaluate N's overall proposal, including this idea, against the other proposals as they stand. If bidder Q's overall proposal is the best, you may, after making clear that you have accepted bidder Q's proposal, ask N for permission to use that idea. (Special circumstances may call for special arrangements. For example, if N's idea is crucial but N's proposal is for some other reason truly unacceptable, some sort of special arrangement seems called for. In some cases, for example, it may be feasible to ask N and Q to collaborate.) It should be noted that a more thorough design study prior to asking for bids might have eliminated the problem.

Case 42: Owning Stock

Is it ethical to own stock in ChEQ, a chemical equipment manufacturer? Consider these four scenarios:

a. You already own stock in ChEQ, a company with which you personally have had no dealings. However, you are assigned to a new project, and as part of your new duties must take bids on a type of item ChEQ, among others, manufactures.

b. Your duties include, from time to time, taking bids on various sorts of chemical equipment. Your stock broker suggests to you that ChEQ would be an excellent investment. (Does it matter whether you personally have, or whether your company has, ever received a bid from ChEQ?)

c. ChEQ is a competing company.

d. Your duties include working with several officers of ChEQ. As a result of this collaboration, you become aware of several features of ChEQ that suggest to you that ChEQ will grow significantly over the next several years. Is it ethical to purchase stock in ChEQ?

Discussion: Owning stock in a chemical equipment company, when your duties even tangentially involve assessing bids from such companies, creates the appearance of impropriety and may in fact create a conflict between your personal interest and exercising your best judgment on your company's behalf. For these reasons, it is to be avoided. Of course, not all such conflicts can be avoided: it would be absurd to insist that you may not become a chemical engineer if your older cousin already works for a chemical equipment company. Some situations in which there may be psychological pressure to make biased choices are inevitable. Nonetheless, one should avoid those situations whenever possible and it is generally a simple matter not to purchase stocks in companies that supply items of a sort your company bids upon. Even if you yourself have never received a bid from ChEQ, the possibility of conflict of interest exists. (You or your company may receive bids from ChEQ in the future and you may be more inclined to find fault with other company's products. In fact, if you buy ChEQ stock, ethical considerations may inhibit you from investigating as thoroughly as you otherwise might what ChEQ can do for your company.) If you already have a significant holding in such a company, you should sell. Many companies do permit engineers to own stock in supplier companies, as long as the stock owned constitutes less than a certain percentage of the total stock of the company. Owning a few shares of a large mutual fund .1% of whose holdings consists of 0.00001% of ChEQ's stock is quite different from personally owning a 20% stake in ChEQ. Next, buying stock on the basis of certain kinds of insider information is illegal, and even when not specifically prohibited by law, it is a bad practice to use your position within the company for personal advantage in the stock market. Finally, buying stock in a competing company not only gives one a financial incentive to do less than one's best for one's own company but also undercuts loyalty and community spirit, and so is a bad idea.

Case 43: Personal Use of Company Facilities

Is it ethical for corporate officers to have their cars repaired or serviced by the company's repair shop? To have repair work on their homes done by company workers on company time? Is it ethical for an executive to have her secretary run personal errands for her? Is it ethical to ask idle workers to do repairs on one's home? Is it ethical to ask subordinates to hang decorations in one's home for an upcoming company party?

Discussion: There is a line to be drawn between fringe benefits and abusing one's position. If you ask company employees to do things for you on company time to which you are not entitled, you are stealing. The company, after all, is paying for the employee's time, and you are taking that time for your own personal use. If you take an hour of a secretary's company-paid time to buy a birthday present for your daughter, that is no different from simply taking the amount of the secretary's hourly wage from the cash drawer. This is true even if the secretary would otherwise be idle. Making illicit personal use of company personnel is just another form of embezzling from the company. Even if you own the company, there may be tax issues (since the employee's wages are deducted as a business expense). However, the use of a company car can be a legitimate fringe benefit. Questions to ask oneself include: what are the rules of the company? Is it lawful? Is this an explicit fringe benefit, and how is it carried on the company's books? What is the accepted business practice? How would it look if it were made public? Is there a legitimate business function involved? (Is your house used for company meetings?) Finally, it goes without saying that putting pressure on employees to do personal errands or favors on their own time is highly improper and unethical.

Case 44: "Rescuing" from the Garbage Dump

Is it ethical for a corporate engineer to "rescue" something from a company garbage can or dump site to be used for private purposes?

Discussion: It may be tempting to regard the item in the dump site as "abandoned property," and so decide that taking it would not be stealing. However, if the item discarded is still usable, loyalty to the company suggests that one has a duty to call the matter to the company's attention. In any case, the ethical thing to do is clear: ask permission before taking the item. Simply taking the item is, at best, not being straightforward and may violate company policy.

Case 45: Copying Company Software

Is it ethical for engineers to make copies of software purchased by the company for (a) their own enjoyment or family use (such as a diagnostic program used in a recreational project or a word processing program used by a son or daughter in school), (b) use in helping friends or civic groups, and (c) one's own profitable projects?

Case 46: Taking Home a Pencil

Is it ethical for an engineer to take home a pencil for use on company business to be done at home? Is it always easy to distinguish between company and personal use? Consider (a) someone with vision problems who uses a magnifying glass when reading professional periodicals, and (b) a "refresher" engineering course taken at night. These are not strictly company activities but they do benefit the company by helping to make one a better engineer.

Case 47: Publicizing Your Work

You are invited by the Left-Handed Engineers of America to give a paper to the LHEA on your recent work in the manufacture of synthetic leather.

Discussion: The values of advancing human knowledge and collegiality within the engineering community mean that participation in professional meetings is an important aspect of professional life. Duties of confidentiality mean that one should not disclose secrets whose dissemination would hurt the company. The way to resolve this conflict is to clear your presentation with your company. Since few companies could operate without considerable co-operation in sharing information, the principle of universality suggests that companies must do their part in advancing the general knowledge shared by engineering as a field. Thus, co-operation is worth a minor risk of inconvenience or loss of competitive advantage to the company, though of course no company ought to part with information that would hurt it significantly. Thus, your company should strive to assist you in sharing as much information as feasible, short of significantly hurting the company.[3]

NOTES

1. Cf. "All employees of the federal administration are in principle prohibited from accepting gifts, hospitality or other benefits," except "an employee may accept a gift if the responsible administrative authority gives its explicit or implicit approval" (Bundesministerium des Innern nd).
2. Cf. Anon. (1961): examples of conflict of interest situations include "owning a financial interest in or holding a position with" or "accepting fees, gifts, entertainment, loans or other favors from" a "supplier, agent, customer or competitor," "acquiring an interest in a business, real estate, or other facilities in which the company may be interested," and "speculating on the basis of inside information."
3. Cf. Alger et al. (1965, 114): "most reputable organizations allow their engineering employees to publish papers covering important new discoveries or advances in the art after patent protection has been received, if possible, and after reasonable time has been allowed for initial use of the new ideas by those responsible for them."

REFERENCES

Alger, Philip, Christensen, N.A. and Olmsted, Sterling P. (1965) *Ethical Problems in Engineering* (John Wiley and Sons).

Anon. (1961) "Conflicts of Interest Pose No Big Problem," *Chemical & Engineering News* 39:43.

Australian Code for the Responsible Conduct of Research (2018) *National Health and Medical Research Council* https://www.nhmrc.gov.au/about-us/publications/australian-code-responsible-conduct-research-2018#block-views-block-file-attachments-content-block-1

Baram, Michael S. (1989) "Protecting Trade Secrets," in Peter Windt et al., eds., *Ethical Issues in the Professions* (Prentice Hall), 303–310.

Bundesministerium des Innern (2014) "Rules on Integrity," https://www.bmi.bund.de/SharedDocs/downloads/EN/publikationen/2014/rules-on-integrity.pdf?__blob=publicationFile

Bundesministerium des Innern (nd) "Private Sector/Federal Administration Anti-Corruption Initiative," https://www.bmi.bund.de/SharedDocs/downloads/EN/themen/moderne-verwaltung/anti-corruption-initiative-questions-and-answers.pdf?__blob=publicationFile&v=1

Engineers Canada (2014) "Public Guideline: Conflict of Interest," https://engineerscanada. ca/public-guideline-conflict-of-interest#-how-to-manage-conflicts-of-interest

Frederick, Robert E. (1983) "Conflict of Interest," in Milton Snoeyenbos, Robert Almeder and James Humber, eds., *Business Ethics* (Prometheus), 125–134.

Gregory, Don and Travers, Eric (2010) "Ethical Challenges of Bid Shopping," *The Construction Lawyer* https://www.keglerbrown.com/publications/ethical-challenges-of-bid-shopping/

Kohn, Philip and Hughson, Roy (1980) "Perplexing Problems in Engineering Ethics," *Chemical Engineering* 87, 100–107.

Martin, Mike and Schinzinger, Roland (1983) *Ethics in Engineering* (McGraw-Hill).

NSPE (2019) "Conflict of Interest—Reviewing and Approving Engineer Offering Redesign Services," https://www.nspe.org/resources/ethics/ethics-resources/board-ethical-review-cases/conflict-interest-reviewing-and

Office of Government Ethics (2007) "Ethics and Procurement Integrity," https://www.fai. gov/sites/default/files/pdfss/OGEprocurementintegrity_07.pdf

Png, I.P.L. and Samila, Sampsa (2013) "Trade Secrets Law and Engineer/Scientist Mobility: Evidence from 'Inevitable Disclosure'," https://citeseerx.ist.psu.edu/viewdoc/download?doi=10.1.1.308.5620&rep=rep1&type=pdf

Quality Mechanicals Inc. (2020) "A Brief Explanation of Ethics," https://qualitymechanicals. com/blog/f/a-brief-explanation-of-ethics

Schlossberger, Eugene (2003) "Trade Secrets and Patents in Engineering: Ethical Issues Concerning Professional Information," in John Rowan and Samuel Zinaich, eds., *Ethics for the Professions* (Wadsworth), 224–227.

Shaw, William and Barry, Vincent (1989) *Moral Issues in Business*, 4th ed. (Wadsworth), 305–306

U.S. Dept. of Justice (2021) "Price Fixing, Bid Rigging, and Market Allocation Schemes: What They Are and What to Look For. An Antitrust Primer," https://www.justice. gov/atr/file/810261/download

XD Agency (2018) "Ethics in the Bidding Process," https://xdagency.com/ethics-in-the-bidding-process/

9

Employee-Employer Relations

Engineers rarely work alone. Engineering operations generally require a team of engineers, support personnel, and people with business/management skills. Thus, most engineers are employed or employ others, and most engineers have superiors or subordinates. As an engineer, you will most likely make workplace relationship decisions. This chapter will help you make those decisions fairly and ethically.

TYPES OF WORK RELATIONSHIPS

There are three kinds of work relationships, according to Gordon and Ross (1983). In an artisan-master relationship, the employee does what the employer wants. In a professional status relationship, the employee gives the employer what the employer ought to have. Finally, in a protégé-patron relationship, the employee gives the employer what the employee wants (for example, researchers who determine their own research programs). Gordon and Ross suggest that a modern research organization may employ different individuals in all three categories. I would suggest, instead, that engineers belong, to a greater or lesser extent, in all three categories. All engineers are bound by organizational and legal rules and policies, and must be guided, to some extent, by the needs of their organization (company, firm, agency, etc.). Ultimately, law and ethics have the final word. In this sense, all engineers are "artisans." But engineers also have a duty to use their brains, to be innovative, and to speak up on the organization's and/or client's behalf. Most organizations welcome suggestions from engineers. Indeed, because engineers must take some responsibility for what goes on in the company, they have a duty not to follow orders mindlessly, without thinking about what ought to happen. (See "Two Principles of Institutional Responsibility" in Chapter 6.) In this sense, all engineers have "professional status." Finally, every engineer is an artist, engaged in the great social task of advancing human welfare, and so, within the limits set for them, should be giving the employer what they, the engineer, want to give, namely their best ideas and effort. In this sense, the engineer is an artist sponsored by the company (protégé-patron relationship).

Balancing these three elements of the work relationship is not always easy. There are some questions you might want to ask yourself when trying to strike a proper balance. First, what is the nature of the work I am asked to do? Some tasks leave more room for freedom than others. For example, an R&D engineer with the primary idea for a new product has more room for "artistic freedom" than does an engineer writing specifications for a cooling system. Second, what is at stake if I give my employer what she ought to have, instead of what she wants? For example, an engineer responsible for the safety of a nuclear reactor has a duty to give the employer what the employer ought to have, rather than what the employer wants, since an unsafe nuclear reactor creates an unacceptable risk to the public. However, while a manufacturing engineer who sees a more efficient way to do what he is asked to do should certainly point out his idea to his supervisor, if the supervisor insists, it should be done another way. When safety or other ethical concerns are not at stake, manufacturing engineers should normally do what the employer wants, since what is at stake here is not the public's safety, but the employer's money. (It is a bit more complicated when the employer is the public and a manager demands a wasteful solution.) In the case of consulting engineering, it is ultimately the firm rather than the individual engineer who is responsible for seeing to the best interests of the client, and it is, ultimately, the client rather than the firm that is responsible for the successful use of the project. Thus, when safety and ethics are not involved, the firm must ultimately give way to the client, and the individual engineer employed by a consulting firm must ultimately give way to the firm. Finally, what particular understanding do I have with my company or firm? A company may give some researchers a freer hand than others, and some companies are less comfortable with "giving researchers their head" than others.

In all three kinds of employment relations, fair treatment and a community atmosphere are crucial for employee motivation. Supervisors and employers can do much to help generate this kind of work environment.

LEADERSHIP AND HEALTHY WORK ENVIRONMENTS

Books on leadership and management fill the rows of libraries and bookstores. Within these books, you will find many theories of good management. Two broad views of supervision deserve special mention. Supervisors often view themselves as the lord of the castle: their role, as they see it, is to give orders, make decisions, and see to it that their subordinates obey. Other supervisors see themselves as facilitators more than order-givers: their role is to help their subordinates function efficiently and co-operatively. In general, facilitators are better supervisors than order-givers, since a facilitator permits subordinates to feel that they are valuable and productive members of a team, doing their best to contribute in the way they best understand. Facilitators help create a community atmosphere and show respect for persons. As a result, subordinates do better work and feel more job satisfaction and company loyalty. Order-givers, who hand down decisions, may lack the experience and perception of subordinates who actually do the work. While many American companies attempt to emulate Japanese methods of involving subordinates in company decisions (such as "total quality management"), their attempts are doomed to fail if supervisors still perceive themselves as lords of the castle. A supervisor who tells a subordinate "get your butt in my office" cannot expect the same subordinate, an hour later, to participate in a meeting as a valued member of the team. To be a good facilitator, a supervisor must bring the appropriate frame of mind to everything she does. An order-giver shows authority by barking commands and making threats. A facilitator's authority shows in her dedication, her self-assurance, her clearness of purpose, her competence, knowledge, and perception (the sense that one cannot "put one over" on her), her fairness, and her ability to make tough decisions when necessary. Establishing a working team environment within a department takes hard work, skill, and care. The results justify the effort.

Facilitators and order-givers use different motivational strategies. Facilitators use incentives that promote teamwork and ethical behavior. For example, a manager who offers a bonus to the individual with the most sales in his department turns that department's sales staff into competitors; each salesperson has a strong reason to hope that their colleagues fail. How can a manager expect one salesperson to assist another when doing so threatens their bonus? Facilitators, who try to promote teamwork, are creative in providing incentives that promote co-operation instead of competition. Offering a bonus to everyone on the sales team if the team as a whole reaches a certain sales goal makes sales staff invested in each other's success. Similarly, offering oil tanker captains a bonus for arriving early gives them an incentive to cut corners, cheat on safety drills, and so forth. Facilitators, who wish to promote ethical conduct, *avoid incentives that reward unsafe or unethical conduct*. Evaluating subordinates on the basis of bottom-line figures pressures subordinates to enhance their figures by taking shortcuts, taking improper risks, and favoring short-term goals over long-term goals. (A department head has a little reason to invest in research and development, for example, when there may be no pay-off for five years: by that time, the department head will probably have moved on and someone else will get the credit.) Facilitators, who want a team dedicated to the company's long-term interests, will *evaluate subordinates on the basis of their long-term contribution to the company*. Since fear is a powerful short-term motivator, order-givers often try to motivate subordinates through fear. Since fear is a poor long-term motivator, facilitators use fear only as a last resort.

Good facilitators concentrate on two key features of supervision: communication and leadership. It is harder to establish good communication than one might think. R.P. Cort found that, in 100 businesses studied, 80% of the information sent down from top management was not understood by workers (Humber 1983). Humber (1983) points out several reasons for the lack of communication. Top management tends to isolate itself to avoid interruption and confrontation. Subordinates are reluctant to pass up unfavorable information. Superiors avoid passing on unfavorable information because they do not wish to lower morale. In large firms and companies, top and local management rarely meet, and it is often difficult to determine who should receive the information. Wilkinson (2021) suggests that good facilitators should guide the process, motivate, establish a foundation for consensus, "respond in advance to avoid dysfunctional behavior," be generous with praise, guide the group to a constructive resolution of conflict, keep projects on track, and be a constructive listener.

Leadership is, in many ways, the key to a good work environment. According to Peters and Austin (1986, 6), "leadership means vision, cheerleading, enthusiasm, love, trust, verve, passion, obsession, consistency, the use of symbols, paying attention as illustrated by the content of one's calendar, out-and-out drama…creating heroes at all levels, coaching, effectively wandering around, and numerous other things." They recommend what they call "naive listening," that is, sending everyone in the company (from executives to R&D people to hourly wage earners) to vendors and customers, and bringing vendors and customers to the company, so that customers and vendors come alive for employees and executives (and vice versa). The key point is listening to what is actually being said, rather than hearing what one expects to hear.

For example, according to Peters and Austin (1986), Apple has its entire executive staff regularly listen in to the customer call-in 800 number (p. 10); a Levi Strauss executive spent a weekend selling bluejeans (p. 12); Campbell Soup regularly sends executives to the kitchens of 300 families (p. 13); Milliken & Co. invites customers to visit the specific factory at which their product is made, and forms joint problem-solving teams with customers that include hourly wage earners from both sides (p. 18); People's Express puts letters from customers (including complaints) in a centralized bulletin board (p. 20); 3M requires its R&D people to make regular sales calls; and every Tandem facility has a weekly "beer bust" to which all employees are invited. "Many say that more business is done in those couple of hours than during the rest of the week combined" (p. 35). "At Hewlett-Packard, each engineer leaves the project he or she is working with out 'on the bench.' Other engineers...take a look at it, play with it, comment on it" (p. 27). (See the discussion of workbenching in Chapter 1.) Peters and Austin suggest that "if you're an R&D VP, make sure you have a second (non-headquarters) office in a lab" (p. 39).

Peters and Austin point out that these techniques foster innovation, since many ideas come from customers or those who deal directly with products or processes, such as linemen. I want to point out that these techniques also help foster a sense of community. Without that sense of community, engineers cannot get satisfaction from being part of a well-functioning team.

Blotnick (1984) points out what happens to the person who doesn't take satisfaction from being part of a team doing a good job. Such a person tends to think about beating the system, and shows an inability to accept authority. He needs more and more money, since work doesn't satisfy, but does less and less to earn it. The effects of this pattern are bitterness, goofing off, and dishonesty. He may feel an excessive need to stand out, feeling that teamwork will make him anonymous, and he will have no credentials to take with him if he needs a new job. As some of the people in Blotnick's book say, "just doing my job isn't getting me the notice I need" (p. 92), and "you have to make people see what you personally can do" (p. 90). Unfortunately, once one person starts hogging credit, it forces everyone

232 • *Ethical Engineering*

else to be "me-oriented" in self-defense. By contrast, Blotnick points out, the best kind of teamwork arises when each person is "busy, day after day, doing what he or she does best" (p. 103).

Blotnick mentions two further harmful patterns: anxiety about becoming obsolete and deliberate amnesia of past accomplishments. Anxiety about becoming obsolete is often a sign that something else is wrong, that the employee has gotten off the right track. Sometimes, this is because the person doesn't recognize that they now have the authority to institute projects and changes, or doesn't realize that praise is sparser for a higher-level person than it used to be, since superiors think it is less needed. Thus, the diminishing of praise is just a normal sign of advancement. Such a person may feel fear of pressure from below, engage in lobbying efforts to prove they are needed, sandbag subordinates, and tend to show indecisiveness.

The second pattern is a strategy for motivating oneself. Some people make a practice of forgetting their past accomplishments. But, says Blotnick, it is important to savor victory when you get it, then move on. (It loses its flavor if you wait till the next project has started.) If you downgrade your past accomplishments, you will have nothing to take with you into the later part of your career. What counts as an "achievement" keeps going up. This may lead to resenting subordinates and to getting involved in overly ambitious undertakings (lacking a sense of reality). The pressure to achieve more and more may actually make you less productive.

PROTÉGÉS

Blotnick (1984) also has some insightful remarks about conducting a protégé relationship. There are several reasons protégé relationships are often unsatisfying. Executives often are fearful of giving away anything important, then complain because the protégé doesn't listen. Too often, the mentor spends a great deal of time sharing complaints about the company. This doesn't help the protégé. After all, if you want people to listen to you, you must give them something worth listening to. Again, executives sometimes look for someone who will humor them instead of a serious protégé who wants to learn. Keep a loose and flexible relationship, and realize that your protégé will have her own way of doing things. Finally, if the relationship feels like a drain or a chore, this is a strong Signal that something has gone wrong. A mentor-protégé relationship should be a two-way street.

DEALING WITH SUBORDINATES

Allan Firmage (1989) makes the following suggestions: supervisors tend to assign repetitive work to the same employee, thinking he/she will get better and quicker at it. This may work for a while, but boredom quickly sets in (p. 55). Recognition awards "should not be a substitute for a simple 'well done.'" In the case of inadequate performance, the supervisor should inform the employee constructively, with a program of improvement formulated in a "joint conference" between supervisor and employee. Firms should "encourage and reward" membership in and attendance at meetings of technical societies. This may involve paying part of the membership dues, allowing reasonable time off to attend meetings, and encouraging the presenting and publishing of papers. Employers should maintain a neat and attractive office environment. Self-esteem is advanced when each employee has her own work area with proper furnishings (p. 57).

To these may be added W.H. Roadstrum's more general reminders that "a good rule in dealing with others is to work from the highest common purposes possible while not overlooking the economic motive" (Roadstrum 1967, 181). Moreover, says Roadstrum, an employee's need for a good self-image is often as important as her economic needs. "Threats against self-image" are to be avoided when feasible. The fulfillment needs of employees are often as important as their economic needs (p. 180).

Nordli (2019) points out that "managing engineers is a balancing act between when to get involved and when to step back, finding projects that tap into an employee's strength and help them grow, and deftly removing obstacles that impede progress." Nordli's advice, culled from speaking to engineers, is to

Create a prioritized list of things you want to accomplish, and block off time on your calendar to get them done; Put your team members in a position to succeed, both by giving them tasks that play to their strengths and ones that stretch their skills; communicate early and often about what's going well, and what isn't; find ways for employees to learn from each other, through collaboration, discussions or exercises like pair programming; make sure everyone knows what they're supposed to be working on. It's not always as obvious as it seems; [and] refrain from trying to solve every technical problem that arises. You don't have the time, and it'll stifle your team's growth.

Nordli 2019

In addition to these suggestions, there are some rules and policies that employers and managers should generally follow. Get out of the way, but still get your hands dirty (don't micromanage, but it is helpful to keep in touch with the actual work your subordinates do.) "Assess, document, and communicate any assumptions you make when planning a project" (More 2021). "No public or private organization shall discriminate against an employee for criticizing the ethical, moral or legal policies and practices of the organization; nor shall any organization discriminate against an employee for engaging in [legal] outside activities of his or her choice [that do not violate the engineer's duties as an engineer or an employee], or for objecting to a directive that violates common norms of morality" (Ewing 1977, 10). (I added the words in brackets because some outside activities, such as moonlighting for a competitor, might be licit grounds for employer sanctions. See also the discussion of Case 50 for additional caveats.) "No organization shall deprive an employee of the enjoyment of reasonable privacy in his or her place of work, and no personal information about employees shall be collected or kept other than that necessary to manage the organization efficiently and to meet legal requirements" (*ibid.*). "No employee of a public or private organization who alleges in good faith that his or her rights have been violated shall be discharged or penalized without a fair hearing in the employer organization" (*ibid.*). "In a personal interview, the employer should inform the employee of the specific reasons for his/her termination" (NSPE Guidelines 1980, 21).

One piece of advice I give students about dealing with almost anyone is: *try to give people a face-saving way of giving you what you want.* "You screwed up doing x—fix it" is less effective than "knowing how conscientious you are, I'm sure x was just an understandable oversight you'll want to correct right away." This strategy applies to subordinates, superiors, clients, regulators, and almost anyone with whom one must deal. Using it will make your life much easier.

EQUALITY/EQUITY, DIVERSITY, AND INCLUSION (EDI)

Equality, Diversity, and Inclusion, or EDI, focuses on three different aspects of issues surrounding individuals from groups that have been under-represented or otherwise subject to discrimination on the basis of race, ethnicity, country of origin, religion, gender, sexual identity or orientation, age, disability status, or other factors, which may be summarized as "opportunity, presence, and welcome." *Equality or Equity* is about *offering equal opportunities* for everyone. *Diversity* is about the *presence of individuals with different backgrounds*, especially those who have been under-represented. For example, "in the year 2000, approximately 1,788,000 men were in some form of engineering occupations in the U.S., while there were only 210,000 female engineers" (Poulsen 2007, 356). Although there has been some improvement, in 2019 only 15% of those working in engineering professions in the US (and 27% of those working in STEM) were women (Martinez and Christnacht 2021). In Japan, 14% of those graduating from engineering programs are women (Foster 2021). Leevers (2020) suggests that "changing career aspirations in response to coronavirus may deepen the current under-representation of women in the engineering and technology workforce." *Inclusion* is about *ensuring that everyone feels welcomed and able to participate*. For example, a Latina engineer reported that "It took…almost two and a half years [for them] to really take me seriously and start giving me assignments" (Rincon and Yates 2018). *Firstup* (2021) mentions "the employee who's a native Spanish speaker but doesn't feel entirely comfortable speaking any language other than English in workplace common areas. Or the breastfeeding mother just returning to work who has no space to pump her breast milk. Or the Muslim employee who feels insecure about maintaining his daily prayer routine on company grounds."

Issues of EDI have assumed increasing importance in recent years. It is important for engineers individually, engineering organizations, and the engineering profession to take an active role in addressing all three concerns. Individual engineers must ensure both that they are not contributing to the problem in their designs, employment decisions, and other actions, and that they are helping to foster a diverse and welcoming workplace. Within an organization, such as a government agency, engineering firm, or a large corporation employing engineers, "Embracing diversity, equity, and inclusion as organizational values is a way to intentionally make space for positive outcomes to flourish" (National Council of Nonprofits 2018). Finally, with regard to the engineering profession, "diversity, equity, and inclusion are necessary to build a stronger, more unified profession" (Brooks 2020). While "this societal ill is, of course, beyond the reach of traditional engineering problem-solving based on models and computation….professional engineers can, and should, contribute to solutions" (Brooks 2020).

Embracing EDI is not only the right thing to do but also promotes good engineering and good business. As AIChE (2021) notes, "We encourage inclusion and intentional representation of people from diverse backgrounds and experiences because it is ethical and honorable, and it enhances the innovation and creativity necessary to find solutions to current and future challenges." Cole (2020) suggests that EDI confers eight (overlapping) benefits on organizations: increasing creativity and innovation, broadening the range of skills, boosting the business and making for happier employees, increasing productivity, better enabling organizations to understand their customers, expanding the talent pool, and increasing revenues. Hunt et al. (2018) found a strong correlation between gender diversity on the executive team and strong economic performance. Leevers (2020) points out that "workforce diversity improves innovation, creativity, productivity, resilience and market insight." Moreover, EDI is intimately connected to the very purpose of engineering: "Our profession of engineering exists to protect people's wellbeing. That is the purpose of practicing engineering--to ethically create products or services that protect the life of people of all diversities" (Yip et al. 2021).

Addressing EDI "requires the exploration and acknowledgement of inequities and overcoming the defensiveness that comes from conversations on power and privilege" (Engineers Canada 2020). It is a mistake, however, to think that addressing EDI in engineering is solely the problem of white males. Perceptual bias, discussed in Chapter 5, is found in all population groups, as it is fostered by widely broadcast stereotypes. Loven et al. (2011) found that women remember and recognize female faces better than male faces. Another study revealed that "men and women employers alike revealed their prejudice against women for a perceived lack of mathematical ability" (Bohannan 2014). Valla et al. (2018) found that both Black and white subjects were more likely to identify an object (such as a gun) as dangerous after seeing a brief image of a Black person than after seeing a brief image of a white person.

Engineers should address EDI on three levels. Some of the suggestions below are things organizations can implement. Others are things individual engineers can do on their own. Finally, some are things engineers can suggest and vocally support in their various communities: workplaces, personal and professional networks, and professional organizations. (See also Section "Hiring Practices".)

1. Be mindful of EDI concerns in design and products. EDI concerns arise in design, sometimes in ways that may not be immediately apparent. For example, "many studies have shown that automobile design as

well as safety policy are biased toward the average male body, resulting in a greater chance of injury or death to female car occupants" (Brooks 2020). Wilson, Hoffman and Morgenstern (2019) found that machine learning systems used in autonomous vehicles have more difficulty discerning dark-skinned pedestrians, leading, potentially, to a higher rate of injuries for darker skinned individuals.

2. Establish employee resource groups and a system for mentoring under-represented groups.

3. Create and support a diversity training program. The program should be meaningful and include specifics, so it is not perceived as time-wasting fluff.

4. Have teams create diversity videos.

5. Engineers can work to establish centers like WinSETT (Canadian Centre for Women in Science, Engineering, Trades, and Technology), "an action-oriented, not-for-profit organization that aspires to recruit, retain and advance women in science, engineering, trades and technology" (Emerson et al. 2021).

6. *Firstup* (2021) recommends writing gender neutral job ads and descriptions, blind review of applicant resumes, and setting specific diversity goals (and tracking progress on meeting those goals).

7. Schooley (2020) recommends celebrating diverse holidays (such as Kwanzaa, Diwali, Chinese New Year, Dia de los Muertos, Purim, and Eid al-Fitr) as a team; creating a buddy system pairing engineers of different backgrounds; and incorporating the team's diversity into the team's product. (However, in following this advice, keep in mind the issue of cultural appropriation discussed in Chapter 3.)

8. Brzezinski (2017) lists eight things "to strengthen common codes of conduct and disrupt harmful corporate cultures that foster abuse": "Stick to an open door policy....Stand up for people....Think about the purpose of every work event you take part in....Bring other people to meetings or events that take place outside of the office.... Promote people for the right reasons....Look to create diversity at all levels of the company....Don't feel like you need to leverage your looks to get ahead....Push back in real time."

9. The Women in the Workplace Survey 2021 found that women of color found the following actions, in order, to convey meaningful allyship: "advocating for new opportunities for women of color; if seeing discrimination against women of color, actively working to confront it; publicly acknowledging women of color for their ideas and work; educating oneself about the experiences of women of color; mentoring or

sponsoring one or more women of color; actively soliciting the perspectives of women of color when making decisions; actively listening to the personal stories of women of color about bias and mistreatment; taking a public stand to support racial equality" (Thomas et al. 2021).

10. The Grainger College of Engineering (2021) mentions three ways engineering faculty can support EDI: "mentoring/advocacy: teaching, tutoring, or mentoring in programs for [under-represented groups] as well as activities advocating for EDI issues in a previous academic position; education/outreach: outreach efforts aimed at [under-represented groups], attendance to conferences, seminars, luncheons, etc. aiming at promoting engagement and supporting [under-represented groups]; community/service: volunteering at a particular organization targeting engagement with [under-represented groups]."

11. A survey respondent who wishes to remain anonymous urged three points/suggestions. (A) "There needs to be more early education about engineering opportunities for all individuals. An engineering degree can open doors in many fields, and does not focus solely around math." (B) "The primary challenge encountered as a female engineer is the existence of sexual innuendo in marketing materials and human interaction.… Sexual innuendo and gender connotation does not belong in a professional environment." (See also Case 48 below and "Advertising" in Chapter 14.) (C) "Politely and privately let someone know if their words or actions are not representative of equality in the workplace. Lead off with a compliment. Specifically identify the action that was offensive or in bad taste. Conclude with a professional explanation of why women find it offensive."

12. Incorporate concepts, technologies, and practices from marginalized or non-Western communities. For example, John Desjarlais points out that "indigenous design and ethics demonstrate integration and historical examples of problem-solving using a reciprocity and virtue-based mindset" (Irving 2021).

SEXUAL HARASSMENT, FAVORITISM, AND PROFESSIONAL RELATIONS

The #MeToo movement has focused attention on sexual harassment in the workplace and, more generally, the treatment of women. LeanIn UK's (2019) survey found that "64% of women report that they've experienced some form of sexual harassment in the workplace, from hearing sexist jokes to being touched in an inappropriate way. And 19% of women say harassment is on the rise." Moreover, sexual harassment is not an isolated ill. An atmosphere in which sexual harassment is normalized is prone to discrimination and other forms of unethical behavior. "Sexual misconduct is not a stand-alone issue....when incivility is tolerated in the workplace, it sets the stage for escalating misconduct" (Geisler 2018).

There is more or less universal agreement that sexual harassment has no place in the workplace. However, while there is a considerable overlap, different cultures (and individuals) may disagree about what precisely constitutes sexual harassment. For example, quid pro quo (demanding sexual favors as a condition of employment, advancement, or favorable treatment) is universally proscribed. However, different cultures have different conceptions of violating personal space or of what is considered inappropriate touching. "Frequently, behavior that is tolerated in Japan is considered inappropriate in the U.S" (Kopp 2014).

India's 2013 Prevention of Sexual Harassment Act defines sexual harassment as "unwelcome sexually determined behaviour, whether directly or by implication, such as: physical contact and advances, a demand or request for sexual favours, sexually coloured remarks, showing pornography, and any other unwelcome physical, verbal or non-verbal conduct of sexual nature." Chile's 2015 Law 20.005 prohibits "Non-consensual sexual requests which threaten the employment situation or opportunities of a worker," including "non-consensual physical contact; verbal proposals of a sexual nature; unsolicited gifts with romantic or physical connotations; and emails or personal letters containing sexual requests." Ethiopia's Labour Proclamation No. 1156/2019, Article 2 (11) states "'sexual harassment' means to persuade or convince another through utterances, signs or any other manner, to submit for sexual favor without his/her consent" (Federal Negarit Gazette 2019). The European Parliament (2006) avers that sexual harassment occurs when "any form of unwanted verbal, non-verbal or physical conduct of a sexual nature occurs, with the purpose or effect of violating the dignity of a person, in particular when creating an intimidating, hostile, degrading, humiliating or offensive environment." In the US, the EEOC's Guidelines recognize two forms of sexual harassment: quid pro quo and creating a hostile work environment. "Unwelcome sexual advances, requests for sexual favors, and other verbal or physical conduct of a sexual nature constitute sexual harassment when (1) submission to such conduct is made either explicitly or implicitly a term or condition of an individual's employment, (2) submission to or rejection of such conduct by an individual is used as the basis for employment decisions affecting such individual, or (3) such conduct has the purpose or effect of unreasonably interfering with an individual's work performance or creating an intimidating, hostile, or offensive working environment" (Code of Federal Regulations 2016). Sachi Barreiro says:

> any of the following actions can be sexual harassment if they happen often enough or are severe enough to make an employee uncomfortable, intimidated, or distracted enough to interfere with their work: repeated compliments of an employee's appearance; commenting on the attractiveness of others in front of an employee; discussing one's sex life in front of an employee; asking an employee about his or her sex life; circulating nude photos or photos of women in bikinis or shirtless men in the workplace; making sexual jokes; sending sexually suggestive text messages or emails; leaving unwanted gifts of a sexual or romantic nature; spreading sexual rumors about an employee, or; repeated hugs or other unwanted touching (such as a hand on an employee's back).

> *Barreiro 2021*

Womenwatch (2019), a United Nations project, includes as possible forms of sexual harassment "unwanted deliberate touching, leaning over, cornering, or pinching....unwanted sexual looks or gestures....hanging around a person....standing close or brushing up against a person.... referring to an adult as a girl, hunk, doll, babe, or honey." With respect to sexual orientation and gender identity, examples given by the EEOC include "repeated, deliberate use of the wrong name or gender pronouns (e.g., he or she); shaming an employee for not acting or dressing in a way that reflects the sex the employee was assigned at birth; refusing to allow an employee to use the restroom associated with the gender the employee identifies with" (Peirce 2021).

Drawing the line is a controversial undertaking. Few would defend workplace groping, for example, but some other behaviors are debated. Some believe that, for example, hugging, standing close, and "hanging around" are friendly more often than sexual indications of personal connection. Jokes and respectful personal compliments, both of which can be problematic in the US, are considered normal in some other societies. Disagreements about particular behaviors sometimes reflect a broader disagreement about the nature of a workplace. Some argue that the workplace should be exclusively focused on business and professional interactions: the workplace is for work, not socializing. Others argue that removing the personal from the workplace dehumanizes the work environment, which is typically about half of a person's waking hours. (This issue is discussed below.) Nonetheless, although there can be reasonable disagreement about exactly what should be considered appropriate in a workplace, *engineers should follow applicable legal guidelines as well as their workplace's rules.* The rule should be, for persons of any gender, both out of consideration for others and out of personal prudence: *if there is any doubt, don't do it.*

Preventing sexual harassment calls for a variety of measures. "Ensure that employees know about and have access to confidential reporting channels through [which] they can report cases of sexual harassment" (Hormann 2021). Volkov (2019) recommends a proactive response consisting in revisiting company policies and procedures; promoting diversity initiatives; instituting innovative training programs; fostering a culture that encourages speaking up and reporting concerns; "a robust investigative protocol" that is "documented, communicated, and scrupulously followed," including an independent committee of senior executives; and recognizing "the need for comprehensive remediation and resources needed to address the problem." Bille (2021) suggests making sure employees understand what sexual harassment is and that preventing it is a company priority; instituting training that goes easy on "legalese" and is positive rather than accusing in tone; and enlisting employees to be "active bystanders" and "social influencers," that is, to be a trusted source of information and attitudes for fellow employees (as well as those who run training programs), to intervene in and formally report sexual harassment when they see it, and to give support to victims. In general, "companies have to recognize that promoting and managing their cultures is one of several important measures to prevent and detect improper sexual harassment and assault incidents from occurring" (Volkov 2019).

For a variety of reasons, many individuals do not feel free to voice their concerns or comfortable doing so. When feasible, communication is helpful and being straightforward is generally the best policy. (That is one reason workplaces should strive to create an atmosphere in which speaking up feels safe and comfortable.) In addition, when someone does indicate that a given behavior makes them uncomfortable, that is normally a strong and sufficient reason to refrain from that behavior.

Non-sexual personal relationships also raise issues of favoritism. It is important to protect subordinates from personal pressures, both explicit and subtle. There are two ways to achieve this. One way is to insist on rigid non-fraternization rules: supervisors should never socialize with subordinates, physicians should never socialize with patients, and faculty should never socialize with students. The other way is to maintain the ideal of integrity in work decisions.

These two methods, the ideal of integrity and non-fraternization rules, lead to very different kinds of work communities. "Fraternization" has come, especially in military contexts, to mean sexual or romantic relationships but it is used here in its original meaning of any personal social connection, especially friendship. (The root of the word, after all, is Latin for brother.) *The rule of non-fraternization insists that I have no personal dealings with subordinates* (or, in some cases, with anyone in the workplace). *The ideal of integrity insists that my personal feelings for or against subordinates not influence my professional decisions.* For example, the rule of non-fraternization says I shouldn't take non-working lunches with subordinates. The ideal of integrity permits and even encourages social lunches, but insists that my subordinates must have absolute confidence that, when I award promotions, I do so entirely on the basis of job-related merit, that those who join me for lunch will not, on that account, be favored, and that those who don't will not, on that account, be penalized.

Again, it is important that these social interactions be entirely free of sexual harassment. If there is any chance a subordinate will perceive a lunch invitation or other form of socializing as sexual harassment, don't do it. More broadly, engineers should keep firmly in mind that a social interaction may appear or feel very different from the other side, especially when the other person does not feel comfortable saying "no." Telling a subordinate about one's marital problems could make the subordinate very uncomfortable but afraid or uneasy about showing or expressing that discomfort. More generally, it goes without saying that engineers must follow the law and adhere to company policy. The ethical engineer will keep within the limits of social behavior between supervisor and subordinate that the law requires and will not violate company policy.

Several reasons suggest that non-fraternization rules are less than ideal. They produce a class-dominated society. Forbidding a manager to talk about personal issues with her assistant sends a strong message to the assistant: you are not the social equal of your boss. You are not merely a subordinate in terms of work responsibility but in your personal life as well. And we send a message to the executive: your assistant is not a person but another business machine. The result is that work relations become purely instrumental. Employees become not a community of mutually concerned persons with common goals and values but a well-oiled business machine.

Fellow employees become not team-mates who care about each other as people, but impersonal cogs and wheels. Non-fraternization rules establish an atmosphere of impersonal efficiency, rather than an atmosphere of fellowship based on pursuing a common goal. This further alienates people from their work. In many fields, non-fraternization rules lower the quality of work. Depersonalization leads to bad engineering, because the element of comradeship and joint innovation disappears. Depersonalization leads to bad medicine, because physicians are prompted to view patients simply as raw material for their skills, not as persons with unique goals and concerns. They are prompted to view their work not as a calling to which they bring the whole of their humanity, but as an impersonal technical exercise. Depersonalization leads to bad education, because faculty are prompted to view students not as members of a community in whose development as thinking human beings they have a personal concern, but as data banks to be filled with information or programs (skills). Finally, non-fraternization rules lead, ironically enough, to a lack of ethics in the work world, since efficiency replaces humanity, and propriety replaces values. When rules replace integrity, what becomes important is not the values associated with integrity but simply keeping to the rules. The ethical dimension drops out and what is left is a game with arbitrary rules to be manipulated to one's advantage.

For these reasons, as well as the value of a community workplace discussed in Chapter 1, the ideal of integrity is attractive. My suggestion is that to the extent that engineers have some latitude, they should generally favor the ideal of integrity over rigid non-fraternization rules (exceptions are discussed below). Integrity calls for special efforts from everyone involved, because it involves trust, and trust must be built and nurtured. It is much easier to keep to a set of rules than it is to use sensitivity and good judgment to build trust, but the extra effort is good business practice and generally improves the lives of employees.

There are, of course, circumstances in which stringent non-fraternization rules are appropriate or even required. In particular, *stringent non-fraternization rules are appropriate when the appearance of propriety is itself a crucial aim of the activity*. In a legal system, for example, assuring the public of the impartiality of judicial proceedings is as important as arriving at just results, for the law's standing depends upon public perception of the law as absolutely impartial. The rule of law is centrally about impartiality. Impartiality is the keystone of law, and respect for the authority of law within a society depends critically upon public perception of impartiality. So, there is some sense in a rule that judges should not socialize with practicing attorneys who might come before them. Bidding is a highly competitive practice, in which, typically, huge sums of money are involved, and the very process of submitting bids depends critically upon impartiality. Bid-taking is centrally about impartiality, and the perception of impartiality is crucial to the practice. Moreover, bid-takers and suppliers generally work for different companies; bid-takers are not on the same "team" as suppliers, and so the ideal of community is not compromised by building walls between bid-takers and suppliers as persons. Because suppliers are not subordinate to bid-takers, non-fraternization rules do not create a class structure. Thus, building walls between suppliers and bid-takers does not dehumanize suppliers. There are good reasons, therefore, for bid-takers to adhere to strict rules about accepting gifts, meals, and so forth from suppliers.

By way of contrast, impartiality of grades is not what the university is centrally about: the university is devoted to the cultivation and advancement of wisdom and learning, a goal to which testing and evaluation of students is secondary. Medicine is about healing, about advancing the well-being of people as people. Building walls between students and faculty, between physicians and patients, does undermine the central tasks of education and medicine, and does dehumanize patients and students.

The application of these remarks to engineers is fairly clear. *Follow faithfully all legal requirements and company policies. Within those limits, follow strict rules of non-fraternization when the context is highly competitive, the perception of impartiality a central goal of the activity, or building walls between the people involved does not undermine a central task. Otherwise strive for a community atmosphere, taking special care to establish trust in your integrity and being sensitive to and respecting the ways in which subordinates might feel pressured or uncomfortable.*

HIRING PRACTICES

Ethical hiring requires a balance between judgment and fair rules. Many important factors are hard to measure, such as integrity, loyalty, motivation, and ability to work with others. These are legitimate and important criteria. However, the hirer must be careful not to let biases, prejudices, and superficial appearances shape her assessment of these qualities.

It cannot be stressed enough that engineers involved in any aspect of hiring must strictly conform to any relevant laws. This certainly includes conducting interviews and making hiring decisions, but it also includes showing applicants around, answering applicant emails, writing or discussing job notices, etc. Whether an engineer is directly or only tangentially involved, they are required to adhere to all applicable legal guidelines concerning hiring and strongly advised to follow both the spirit and letter of guidelines from relevant organizations. As these guidelines change with time and are different in different jurisdictions, I will not try to enumerate them here. Engineers involved in any way in the hiring process must be sure to keep current about them. Here, as in several other sensitive areas, err on the side of caution. For example, if there is any doubt at all about whether an interview question is appropriate, don't ask it.

Constructing a detailed job description is one tactic that helps ensure fairness in hiring. Snoeyenbos and Almeder (1983, 199) cite two reasons for drawing up a detailed job description. First, job descriptions are often required by law. "More fundamentally, utility for the firm and fairness to the prospective employee are enhanced when the employee knows what is expected of him."

However, the value of a job description must be balanced against the more flexible demands of community. Providing job interviewers with a detailed job description of the position to be filled helps to ensure that the factors considered by interviewers are relevant to the job and makes it easier for engineers to be clear about they must do to succeed. But in a community atmosphere, engineers do what they can to help the project, rather than limiting their contributions to what is in their job descriptions. In short, job descriptions may tend to promote rigidity and compartmentalization instead of promoting teamwork, flexibility, and initiative. Perhaps the best compromise would be drawing up a job description that leaves room for flexibility without being overly vague. For example, the job description might spell out clearly some duties, but include a clause that the engineer is expected to contribute in appropriate ways to the success of the company or firm. This kind of job description tends to favor the employer. Since job descriptions should be mutually beneficial, job descriptions should also include some things the company or firm owes the employee.

Dessler (1981) suggests that interviewers be properly trained and carefully selected, possess a detailed job description, draw their questions from a structured set of guidelines, take care to avoid premature decisions, and supplement interviews with documentable tools such as reference checks and tests. In addition, the tools and techniques the interviewers employ should be validated (documented to be valid predictors of job success).

Again, it must be remembered that hiring decisions demonstrate a conflict between rules and judgment on the one hand, and fairness and community on the other. While testing does serve as a check against interviewer bias and is helpful in protecting hirers against legal action, too much reliance on testing results in poor hiring decisions.

INTERDEPARTMENTAL DEALINGS AND HIRING AWAY FROM ANOTHER FIRM

In dealing with another department within the firm or company, a balance must be struck between mutual concern and respect for autonomy. On the one hand, loyalty means that every department has a stake in the flourishing of every other department: all departments are members of the same team. On the other hand, no department will function smoothly unless channels within a department are respected by other departments. The question to ask yourself is "would doing this undermine the orderly functioning of the other department?"

In engineering, there is a kind of gentleperson's agreement that companies and firms will not hire engineers away from other companies and firms. To some extent, this serves the companies and firms, since they do not have to compete for the services of top engineers, and cannot be as easily manipulated by engineers seeking to play one company against another. There is, however, a legitimate ethical reason for this agreement. The impossibility of drawing a clear line between trade secrets and an engineer's general knowledge raises ethical problems when engineers change jobs. It is difficult for a Human Resources Officer, however ethical, to be sure that she is recruiting engineer Smith of P Corporation for their general knowledge and ability, rather than for their familiarity with P Corporation's trade secrets and current research. Thus, it is preferable not to recruit engineers from another firm. (See "Confidentiality and Trade Secrets" in Chapter 8.)

However, there may be emergencies when a particular kind of skill (not a trade secret) is desperately needed, or when an engineer who is under-utilized (or underpaid or ill-treated) at his current firm would be perfect for a more responsible position in another firm. In such cases, there are two approaches that companies have used. When Michaels of Z Company wishes to hire Smith of X company, Michaels might speak to Smith's supervisor at X company, rather than only to Smith himself. The other possibility (more commonly encountered) is for Michaels to mention to a common acquaintance, Jones, that the position at Z company is available and leave it to Jones to inform Smith. This approach preserves the fiction that Michaels has not made contacted with or recruited Smith. (This case illustrates that flexible rules can still be very useful. An overly strict rule against "hiring away" would not be as useful, since firms could not meet urgent needs. However, an overly lax rule would result in constant raiding and hiring away, creating problems of instability and confidentiality. The actual practice, which makes it difficult and cumbersome but not impossible to recruit engineers in other firms, insures that hiring away does not occur except in cases of special need, the arrangement that best suits most firms.)

Of course, engineers have a right to seek jobs and change employers, for a variety of legitimate reasons. While issues about trade secrets mean that engineers should be conservative about changing jobs, they are not slaves bound for life to a single employer.

Case 48: Sexual Innuendos in Advertising and Session Titles

With regard to advertising (and conference session titles) and sexual innuendo, where is the line between harmlessly intriguing and inadvisable or impermissible? For example, is it ok for a drilling company to use a poster ad that says, in large letters, "Harder, Faster, Deeper"? Should one avoid entitling a talk on industrial vibrators "Does Size Really Matter?" Two questions are relevant here. First, should engineers avoid deliberate innuendo? Second, to what extent should engineers think about and take steps to avoid their legitimate titles, which they did not intend to be sexual, being given an unintended sexual meaning by someone else?

Case 49: Moonlighting

What guidelines govern the acceptance of additional employment outside of office hours? Consider these factors: (a) the potential conflict of interest between the two employers, (b) the effects of informing both employers, (c) the extent to which having a second job interferes with whole-hearted performance of the first (for example, by tiring the employee, leading to burn-out or reduced effectiveness; by cutting into time for further study, community involvement, and dreaming up new ideas; and by mitigating social and professional allegiances), and (d) the degree to which the employer's perception of a "moonlighting" employee will harm that employee's career.

Case 50: Dismissal for Non-Work Related Conduct

To what extent is personal conduct not directly related to employment duties a valid reason for dismissal? Consider:

1. conduct that embarrasses the company, such as adultery; taking an unpopular public stand on issues outside of engineering; domestic violence; gambling, drinking or using drugs outside the office; racist posts on social media; and commission of crimes not related to job activities, such as non-job-related insider trading or carrying a gun without a license, and
2. conduct in the workplace that is not specifically job-related, such as a conceited manner toward other employees, or borrowing and not returning money or articles from other employees.

Discussion: Two opposing factors operate here. The communal character of the profession and the company pulls in one direction, while due process and the autonomy of the individual pull in the other. A few remarks might prove helpful. In our discussion of dealing with subordinates, we said that employers must not punish employees for outside (legal) activities of their choice. Recall also the importance of respecting rights. No employer should act so as to place a "chilling effect" on the exercise of basic freedoms such as freedom of speech, political association, and assembly. These freedoms are essential to a democratic way of life, and so employers must be sensitive to any pressure they may place upon employees not to exercise these rights freely. An employee's taking an unpopular stand, or engaging in a lifestyle found repugnant by supervisors or other employees, must not be allowed to affect adversely, or be seen by the employee as adversely affecting, the employee's career. If, as a result, the profession or the company loses some public esteem, that is, the price it must pay to operate in a free society. A company may not, without further evidence, assume that an employee who jaywalks will be careless at work, or fail to observe the law scrupulously. To do so would be to convict the employee of crimes she did not commit, on flimsy evidence, in a way that denies the employee the opportunity to speak in her own behalf. However, a successful and ethical company is founded on mutual trust and community. The employee is in many ways the legal agent of the company, and the company is responsible to insure and oversee the employee's probity in job-related matters. A responsible company will not endanger the public by giving responsibility for a project (or aspect of a project) to an employee whose judgment, the company has reason to believe, is distorted. So, it is a matter of legitimate concern to a company if an employee with a responsible position is often drunk outside of office hours or has a drug problem. Indeed, the community model suggests that the company might be a partner in dealing with these problems. Thus, when basic freedoms are not involved, the company may reprimand the employee for, and require the employee to correct, situations that compromise professional values or the community atmosphere, such as not returning borrowed items or failing to treat other employees with respect and comradeship. The company may take special steps, if it seems appropriate, to monitor the activities of those employees whose outside behavior casts doubt on their probity, warn those employees that strict observance of the law in work-related matters is mandatory, and provide assistance and counseling where appropriate. When the outside behavior is both well-documented and of such a nature as to cast severe doubt on the engineer's ability to act safely and responsibly, suspension or dismissal may be necessary. (Of course, it is necessary to give the employee a fair hearing before taking detrimental action.)

Finally, an employee's non-job behavior that violates foundational values raises sensitive issues. For example, an ethical company or organization will oppose racism and be vigorous in ensuring its activities are not racist. How should it respond, then, when an employee publicly posts blatantly racist statements? Can the public and fellow employees trust that the poster will not display racial bias in their job activities? If an employee posts that Jews are agents of Satan, for example, can Jewish customers and co-workers be confident that the employee will treat them fairly and faithfully? If not, that is a compelling reason for company sanction of the employee. By contrast a merely unpopular post, such as a post by a chemical engineer that the legal age of consent for sex should be changed to 12, would not be. However unsavory many will find that post, the post does not give grounds for doubting that the poster, a chemical engineer (and not a baby-sitter), will perform their work activities fairly and scrupulously. (Of course, if the engineer in any way presents himself as a representative of the company, that changes the case entirely.)

Case 51: Giving Reasons for Dismissal

Must a discharged engineer be apprised of the reasons for her dismissal? How frank and detailed must be employer be, and how should the matter be handled?

Discussion: One guideline in "Dealing with Subordinates" specifically says she must be given a frank answer. Treating others fairly, the values of the engineering profession and creating a community atmosphere all require that the employer should be as frank as possible in informing employees of the reasons for their dismissal. Vagueness fosters mistrust, and only by being frank can an employer help an employee improve (make the most of themselves). In a few cases, frankness may even lead to the elimination of a misunderstanding that renders the dismissal unnecessary. "In all justice, an employee is entitled to know in what respect or respects he fails to be satisfactory. A proper explanation may well assist him in altering his attitude to the end that he becomes a valued employee" (Alger et al. 1965, 109).

Case 52: Hiring Away

You need an engineer with knowledge, skill, and experience profile P. A friend of yours tells you that Smith, an engineer with firm Y, fits profile P quite well, and that Smith is somewhat unhappy with Y. Do you make use of this information? Do you approach Smith, and if so, how?

Case 53: Complaints about One's Successor Part 1

Garcia, before her transfer to Department Q, was the well-liked and respected head of Department P. Of late, the employees in Department P have been telling her that her successor, Adams, routinely puts down employees, does not listen to suggestions, makes arbitrary decisions, and doesn't let anyone else in P know what's going on. What, if anything, should Garcia do?

Discussion: Although it is no longer her department, if Garcia does have the interests of the company at heart, she will not simply ignore the problem. Insofar as she is able to do so, she ought certainly to be a "sounding board" for her former employees. However, it would be not only bad management, but also a violation of community and fairness, for Garcia to foster further discontent in Department P, or to complain about Adams to Adams' superior. (For example, Garcia has not heard Adams' side of things.) Rather, Garcia ought to be a voice for a community atmosphere and for treating people fairly and well. The proper course is (a) to be a sympathetic listener without suggesting that Adams is no good, (b) to remind the employees in Department P that they have to work with Adams, and that a constructive attitude is more helpful than resentment, (c) suggest that the employees in Department P speak frankly with Adams about the problems and try to work them out, reminding them of the importance of giving Adams a face-saving way of responding to their needs. If this is not successful, then (d) Garcia may volunteer to take Adams to lunch and, as sympathetically as possible, explain the problems to Adams and indicate a willingness to help. If, after all this is done, Adams continues his ways, Garcia may counsel the employees in Department P about how to proceed. (See the next case.)

Case 54: Complaints about One's Successor Part II

Adams, the head of Department P, routinely puts down employees, does not listen to suggestions, makes arbitrary decisions, and doesn't let anyone else in P know what's going on. The employees in P have made several good faith attempts to speak with Adams about these problems, to no avail. The former head of Department P has spoken to Adams informally about these problems, but Adams insists that "it's my department, and nobody is going to tell me how to run it." What should the employees do?

Discussion: Every good faith effort has already been made to treat Adams fairly and well, and to handle the matter informally in a co-operative way. If the company is generally trustworthy, the employees might sign a statement articulating their discontents and noting the steps previously taken to correct the problem. They should give a copy of this statement to Adams, indicating that if nothing is done, or if Adams retaliates against those who signed the statement, a copy of the statement will be sent to Adams' superiors. This should not harm the employees: if the company is well run, Adams' superiors will not regard the signers as "troublemakers," since (a) many members of the department signed the statement, which indicates that the problem is real, (b) the statement points out that many informal and constructive steps were taken before the employees complained to Adams' superiors, and (c) if the complaint is true, it is in the company's interests to do something about the situation. Of course, the best way to handle the matter depends on the particular situation. For example, if Adams is the brother of the CEO, another approach might prove more fruitful. That is why the five-step process is useful.

REFERENCES

AIChE (2021) "AIChE Equity, Diversity, and Inclusion Statement," https://www.aiche. org/Equity-Diversity-Inclusion

Alger, Philip, Christensen, N.A. and Olmsted, Sterling P. (1965) *Ethical Problems in Engineering* (John Wiley and Sons).

Barreiro, Sachi (2021) "What Kinds of Behavior Are Considered Sexual Harassment"? *NOLO* https://www.nolo.com/legal-encyclopedia/what-kinds-of-behaviors-are-considered-sexual-harassment.html

Bille, Elizabeth (2021) "7 Strategies for Preventing Sexual Harassment at Work," *Everfi* https://everfi.com/blog/workplace-training/strategies-to-prevent-sexual-harassment-at-work/

Blotnick, Srully (1984) *The Corporate Steeple Chase* (Penguin).

Bohannan, John (2014) "Gender Bias Seems to Affect Men's—And Women's—Perception of Women's Math Skills," *Washington Post* https://www.washingtonpost.com/national/health-science/gender-bias-seems-to-affects-mens–and-womens–perception-of-womens-math-skills/2014/03/17/e26f8aee-aad1-11e3-98f6-8e3c562f9996_story.html

Brooks, Jacquelyn (2020) "Why Should I Care about Diversity in Engineering?" *NSPE PE Magazine* https://www.nspe.org/resources/pe-magazine/july-2020/why-should-i-care-about-diversity-engineering

Brzezinski, Mika (2017) "With #MeToo, We Need a Serious Talk about Workplace Ethics," NBC News https://www.nbcnews.com/know-your-value/feature/metoo-we-need-serious-conversation-about-workplace-ethics-ncna830101

Code of Federal Regulations (2016) "Guidelines on Discrimination Because of Sex," 1604.11 https://www.govinfo.gov/content/pkg/CFR-2016-title29-vol4/xml/CFR-2016-title29-vol4-part1604.xml

Cole, Bianca Miller (2020) "Eight Reasons Why Diversity and Inclusion Are Essential to Business Success," *Forbes* https://www.forbes.com/sites/biancamillercole/

2020/09/15/8-reasons-why-diversity-and-inclusion-are-essential-to-business-success/?sh=4031edc91824

Dessler, Gary (1981) *Personnel Management*, 2nd ed. (Reston).

Emerson, Carolyn J., Lefsrud, Lianne M., Robinson, Jane and Hollett, Susan (2021) *Canadian Journal of Chemistry* 99, 8 https://cdnsciencepub.com/doi/full/10.1139/cjc-2020-0327

Engineers Canada (2020) "EDI in the Engineering Profession: The Importance of Maintaining Momentum on Diversity and Inclusion Initiatives," https://engineerscanada.ca/news-and-events/news/edi-in-the-engineering-profession-the-importance-of-maintaining-momentum-on-diversity-and-inclusion-initiatives

European Parliament (2006) "Directive 2006/54/EC," Article 2 https://eur-lex.europa.eu/legal-content/EN/TXT/?qid=1582119207866&uri=CELEX:32006L0054

Ewing, David (1977) "What Business Thinks About Employee Rights," *Harvard Business Review* 55, 234–235, quoted in Martin, Mike and Schinzinger, Roland (1983) *Ethics in Engineering* (McGraw-Hill).

Federal Negarit Gazette (2019) 25th Year No. 89 https://www.mtalawoffice.com/images/upload/Labour-Proclamation-No_-1156-2019.pdf

Firmage, D. Allan (1989) "Management/Employee Ethics in Engineering Offices," *Professional Issues in Engineering* 115, 53–58.

Firstup (2021) "15 Ways to Improve Diversity and Inclusion in the Workplace," https://firstup.io/blog/15-ways-to-improve-diversity-and-inclusion-in-the-workplace/

Foster, Malcolm (2021) "Japan's Tech Push Faces Gender Gap," *New York Times* September 2 Section B page 1. https://www.nytimes.com/2021/09/01/business/japan-tech-workers-women.html

Geisler, Jill (2018) "Ten Ethics Lessons from the #MeToo Movement in Media—And Beyond," *Freedom Forum Institute* https://www.freedomforuminstitute.org/2018/03/19/ten-ethics-lessons-from-the-metoo-movement-in-media-and-beyond/

Gordon, Bruce F. and Ross, Ian C. (1983) "Professionals and the Corporation," in James Schaub and Karl Pavlovic, eds., *Engineering Professionalism and Ethics* (John Wiley and Sons), 149–157.

Grainger College of Engineering (2021) "Guidelines to Write an Equity, Diversity, and Inclusion Statement for Faculty Candidates," https://grainger.illinois.edu/about/diversity/guidelines

Hormann, Moritz (2021) "How to Prevent Sexual Harassment in the Workplace," *EQS Group* https://www.eqs.com/compliance-blog/sexual-harassment-workplace/

Humber, James M. (1983) "Honesty in Organizational Communication," in Milton Snoeyenbos et al., eds, *Business Ethics* (Prometheus), 175–184.

Hunt, Vivian, Yee, Lareina, Prince, Sara and Dixon-Fyle, Sundiatu (2018) "Delivering through Diversity," *McKinsey & Co.* https://www.mckinsey.com/business-functions/organization/our-insights/delivering-through-diversity

Irving, Tyler (2021) "'Engineering is Not a Western Construct': Lecture Examines the Role of Indigenous Design and Ethics in the Profession," *University of Toronto Engineering News* https://news.engineering.utoronto.ca/engineering-is-not-a-western-construct-lecture-examines-the-role-of-indigenous-design-and-ethics-in-the-profession/

Kopp, Rochelle (2014) "What's OK in Japan Can Be Sexual Harassment in the U.S.," *Japan Close-up and Japan Intercultural* https://japanintercultural.com/free-resources/articles/whats-ok-in-japan-can-be-sexual-harassment-in-the-u-s/

LeanIn UK (2019) "Working Relationships in the #MeToo Era," https://leanin.org/working-relationships-survey-results-uk

Leevers, Hilary (2020) "Equality, Diversity, and Inclusion Strategy 2019-2022," *Engineering U.K.* https://www.engineeringuk.com/media/232364/edi-strategy-final.pdf

Loven, Johanna, Herlitz, Agneta and Rehnman, Jenny (2011) "Women's Own-Gender Bias in Face Recognition Memory," *Experimental Psychology* 58:4, 333–340.

Martinez, Anthony and Christnacht, Cheridan (2021) "Women Are Nearly Half of U.S. Workforce but Only 27% of STEM Workers," *United States Census Bureau* https://www.census.gov/library/stories/2021/01/women-making-gains-in-stem-occupations-but-still-underrepresented.html

More, Vidya (2021) "Essential Skills for Effective Engineering Management," https://medium.com/@vidya.more/engineering-management-22644da13c5d

National Council of Nonprofits (2018) https://www.councilofnonprofits.org/tools-resources/why-diversity-equity-and-inclusion-matter-nonprofits

Nordli, Brian (2019) "Tips for Being an Effective Engineering Manager," *BuiltIn* https://builtin.com/software-engineering-perspectives/career-advice-for-engineering-managers

NSPE (1980) "NSPE Guidelines to Professional Employment for Engineers and Scientists," *Professional Engineer* 50, 15–21.

Peirce, Brooks (2021) "EEOC's New Guidance Regarding Sexual Orientation and Gender Identity Workplace Discrimination," *JD Supra* https://www.jdsupra.com/legalnews/eeoc-s-new-guidance-regarding-sexual-4669355/

Peters, Tom and Austin, Nancy (1986) *A Passion for Excellence* (Warner).

Poulsen, Eva Sorum (2007) "Female Engineers in Europe," in Steen Hyldgaard Christensen et al. *Philosophy in Engineering* (Academica), 353–367.

Rincon, Roberta M. and Yates, Nicole (2018) "Women of Color in the Engineering Workplace," *SWE* https://alltogether.swe.org/wp-content/uploads/2018/02/Women-of-Color-Research-2018.pdf

Roadstrum, W. H. (1967) *Excellence in Engineering* (John Wiley & Sons).

Schooley, Sky (2020) "7 Ways to Improve Workplace Diversity and Inclusion," Business.com https://www.business.com/articles/diversity-and-inclusion-examples/

Snoeyenbos, Milton and Almeder, Robert (1983) "Ethical Hiring Practices," in Milton Snoeyenbos, et al., eds., *Business Ethics* (Prometheus), 194–207.

Thomas, Rachel et al. (2021) "Women in the Workplace 2021," *McKinsey & Co.* https://wiw-report.s3.amazonaws.com/Women_in_the_Workplace_2021.pdf

Valla, L. G., Bossi, F., Calì, R., Fox, V., Ali, S. I. and Rivolta, D. (2018) "Not Only Whites: Racial Priming Effect for Black Faces in Black People," *Basic and Applied Social Psychology* 40:4, 195–200.

Volkov, Michael (2019) "The Urgency of Now: Corporate Ethics and the #MeToo Movement," *JD Supra* https://www.jdsupra.com/legalnews/the-urgency-of-now-corporate-ethics-and-29599/

Wilkinson, Michael (2021) "The 8 Roles of a Facilitator," https://www.leadstrat.com/8-roles-of-a-facilitator/

Wilson, Benjamin, Hoffman, Judy and Morgenstern, Jamie (2019) "Predictive Inequality in Object Detection," https://arxiv.org/pdf/1902.11097.pdf

Womenwatch (2019) "What Is Sexual Harassment," https://www.un.org/womenwatch/osagi/pdf/whatissh.pdf

Yip, Chris, Sterling, Marisa, Pepier, Daniel, Koursiniouris, Christopher and Mahmood, Safdar (2021) "Our Shared Values of Diversity, Inclusion, and Professionalism," *University of Toronto Dept. of Engineering* https://www.engineering.utoronto.ca/about/equity-diversity-and-inclusion/

10

The Environment, Climate Change, and Sustainability

In recent years, the engineering community as a whole has paid increasing attention to environmental concerns. Chapter 4 discussed the importance of a partnership with nature and different views of humanity's relationship to nature, such as Gaia Theory and Deep Ecology. To which specific environmental issues must engineers attend? The answer falls into four (overlapping) categories: climate change, pollution (besides greenhouse gases), sustainability, and conservation/preservation. (It should be noted that, as technology and measurements are rapidly changing, the information in this chapter may become outdated.)

CLIMATE CHANGE

While there is some disagreement about the exact nature and timetable of global warming, it is firmly established that human activity, especially the burning of fossil fuels, has significantly fueled an unprecedentedly rapid rise in global temperatures. Antonio Guterres, Secretary-General of the United Nations, called the IPCCs 2021 Sixth Assessment Report a "code red for humanity" (UN News 2021). The report, authored by 234 scientists from 66 countries, claimed that continued sea rise is already irreversible for centuries to come. While our climate had been relatively stable for the last 12,000 years, the average temperature of the earth rose roughly 2 degrees C between 1900 and 2000 and is projected to continue rising. This rise in temperature produces significant environmental changes. If current trends continue, dramatic changes will ensue. Predictions vary in detail but possible results include flooding of coastal regions, more

DOI: 10.1201/9781003242574-13

frequent and uncontrollable wildfires, increasingly powerful storms, drought in some areas and flooding in others (with resulting famine), widespread extinctions, and, in the worst case, runaway warming making the planet uninhabitable. Without serious responsive action, some or all of these are inevitable.

The three strategies for dealing with climate change are *mitigation, adaptation, and geoengineering*. Mitigation attempts to slow or reverse the increase in global temperature, adaptation tries to lessen its effects, while geoengineering seeks to directly change our planet's natural systems. Adaptation efforts include building flood defenses; laying down water-permeable surfacing; increasing water storage (and using water more efficiently); and creating new varieties of crops better suited to drought and/or high temperatures. Mitigation efforts include switching from high-carbon to low-carbon energy technologies; capturing carbon from the atmosphere, either by natural means (maintaining and expanding forests or using seaweed farms to remove carbon from the oceans) or via technology (carbon mineralization, scrubbers, and direct air capture); reducing energy demand; and moving away from animal-based to plant-based food. Geoengineering approaches include a system of solar reflectors (most likely in space) and injecting sulfate particles into the upper atmosphere, mimicking the action of a volcanic eruption, or dispersing sea salt into the lower atmosphere to seed clouds.

Mitigation programs are urgently needed. Geoengineering programs that would make a large dent in climate change are some years away, and "geoengineering is not a cure. At best, it's a Band-Aid or tourniquet; at worst, it could be a self-inflicted wound" (Conway 2014). Adaptation efforts are required since climate change is already underway. However, although adaptation can help soften the impact, the impact without mitigation will remain severe.

Mitigation involves changes in several aspects of energy technology. In discussing energy technologies, we must distinguish between *sources of energy* and *methods of use*. For example, electric cars that use batteries instead of combustion engines do not include a source of energy: the energy for the batteries must be generated by some other source, such as solar panels or coal-burning electrical generators. The batteries are a method of use, not a source of energy. Next, we must distinguish between *renewable* and *nonrenewable* sources. Nuclear fission is nonrenewable (at present), since the amount of radioactive material is limited, while solar energy is renewable, since the sun keeps shining (at least for the next five billion years). Nuclear fusion is generally classified as renewable: while technically it is not, since the hydrogen fused into helium is not replaced, the amount of hydrogen available in water vastly exceeds the amount needed for fusion. ("In practice nondepletable" is perhaps a more accurate term.) In addition, sources and methods of use can be divided into *carbon-intensive* and *carbon-neutral* (or low-carbon) technologies: burning biomass releases a large amount of CO_2 into the atmosphere, while burning hydrogen does not (it releases water). Energy technologies can be *dirty* or *clean* in other ways besides releasing CO_2: nuclear fission produces radioactive waste, for example. Finally, energy technologies can be relatively *environmentally troublesome* or *environmentally gentle* in other ways. For example, hydroelectric dams are renewable, low-carbon, and produce little pollution but can create environmental harms such as blocking salmon runs.

With regard to *sources of energy*, figures vary somewhat on the percentage of each source's current contribution to the world's energy mix. Rapier (2020) reports that in 2019, roughly 84% of the world's energy came from oil, coal, and gas. IEA (2020) says that in the first quarter of 2020, 28% of energy worldwide was produced by renewable sources. The World Nuclear Organization (2021) estimated that nuclear power contributed 10% of the world's electricity (and 29% of low-carbon power). Ritchie and Roser (2020) state that 4% of the world's total energy comes from nuclear. REN21 reports "the share of fossil fuels in the global energy mix was 80.2% in 2019, compared to 80.3% in 2009, while renewables such as wind and solar made up 11.2% of the energy mix" (Chestney 2021). What seems clear is that the preponderance of energy, worldwide, still comes from fossil fuels but that renewables and nuclear contribute a significant and growing percentage of the total. Nuclear is holding steady or declining, while wind and solar are growing.

Carbon-intensive energy sources include fossil fuels (oil, natural gas, and coal) and biomass. Fossil fuels are carbon-intensive, nonrenewable, and dirty. Burning fossil fuels produces CO_2. It generally takes on the order of 60 million years for buried dead plants and animals to form hydrocarbon liquids. Oil, gas, and coal also produce large amounts of other pollutants, some of which may be lessened by on-site amelioration such as smokestack scrubbers but they are all classified as dirty. The extraction and processing of these nonrenewables also has the potential for environmentally damaging accidents such as massive oil spills and environmental destruction, such as strip mining. There is wide agreement among scientists that these nonrenewable, carbon-intensive, dirty, and environmentally harsh technologies need to be phased out. Biomass technologies are renewable but because they are carbon-intensive and dirty, are not a viable, long-term replacement.

Low-carbon technologies include wind, solar, geothermal, hydraulic, and several potential technologies it is still too early to evaluate, such as nuclear fusion, and harnessing lightning or waves/tides, though the latter shows early promise.

Nuclear fission, unlike the other low-carbon sources discussed, is not renewable in the sense that the earth's available supply of radioactive material is limited. (However, advances in reprocessing technology may mitigate this concern.) It is also classified as 'dirty because of the potentially widespread and dangerous pollution from nuclear accidents and, even if reactors could be made accident-proof, because of the dangerous radioactive wastes it produces. About 10% of radioactive waste has a half-life of 29 years or more. At the lower end of this range are Sr90 and Ce90, which are of particular concern because the body mistakes Ce90 for potassium and Sr90 for calcium (so it is incorporated in bones and teeth). Plutonium and other waste products remain radioactive for thousands or even millions of years. It takes roughly 200,000 years for the waste to match the radioactivity level of natural uranium. No current technology comes close to safely maintaining wastes on that time scale, although some progress is being made in recycling the radioactive waste. Other concerns include CO_2 emissions from the construction and maintenance of nuclear plants and warming of local water by the cooling system.

Wind energy is renewable but raises several concerns apart from the fact that wind speeds are variable (and so energy storage systems are needed): land use, damage to birds and bats, pollution created by producing and transporting materials for wind turbines as well as their assembly, operation, and dismantling. Wind technology also requires maintenance due to moving parts.

Solar energy is renewable and requires little maintenance since there are no moving parts. Sunlight is not always available, so energy storage is needed. However, if photovoltaic cells sensitive to the full spectrum of light are developed and used, efficiency would greatly increase and electricity could be generated from local heat even during night. Concerns include land use, use of water and scare materials in manufacture, effects of mining materials for manufacture, and hazardous materials that may be used in cleaning and purifying surfaces, such as hydrochloric acid, sulfuric acid, nitric acid, hydrogen fluoride, 1,1,1-trichloroethane, and acetone. In addition, thin-film pv cells may contain gallium arsenide, copper indium gallium diselenide, and cadmium-telluride.

Geothermal energy is renewable and constant. Concerns include gases released into the atmosphere during digging and surface instability caused by the removal of water and steam.

Hydraulic energy is renewable, although there are a limited number of feasibly damnable sites. Concerns include harming aquatic wildlife (e.g., blocking salmon runs), increasing incidence of schistosomiasis, destroying wildlife habitat and displacing people (as a result of flooding to create a hydraulic reservoir), changing water temperature, accumulating pollutants trapped behind the dam, and changing riverbed patterns below the damn. Damns that prevent silt from being deposited downriver increase coastal erosion. Reservoir flooding in tropical areas can also produce methane and CO_2 from decomposition of peat (Union of Concerned Scientists 2013). Some of these effects may be partly mitigated, e.g., building fish ladders, instituting short periods of high velocity water release to mimic natural storms, removing trapped pollutants, and even constructing Standard Modular Hydropower plants (which can pass water, fish, and small craft) to preserve river function. Nonetheless, for these reasons hydroelectric is often classified as dirty.

In addition to shifting from fossil fuels to other sources of energy, alternate *methods of using* the energy created can prove helpful. Hydrogen fuel cells and electric batteries (e.g., in electric cars) are two prominent alternatives. At present, while the manufacture of electric vehicles produces higher emissions than the manufacture of standard vehicles, electric vehicles nonetheless produce less over their lifetimes than do fossil fuel powered vehicles. (Moreover, Chinese facilities producing batteries could cut their emissions by more than half if they adopted American and European techniques.) In addition, "more than half of the world's lithium resources lies beneath the salt flats in the Andean regions of Argentina, Bolivia and Chile, where indigenous quinoa farmers and llama herders must now compete with miners for water in one of the world's driest regions" (UN Conference on Trade and Development 2020). The excavation dust from cobalt extracted from artisanal mines under dangerous conditions, often by children, may contain uranium and other toxic materials. Acid mine drainage containing sulfuric acid can affect rivers and streams for centuries (UN Conference on Trade and Development 2020). Many of these concerns can be alleviated by the use of better mining and disposal practices. In addition, new battery technologies currently being researched may use less problematic materials.

Reducing demand is also an essential part of the strategy. Car-pooling makes individual transportation more energy and carbon efficient, while public transportation, especially in reasonably high population density areas, is generally more efficient than individual transportation. The three Rs (Reduce, Reuse, Recycle) are certainly relevant here. Many have argued that radical changes in our practices and mindset are needed.

Hydrogen fuel cells burn clean, producing water rather than CO_2, ash, sulfuric acid, or other pollutants. However, the hydrogen is typically extracted by hydrolysis, which uses electricity generated by other means, and from natural gas by separation. A possible clean and renewable source is freshwater algae, which produce hydrogen when starved. A viable algae-to fuel-cell technology would move hydrogen from method of use to source of energy.

Carbon-capture technologies are important even if we achieve low-carbon emissions. Even if we switched today entirely to low-carbon or zero-carbon sources and methods of use, global temperatures would remain elevated for centuries. In addition to replacing fossil fuel with low-carbon energy technologies, other methods of carbon remediation are called for. Carbon capture typically involves removing carbon from the atmosphere, either at the point of emission or by atmospheric scrubbing. Exxon's Hawiyah plant directly captures and processes up to 45 million cubic feet of CO_2 daily. Artificial trees developed by Klaus Lackner are another potential avenue for atmospheric scrubbing. A natural method is increasing photosynthetic biomass. Planting trees, allowing land to revert to forest or grassland, and promoting and protecting aquatic plant life all increase the amount of CO_2 converted by plants to O_2. A less "natural" plan calls for the bioengineering of plants that either grow in harsh conditions or more efficiently capture carbon.

Once captured, carbon must be *stored*. The two basic strategies are sequestration and repurposing. Proposed sequestration efforts include storing carbon, perhaps as a supercritical fluid, in granite, deep caves, or salt beds, via structural, residual, solubility, or mineral trapping. Other proposals include storing carbon under the ocean and launching carbon into space. Repurposing possibilities include using captured carbon in other products, such as high-quality plastics, olefins, cement, and bio-mass growth. Solidia is a cement whose carbon footprint is claimed to be as much as 70% lower than typical Portland Cement. (For updates and more information on carbon capture and storage, see the National Energy Technology Laboratory website: https://netl.doe.gov/.)

POLLUTION

There are at least seven (overlapping) types of pollution: air pollution (including greenhouse gases discussed above); water pollution; soil pollution; thermal pollution; radioactive pollution; light pollution; and noise pollution.

According to the WHO (2016, 49), "Ambient air pollution kills about 3 million people annually and is affecting all regions of the world, although Western Pacific and South East Asia are the most affected. About 90% of people breathe air that does not comply with the WHO Air Quality Guidelines." Acid rain can occur naturally but burning coal or gasoline may produce SO_2, which becomes SO_3 in the air, and then joins with water droplets to form SO_4, or sulfuric acid.

Water pollution is caused by rapid urban development, improper treatment of sewage, run-off from fertilizers and pesticides, oil spills, dumping of chemical wastes, and release of radioactive wastes.

Chemical waste results from dumping and accidents, such as the Bhopal methyl isocyanate gas leak from a pesticide plant in 1984 (see Case 74 in Appendix III), and benzene leaks from the 2005 explosions in the Jilin City, China petrochemical plant.

The extraction and refining of fossil fuels is one major source of pollution. Oil spills from wells (such as Deepwater Horizon in 2010 – see Case 82 in Appendix III), transportation (such as the Exxon Valdez in 1989), and processing and storage (the Kolva River spill in 1994) have wrecked environmental havoc on local (and not so local) ecosystems. The Ixtoc spill in 1979 resulted in over three million barrels of oil spilling into the Gulf of Mexico, affecting 162 miles of beaches. The Gulf War spill in 1991, allegedly caused by Iraqi armed forces, caused 10–20 million tons of oil to pool in lakes and rivers. In addition, 700 km of Saudi Arabian coastline was contaminated. In 1992, gasoline from a corroded pipe leaked into sewer pipes in Guadalajara, causing nine explosions downtown in the Analco district. Estimates of those killed ranged from 162 to 215, with 800–1,500 injured.

In addition to oil spills, fracking and strip mining affect the environment. Fracking (hydraulic fracturing) is a process of extracting oil or natural gas by drilling vertically or horizontally into shale. The well is encased in steel and/or cement. A high pressure (sometimes more than 9,000 pounds per square inch) blast of slickwater, a mix of water, sand, and chemicals, is unleashed, fracturing the rock and releasing the oil or gas, which can then flow to the wellhead, along with flowback liquid.

Fracking significantly increases available oil and gas reserves at a reasonable price. Proponents say it provides jobs and provides energy security. However, critics claim that fracking uses large amounts of water. Transporting the water to the site can have negative environmental consequences. In addition, critics charge, the chemicals released can seep into the water table. Flowback liquid can contain radon, heavy metals, hydrocarbons, and other dangerous materials. Air and water pollution can result from accidents. For example, a malfunction in a Pennsylvania fracking well in 2011 caused toxins to gush forth for over 12 hours. Treating the vast amount of flowback liquid is expensive and can be more than local facilities can handle. Some researchers claim that fracking may cause tremors (Watson 2021; Bressan 2021), although the USGS (2019) says, "reports of hydraulic fracturing causing felt earthquakes are extremely rare. However, wastewater produced by wells that were hydraulic fractured can cause 'induced' earthquakes when it is injected into deep wastewater wells." In any case, the oil and gas produced are burned, releasing carbon dioxide into the environment.

Engineers who are involved in projects any aspect of which involves these forms of pollution need to pay close attention to the potential harms of those projects and to the ways in which those harms can be eliminated or minimized.

SUSTAINABILITY

The need for sustainable practices is widely recognized, in engineering and elsewhere. "From the depletion of natural resources, to political and economic systems that favor short-term benefit for the few over long-term health for the many, to the accelerating crisis of climate change, scientific consensus is that we will need to change our behaviors if we wish to thrive together on a crowded Earth with finite resources" (Council on Sustainability and Social Responsibility 2016). What is sustainability? Most contemporary discussions of sustainability combine two areas of ethical concern: environmental impact and social justice. Sustainable practices are those that can be carried out over time without significantly (a) harming or depleting natural resources or (b) causing long-term social harm. The World Commission on Environment and Development (1987, 54) stresses equity within and between generations: "sustainable development is development that meets the needs of the present without compromising the ability of future generations to meet their own needs. It contains within it two key concepts: the concept of 'needs', in particular the essential needs of the world's poor, to which overriding priority should be given; and the idea of limitations imposed by the state of technology and social organization on the environment's ability to meet present and future needs." Cortese and Rowe (2004) note that "sustainability is a vision for the world in which current and future humans are reasonably healthy; communities and nations are secure, peaceful and thriving; there is economic opportunity for all; and the integrity of the life-supporting biosphere is restored and sustained at a level necessary to make these goals possible. All…dimensions of sustainability must be addressed to achieve this vision."

A series of UN summits on sustainable development, including United Nations Conference on the Human Environment, Stockholm 1972; Earth Summit, Rio de Janeiro 1992 (which produced Agenda 21 and the Rio Declaration); World Summit on Sustainable Development, Johannesburg, 2002; and the United Nations Summit on Sustainable Development, which promulgated Agenda 2030's 17 sustainable development goals (UN 2015), emphasized both development (especially the eradication of poverty) and sustainability. Agenda 2030's 17 goals are often summarized as (1) no poverty, (2) zero hunger, (3) good health and well-being, (4) quality education, (5) gender equality, (6) clean water and sanitation, (7) affordable and clean energy, (8) decent work and economic growth, (9) industry, innovation, and infrastructure, (10) reduced inequality, (11) sustainable cities and communities, (12) responsible consumption and production, (13) climate action, (14) life below water ["conserve and sustainably use the oceans, seas and marine resources for sustainable development"], (15) life on land ["protect, restore and promote sustainable use of terrestrial ecosystems, sustainably manage forests, combat desertification, and halt and reverse land degradation and halt biodiversity loss"], (16) peace, justice, and strong institutions ["promote peaceful and inclusive societies for sustainable development, provide access to justice for all and build effective, accountable and inclusive institutions at all levels"], and (17) partnerships for the goals (quoted material in brackets from UN 2015).

Two central values of the engineering profession, advancing human welfare and partnership with nature, require engineers to work toward sustainable development. Both environmentalism as prudence and environmentalism as an ethic demand sustainable technologies and practices since runaway unsustainable practices threaten both humanity's future and the environment itself.

CONSERVATION AND PRESERVATION

Conservation describes the sustainable use of natural resources, while preservation is about protecting natural resources from being used or destroyed. Limiting water use and restricting logging are conservation measures while outlawing logging and development of a virgin forest is preservation. Both are important components of protecting the environment. As the idea of partnership with nature suggests, some aspects of the natural world should be left alone while in other ways respectful use of and cooperation with nature are called for (see Chapter 4). The most direct and visible preservation efforts are government or NGO established natural parks, wildlife refuges, and nature reserves as well as laws and regulations protecting species and habitats like China's Wild Animal Protection Law which, for example, imposes a minimal ten-year jail sentence for panda hunting or smuggling. In addition, several organizations promote both conservation and preservation, such as The Sierra Club, The Nature Conservancy, The World Wildlife Fund, Conservation International, International Anti-Poaching Foundation, National Geographic Society, Oceana, and the National Resources Defense Council.

SOLUTIONS

Balancing Environmental Risk and Benefit

While engineers should go the extra mile in planning for and avoiding hazards, accidents, and environmental harm, a certain amount of risk and environmental loss is inevitable. No activity is without risk, and so engineers should not be expected to eliminate every risk completely. Similarly, change is the only constant in life, and every change for the better leaves the world different from the way it was before. Some degree of environmental impact is inevitable. Every benefit has its price: in taking a step forward, one always loses something. Engineers and environmentalists must realize that nothing in life is all good, and there is a downside to even the most benign and beneficial decisions. This means that engineers should not be afraid to acknowledge any adverse effects of their projects, and environmentalists must accept some undesirable consequences as the price of a worthwhile and beneficial project.

Thus, for any project, the risk and harm to the environment must be weighed against the benefit of the product or process and the cost of correcting the problem. This can only be done rationally in the light of honest and unbiased information and flexible and fair judgment.

Unfortunately, companies sometimes overlook indirect effects of their projects, and environmentalists often overlook indirect costs of correcting a problem. Thus, it is useful to have some guidance in assessing environmental harm, and in assessing costs of correcting a problem.

In assessing the cost of correcting a problem, one must consider both "resource costs," that is, the money, labor, etc., used to rectify the problem, and "factor income costs," the changes in capital income that result from rectifying the problem. The labor and cost of producing and installing smokestack scrubbers is an example of a resource cost. By contrast, if many companies switch from sulfurous coal to a less polluting fuel, the lower price of sulfurous coal is a factor income cost to coal mining companies. For example, suppose company XYZ withdraws from the market a profitable, long-lived pesticide, in favor of a short-lived pesticide with fewer environmental consequences. XYZ suffers resource costs in developing, testing, marketing and manufacturing the new pesticide, and factor income costs to the extent that the profit earned on the new pesticide is less than that earned on the old pesticide. In addition, the new pesticide increases costs to farmers who use the pesticide, including resource costs such as the cost of buying the more expensive pesticide, the cost of more frequent applications, and the cost of any new equipment necessary to apply the new pesticide, as well as the factor income cost of decreased yield if the new pesticide is less effective in controlling pests.

In assessing the environmental effects of a project, one must consider the consequences, direct and indirect, of developing, manufacturing, using and disposing of the product. A company that produces chemical fertilizer may not ignore the effects on the environment that use of the fertilizer produces (such as algae growth from nitrogen run-off), the effects of mining, drilling for or manufacturing materials used in manufacturing the fertilizer, and the effect of depleting scarce resources used in producing the fertilizer. A firm that builds a dam designed to generate electricity may not ignore the environmental consequences of the dam's construction, such as cutting pathways to the dam for construction materials, the consequences to the river's ecosystem, such as the difficulty salmon may have in returning upstream, and the effects of the use of the electricity produced. Companies must also consider the costs of dealing with risks or environmental harms. These costs include the costs of safety drills, the costs of planning for and putting in place procedures for accidents, the costs of recalls that may be necessary, and the costs of lawsuits.

There are at least three different kinds of solutions to an environmental problem: *technical modifications of a process* (e.g., adding scrubbers to smoke stacks of coal-burning plants); *alternate technologies* (e.g., using geothermal power instead of burning oil or coal); and *non-technological solutions*.

Questions to be asked about alternate technologies and technical modifications include the following: (1) Will it solve the problem with reasonable certainty? (2) Will the solution cause new problems: technically, economically, socially? (3) Could we, instead, substitute another technology that does not cause this problem? (4) Could we do without the technology that causes the problem (or could we do with less of it)?

Questions to be asked about non-technological solutions include the following: (1) Does this represent a real solution. That is, will it work? Is it politically acceptable? (2) Is the idea economically justifiable? Is the cost reasonable compared to benefits gained? (3) Is the idea socially justifiable? Who gains and who loses?

For example, engineers should strive, when feasible, to avoid thermosets, which are hard to recycle. Is the strength and flexibility of thermosets really needed for this application? Are there recyclable alternatives available that will work adequately?

Given the complexity of these factors, it follows that weighing the benefits of a project against its risk or environmental impact is not easy. However, since engineering is devoted to innovation in advancing human welfare in partnership with nature, engineers have a duty to use their abilities to discover ways to minimize the risks and environmental harms of their projects.

Of course, any company's or firm's expertise and budget are limited. It is not mandatory to solve every problem one's product may cause. So, there are limits on what a company should be expected to do by way of protecting the environment. But an ethical firm is committed, to the extent feasible, to take reasonable steps to develop, promote, and encourage solutions to these problems. This means that the ethical engineer's job is not finished when the product is perfected. She must ask herself about the environmental effects of the product or process. The two principles of institutional accountability suggest that she should *take responsibility* for doing what is feasible to devise methods of minimizing harmful effects herself, or to encourage the company to either devise such methods or encourage others to do so. It is not her responsibility alone but she must be a voice for good. This may take several forms, including (a) speaking to supervisors about the need to address the problem, (b) rewarding subordinates who work on or raise the problem, (c) speaking to legislators about the need to address the problem, and (d) publication of the problem in an appropriate forum (whether a trade publication or even a letter to the editor). Obviously, an employee contemplating (c) or (d) should first consult her supervisors, and should approach legislators or the public in a way that will not hurt the company. (It is generally possible to present the company as concerned and seeking solutions to this problem, rather than as an environmental villain.)

Recycle, Reuse, and Reduce

The three Rs (reduce, reuse, recycle) are fundamentals of environmentally conscious engineering. Reducing the amount of waste and reusing discarded materials are the most environmentally friendly options, when available but even after efforts to reduce and reuse, large amounts of waste remain. As much of this waste as feasible should be recycled. While figures on recycling vary, it is clear that much more can be done. According to the EPA (2021b), the United States in 2018 generated 292.4 million tons of municipal solid waste, including 63.1 million tons of food waste, 35.4 million tons of yard waste, 35.7 million tons of plastics, and 2.7 million tons of consumer electronics. 60 million tons of this waste were recycled and 29 million tons were composted (a total of 32.1%). "An additional 17.7 million tons of food were managed by other methods. Other food management includes the following management pathways: animal feed, bio-based materials/biochemical processing, co-digestion/anaerobic digestion, donation, land application and sewer/wastewater treatment" (EPA 2021b). 34.6 million tons of waste were burned to produce energy. However, this method contributes to global warming. Nonetheless, "in 2018, the recycling, composting, combustion with energy recovery and landfilling of MSW saved over 193 million metric tons of carbon dioxide equivalent" (EPA 2021b). Worldwide, over 2 billion tons of waste are produced annually. 300 million tons of this waste are plastic (roughly 150 million tons of which are single use) (Miniwiz 2021). Total global production of plastics in 2017 exceeds 400 million metric tons (Geyer 2020). The World Bank warns that, if current trends continue, worldwide waste production will reach 3.4 billion tons by 2050 (World Bank 2018). Of the 80 million tons of waste Brazil produces, only 3% are recycled. At the other end of the spectrum, Germany boasts the world's highest recycling percentage, 56%.

There are five basic types of recycling: mechanical, energy, chemical, soil incorporation, and salvage. Most common is *mechanical recycling*, in which materials are reshaped without changing their chemical properties. For example, used paper products are gathered, sorted, shredded, pulped by a puller with hydrogen peroxide, pushed through screens to remove contaminants, de-inked with chemical additives and flotation using air bubbles, bleached, run through press rollers and a heated metal roller, dried, and passed through steam heated cylinders. The resulting sheet is then cut to the desired shape and size. The Timsfors Pulp Mill in Sweden flocculated fibrous wastewater with alums and reused the resulting sludge in making low-grade paper (Eckenfelder 1975). Mechanically recycled plastics are fashioned into garbage bags, hoses, flooring, and much else. Typical plastic recycling may involve milling, washing, drying, followed by agglutination, extrusion, and cooling to form pellets, which are then worked into raw material for new products. 25 billion polystyrene containers are made each year in the US. The National Polystyrene Recycling Company (formed by Dow, Amoco, Mobil, Atlantic Richfield, and others) grinds polystyrene into small pieces, washes it with hot water, fires it, and melts it at high temperatures, after which it is filtered, cooled, and chopped into pellets for reuse (Steger and Bowermaster 1990). Although food containers cannot generally be purified enough for reuse in food, PET (polyethylene terephthalate), used in soda bottles, can be made into carpet fibers and backing, cushion stuffing, scouring pads, fiberfill, refrigerator insulation and paintbrushes. HDPE (high-density polyethylene), used in milk jugs, can be turned into construction materials (such as posts and boards), trashcans, and bins (Hynes 1990). In addition, well over a billion tires worldwide are discarded every year (279 million in the United States alone). Tirecycle has combined the old tires with new rubber and various plastics to produce gaskets, storage bins, and railroad crossing pads. *Energy recycling* converts wastes into energy, typically by incineration using catalyzers to reduce emissions. *Chemical recycling* changes the chemical structure of wastes. For example, plastics may be transformed by hydrogenation, gasification, or pyrolysis (reducing solid waste at high temperature to pyrolytic oils, which can be used as fuel). *Soil incorporation* or land farming has been used with sludge from paper mills and fruit canneries, sewage sludge, pharmaceutical wastes, and petroleum refinery wastes (Maugh 1979).[1] Finally, in *salvage or recovery recycling*, materials are extracted for reuse, e.g., gold from printed circuit boards or mercury from thermometers. The slurry in wet scrubbers can be used to produce gypsum, or the calcium or magnesium compound in the slurry may be recycled, producing sulfur or sulfuric acid for sale (Fowler 1984). The goals of wastewater treatment are to purify water, making it safe while safely disposing of hazardous materials in the water, and, secondarily, retrieving resources (such as energy, nutrients, and useable materials) from the water. Water treatment occurs in three stages: primary, including filtration, coagulation, sedimentation and gravity separation, and screens for larger particles like paper; secondary, including conversion by aerobic and anaerobic microbes into gases and removable solids; and tertiary, including crystallization, evaporation, solvent extraction, oxidation (for example, cyanides can be oxidized with sodium hydroxide and chlorine or sodium hypochloride), precipitation (for example, lime applied to liquids from electroplating and steel finishing balances the PH and precipitates out heavy metal ions), ion exchange separation, reverse osmosis (to concentrate solids and recover water), filtration, hydrolysis, and absorption (Ince and Ince 2019). These techniques can, of course be combined. For example, nitric acid wastes, used in etching silicon wafers, can be neutralized with lime. The resulting calcium nitrate can then be reused in fertilizer. Moreover, one waste product can sometimes be used to neutralize another. For example, cyanide wastes can be used to reduce chromium VI wastes (Maugh 1979).

Engineers should design for recycling. This involves more than calling for recyclable materials. For example, in many soda and water bottles, the cap is secured via a plastic ring that remains around the neck of the bottle when the cap is removed. This ring must be removed before the bottle can be recycled, which is one of many reasons that vast amounts of US recycling were shipped to China for processing, instead of being processed locally, until China imposed restrictions on waste imports. A design in which the ring broke off with the cap instead of remaining on the bottle would make bottle recycling much easier.

In designing for recycling, "the first stage is to review all the processes intervening in the design of a product and to find solutions to reduce the impacts on the product's life cycle....In the materials choice phase, one can choose less impacting materials; reduce the quantities of materials; improve process techniques, transport, and the usage phase; and optimize the life cycle and end-of-life of products" (Maris 2014). East West Manufacturing offers the following tips for facilitating disassembly: "Use fewer parts; Use common parts; Reduce the type of number of fasteners used in an assembly; Use common fasteners that don't require specialty tools for removal; Avoid using glue or other adhesives if possible; If you must use glue, consider a soluble adhesive for easier disassembly; Include disassembly instructions with product; Post a disassembly tutorial video on YouTube and share via social media accounts." They also suggest "Use clearly labeled and commonly recycled materials; Avoid paints and other coatings on plastics to avoid contamination; Spell out any special recycling instructions; If product is difficult to recycle (like polystyrene), recommend a recycling center that accepts this product; Include detailed recycling instructions and information on your website; Remind customers to recycle via your social media accounts; Provide lists of local recycling centers in each state or region; Teach customers how to properly pre-sort materials before arriving at recycling centers" (Hassiotis 2015). The RecyClass Design for Recycling Guidelines may be found at https://recyclass.eu/recyclability/design-for-recycling-guidelines/

For examples of designing for recycling, engineers might look at The Institute of Scrap Recycling Industries (ISRI) Design for Recycling Award. Recipients have included The Dell Latitude 5590 Laptop, which "contains a removable battery, is free of harmful substance such as mercury, eliminated the use of glues and adhesives, contains a modular design making easier to access and disassemble, and uses standardized fasteners" and The Latitude 5285 2-in-1, which "uses gold recycled from used electronic products"; The Samsung 2016 Curved Full HD TV, which "incorporates easy-to-disassemble, snap-together parts that are made with minimal chemical content. The snap closures eliminate the use of many screws, making it easier for recyclers to disassemble"; and the Nestle Pure Life 700ml bottle, "made entirely from recycled content and [featuring] a state-of-the-art, pressure-sensitive label that releases easily during the wash stage of the recycling process" (ISRI 2021).

It is worth noting that in addition to benefitting the environment, recycling can save money, produce jobs, and can even improve quality. Waste solvents for electronics are sometimes of higher quality than the new materials used in other processes that tolerate higher levels of impurity. Thus, companies should make use of Waste Exchanges, which list available materials without identifying the source, and handle the exchanges.

"The most beneficial actions that could improve recycling rates are increased collection rates of discarded products, improved design for recycling, and the enhanced deployment of modern recycling methodology" (Reck and Graedel 2012, 690). Corporate and government outreach programs can be useful. For example, Amoco, Rubbermaid, and McDonald started a demonstration project in Brooklyn that sorted and recycled trash from school cafeterias and McDonald restaurants (Steger and Bowermaster 1990). "Metal price is a key driver directly affecting collection and processing efficiencies" (Reck and Graedel 2012, 692). "EOL-RR [is] defined as the fraction of metal in discarded products that is reused in such a way as to retain its functional properties" (Reck and Graedel, 690). Although EOL-RRs for such commonly used metals as iron, copper, and zinc top 50%, most other elements are rarely recycled (Reck and Graedel 2012).

Finally, waste disposal of non-recyclable material also calls for engineering ingenuity. Among the hazardous wastes are heavy metals, radioactive materials, carcinogens, waste oil, herbicides, fungicides, and insecticides, and microbes. Wastes can be industrial, commercial, domestic, or agricultural. The EPA distinguishes between listed wastes (materials that occur on the EPA's F-list, K-list, P-list, and U-list), characteristic wastes (wastes that are flammable, toxic, corrosive, and reactive, e.g., explosive), and mixed wastes containing both hazardous and radioactive materials (EPA 2021a). To this trio is sometimes added universal wastes (commonly produced wastes in batteries, mercury-containing bulbs, etc.).

Nine General Responses to Environmental Problems

In general, engineers should keep in mind these nine responses to an environmental problem:

- When feasible, seek an alternative product or process that is not harmful (or at least less harmful).
- Develop and publicize an additional product, process, or method of use that minimizes the harmful effects.

- Publicize the problem and encourage solutions.
- Take a harmful product off the market.
- When feasible, build biodegradability into the product.
- Develop and publicize processes for reusing, recycling, or at least safely disposing of, the product.
- Minimize use of the scarce resources by finding an alternative, or using as little as possible of the resource.
- Develop a method of extracting and reusing scarce materials.
- Develop a new method of extracting or a new source of scarce materials.

Case 55: Silver Bay and Reserve Mining

In 1955, Reserve Mining built its Silver Bay plant on Lake Superior, 55 miles from Duluth, to extract iron from taconite ore. In the process, up to 67,000 tons of waste tailings were dumped daily into Lake Superior, creating, over 25 years, a 1/3-mile delta of rubble. The company employed 80% of the 3,500 local workers in the Silver Bay, Minnesota region. Much of the town of Silver Bay was built by Reserve Mining, which sold houses to workers with no down payment. The tailings occluded the water, killing fish and affecting the drinking water of 100,000 people in Duluth and surrounding communities. In 1973, the EPA concluded that Duluth's drinking water was contaminated by amphibole (asbestos-like) fibers. Reserve's tailings released the fibers into the water and into the air. Asbestos was known to be a potent carcinogen when inhaled. In addition, asbestos used to polish rice in Japan was linked to cancer. At the time, there was limited evidence that ingested asbestos was dangerous. Animal studies indicated that ingested asbestos was rapidly spread through the bloodstream. Workers in a Paterson, NJ asbestos plant showed three times the expected death rate from gastrointestinal cancer. During several years of litigation, Reserve insisted that the plant would have to close if it were ordered to cease dumping tailings into the lake. In 1974, Judge Lord ordered the plant to cease dumping within 24 hours. Two days later, three Court of Appeal judges reversed Judge Lord's order, allowing the plant to continue operations while Reserve established alternative methods of disposal. In 1980, the company switched to disposal in an inland basin via rail and pipeline.

· *Discussion:* This case is a good example of a high-stakes decision under conditions of considerable uncertainty. On the one hand, if the plant is shut down while the new facility is built, people will lose their jobs, their homes, their lifelong (even multigenerational) network of friends and support systems. On the other hand, airborne asbestos is a potent carcinogen. It seems insane, one might argue, to pump asbestos-like fibers into the water supply of large numbers of people. The fibers, which presumably do not biodegrade, are likely to disperse throughout a wide region of Lake Superior and perhaps the Great Lakes system. Although concentration levels will be very low over such a wide area, OSHA has set a limit of 0.1 fiber per cubic centimeter. The National Cancer Institute (2021) and the Australian Government Department of Health (2013) say that there is no safe level of asbestos exposure, although the latter adds "occasional exposure to low levels of fibers poses only a low risk to your health." However, while the danger of airborne asbestos is well known, much less is known about waterborne (imbibed) asbestos.[2] At the time, there was no conclusive data about that. Lung tissue is quite different from stomach and intestinal tissue, which are protected in any case by a lining. At the time it was not known whether imbibed asbestos remains in the body and is absorbed by cells and what effect asbestos would have on those cells in any given concentration. Moreover, hard evidence about this is hard to obtain and may take generations. Three kinds of tests might be used. Laboratory studies might show whether, in lab conditions (which are not quite the same as the body) certain kinds of cells absorb waterborne asbestos or undergo very short-term DNA mutation when exposed to asbestos. These tests are suggestive but are not likely to reveal anything definitive about the effect in the human body of drinking asbestos. Population studies (comparing incidence of cancer in areas exposed to waterborne asbestos with cancer rates in areas not exposed to waterborne asbestos) cannot control for significant variables and so are often unreliable. A given town whose water is high in asbestos might have a higher cancer rate but that could be due to factors other than waterborne asbestos. Controlled studies of the right sort are not ethically or logistically feasible and would take 20 years. (We may not set aside 10,000 people for 20 years, control everything that might affect cancer rates, such as the amount of sunlight, and give half of them a possibly dangerous level of asbestos to drink.) So, as far as anyone definitively knew at the time, waterborne asbestos may have posed no danger, especially in the relevant concentration. But would you let your children drink a known, highly virile airborne carcinogen whose effect in water is unknown? However, how would you feel if you lost your house and your job and had to move away from the community your family lived in for generations because of something that might not harm anyone at all?

The difficulty of the decision emerges when considering the many ethical factors. For instance, while the risk is severe (death) and widespread (potentially millions of people), the likelihood of harm is unknown. Workers and their families benefit from the risk and assume the risk voluntarily but Great Lake residents further from the plant do not. Keeping the plant open respects the workers' autonomy but not the autonomy of the residents further along the Lake. It is unclear whether closing or keeping open the plant promotes better consequences. Closing the plant creates certain and immediate harm (the company may not survive, workers lose jobs, the local economy tanks), while keeping it open might (but might not) create even greater harm, long term. Environmentalism as an ethic tells against keeping the plant open, since the turbidity and fibers adversely impact the environment. If waterborne asbestos is harmless (in those concentrations), however, the environmental damage, while significant, is not disastrous. While the Precautionary Principle (Chapter 6) certainly suggests the danger should be taken seriously, it does not give a clear answer here. The Rio Declaration discusses cost-effective measures, and closing the plant is at least arguably not cost-effective. The Wingspread Statement calls for considering the entire panoply of alternatives, including taking no action.

Case 56: Environmentally Harmful Products (Toxinal)

The company's product, Toxinal, is toxic to natural ecosystems, or interferes with natural ecosystems in other ways (for example, nitrogen run-off from the use of Toxinal stimulates the growth of algae in rivers).

Discussion: Four responses to this problem are available.

Response 1: When feasible, seek an alternative product or process that is not harmful, or is less harmful than Toxinal. If this is not feasible, then:

Response 2: Develop and publicize an additional product, process, or method of use that minimizes the harmful effects of Toxinal.

For example, when chemical fertilizers are applied to fields where seed has been poked into last year's stubble, rather than to fields that are burned and clear-cut, run-off is minimized.

Response 3: Publicize the problem and encourage solutions. Encouragement of solutions may take several forms, including providing research funding for a university to address the problem, lobbying state or federal agencies to fund research, or simply offering to assist and provide information to anyone working on a solution.

Response 4: If the harm caused by Toxinal, even with the measures mentioned above, is serious enough, the company should consider taking Toxinal off the market. Three questions are helpful in deciding whether to remove the product. First, how bad is the problem? Here the safety factors discussed above come into play. A rough guide to the extent of the problem can be obtained by subtracting the good done by the product, its manufacture and its use, directly and indirectly, from the harm done by the product, its manufacture and use, directly and indirectly. However, as we saw earlier, this kind of purely quantitative analysis is inadequate. For example, severe harm to a few individuals counts more than minor inconvenience to a large number of people. And it makes a difference if the risk or harm is limited to people who use the product voluntarily, knowing the risk. Second, how much would the company suffer by removing the product from the market? Here, too, many factors come into play. How much of the company's revenue depends upon this product? How much overhead is tied up in the product? (This includes advertising, production machinery, inventory, etc.) How much of that overhead could be re-utilized for another purpose? (How much would it cost to convert the Toxinal plant?) How difficult would it be for the company to replace this source of revenue? Does the company have sufficient capital resources to weather the change-over period? How many jobs would be lost? Finally, would removal do any good? That is, what is the likelihood that were Toxinal removed, it would simply be replaced by another, equally harmful product? This factor should not be used as an all-purpose excuse for unethical conduct. A hired killer cannot excuse his killings by saying that if he didn't take the job, someone else would. If the harm done by Toxinal is bad enough, no ethical company would market it, period. However, the likelihood that someone else would market Toxinal (or an equally harmful product) is a mitigating factor. If the decision to remove Toxinal is a close one, this factor may make the difference.

Case 57: Environmentally Harmful Manufacturing

The process of making the product produces harmful effects. For example, cadmium is used in the manufacture of paints, alloys, light bulbs, pesticides and nuclear reactor parts. Cadmium in the environment is accumulated by shellfish, rice, wheat and fish liver. Once introduced into the body, cadmium accumulates in the kidneys, and is associated with kidney disease and high blood pressure. Cadmium stays in the body a long time: it has a half-life of 10–30 years. Again, nitrates, which are added to meat for curing, and are often found in fertilizer run-off, change Hemoglobin from $Fe2+$ to $Fe3+$, which cannot oxygenate. Thus, nitrates cause slow suffocation. Infants are particularly susceptible. Nitrosamines have been found to be carcinogenic. Mercury, used as a preservative in Latex paint, and used in paper mills, fungicides, electrical devices and drugs, is a permanent pollutant: it is recycled within the environment. Bacteria can change inorganic Mercury, which is easily excreted, to more toxic Alkyl Mercury compounds, such as Methyl Mercury. Mercury is a nerve poison: the expression "mad as a hatter" originally described the effects of exposure to Mercury on workers in the felt-hatting industry (Revelle & Revelle 1981).

Discussion: Here again, several responses are available. We can modify the process to minimize the effects, institute ameliorating mechanisms (e.g., scrubbing pollutants, pre-cooling water or pipes that may raise the temperature of rivers), or publicize the problem and encourage solutions.

Case 58: Environmental Accidents

Accidents during the transportation or synthesis of a product may cause harm.

Discussion: Responses include designing transportation/synthesis processes to minimize risk, such as including safety mechanisms and choosing the safest routes or sites; preparing, in advance, clean-up procedures, and making sure clean-up can be rapidly and effectively instituted; instituting periodic safety drills and making sure these drills are taken seriously by employees; inspecting all of above frequently; and publicizing the problem and encouraging solutions.

Case 59: Waste Materials

Some safely manufactured products may do no harm while in use but nonetheless pose environmental problems after their usefulness has expired. Obviously, toxic materials such as nuclear waste pose a problem. But so do long-lived plastics. The long-term effects of turning the earth's resources into garbage are serious and cumulative. The ethical engineer must take responsibility for ensuring that neither the depletion of limited resources nor the production of garbage gets out of hand.

Discussion: Engineers should pay close attention to the three Rs (reduce, reuse, and recycle) discussed above. When feasible, build biodegradability into the product. In any case, develop and publicize processes for reusing, recycling, or at least safely disposing of, the product.

Case 60: Using Non-Renewable Resources

The product or process of manufacture uses scarce, effectively nonrenewable resources.

Discussion: Minimize use of the scarce resource by finding an alternative, or using as little as possible of the resource. Develop a method of extracting and reusing the scarce material. Develop a new method of extracting or a new source of the material. Publicize the problem and encourage solutions.

CONCLUSION

Engineers have a responsibility to address sustainability, pollution, and climate change on four levels:

- as an individual engineer,
- as a member of an organization,
- as a member of the engineering profession, and
- as a citizen/member of society.

Thus, engineers should:

- Stay *informed* and current about issues and technologies pertaining to the environment.

- Be *active* at all four levels, developing contacts and participating in relevant discussions, forums, and projects.

This will enable you to:

- *Consider the environmental consequences* of your projects and *strive to increase the environmental friendliness* of your projects.
- *Be a voice* for the environment within your organization.
- Support the profession's efforts to *draw attention to and promote solutions to* environmental concerns.
- *Be an ambassador* for the environment within your community.

NOTES

1. According to Maugh (1979), there are four steps in soil incorporation: applying wastes to soil, mixing for aeration, addition of nutrients, and periodic re-mixing. The farmed area must be 1.5 meters above the groundwater table and should be 150 meters from any potable water sources. The soil should be periodically monitored down three feet to check for migration of contaminants. The process can be accelerated by aerobic composting but this process requires containment, protection from rain, and bulking agents to keep the material porous. Eckenfelder (1975) mentions two other methods of speeding up soil incorporation: using pure oxygen and multi-staging the process to make use of the higher reaction rates in initial stages.
2. A 1991 study concluded "some epidemiological and laboratory-animal studies have revealed an association of asbestos ingestion with various types of cancer, while other studies have not. Ingested asbestos, usually short fibers, is capable of penetrating the intestinal wall and being eliminated in urine or accumulating in various tissues and organs" (Weber and Covey 1991). The International Chrysotile Association (not a disinterested source) said in 2001, "Epidemiological studies on human health effects related to asbestos levels in drinking water have failed to indicate any increased risk of alimentary tract tumours following the direct ingestion of asbestos fibres" (ICA 2001). More recently, a 2017 paper concluded that, "The presence of asbestos fibres (AFs) in drinking water could be linked with gastrointestinal cancers. However, it is not regulated in several countries due to conflicting evidence" (Di Ciaula 2017).

REFERENCES

Australian Government Department of Health (2013) "Who Is at Risk of Developing Asbestos-Related Diseases," http://www.health.gov.au/internet/publications/publishing.nsf/Content/asbestos-toc˜asbestos-health˜asbestos-risk-diseases

Bressan, David (2021) "New Type of Earthquake 'Triggered by Fracking' Discovered," *Forbes* https://www.forbes.com/sites/davidbressan/2021/12/06/new-type-of-earthquake-triggered-by-fracking-discovered/?sh=708cc05f672f

Chestney, Nina (2021) "Global Fossil Fuel Use Similar to Decade Ago in Energy Mix, Report Says," *Reuters* https://www.reuters.com/business/environment/global-fossil-fuel-use-similar-decade-ago-energy-mix-report-says-2021-06-14/

Conway, Erik (2014) "Just 5 Questions: Hacking the Planet," *NASA* https://climate.nasa.gov/news/1066/just-5-questions-hacking-the-planet/

Cortese, Anthony D. and Rowe, Debra (2004) "Higher Education and Sustainability Overview," *College of Engineering, University of Toledo* [quoted https://www.haverford.edu/sites/default/files/HaverfordSustainabilityPlanAprilDraft.pdf and elsewhere].

Council on Sustainability and Social Responsibility (2016) "Sustainability Strategic Plan," Haverford University https://www.haverford.edu/sites/default/files/HaverfordSustainabilityPlanAprilDraft.pdf

Di Ciaula, A. (2017) "Asbestos Ingestion and Gastrointestinal Cancer: A Possible Underestimated Hazard," *Expert Review of Gastroenterology & Hepatology* 11:5, 419–425 https://www.ncbi.nlm.nih.gov/pubmed/28276807

Eckenfelder, W. Wesley Jr. (1975) "Economic Alternatives for Industrial Waste Treatment," in Richard A. Tybout, ed., *Environmental Quality and Society* (Ohio State).

EPA (2021a) "Defining Hazardous Waste: Listed, Characteristic and Mixed Radiological Wastes," https://www.epa.gov/hw/defining-hazardous-waste-listed-characteristic-and-mixed-radiological-wastes

EPA (2021b) "National Overview: Facts and Figures on Materials, Wastes and Recycling," https://www.epa.gov/facts-and-figures-about-materials-waste-and-recycling/national-overview-facts-and-figures-materials

Fowler, John M. (1984) *Energy and the Environment*, 2nd ed. (McGraw-Hill).

Geyer, Roland (2020) "Production, Use, and Fate of Synthetic Polymers," in Trevor M. Letcher, ed., *Plastic Waste and Recycling* (London: Academic Press), 13–32. https://www.researchgate.net/profile/Mohanraj-Chandran/publication/339905534_Conversion_of_plastic_waste_to_fuel/links/5e982e474585150839e08d12/Conversion-of-plastic-waste-to-fuel.pdf#page=36

Hassiotis, Mary-Kerstin (2015) "How to Design Sustainable Products for Recycling by the End User," *East West Manufacturing* https://news.ewmfg.com/blog/how-to-design-sustainable-products-for-recycling-by-the-end-user

Hynes, H. Patricia (1990) *Earth Right* (Rocklin, CA: Prima Publishing).

ICA (International Chrysotile Association) (2001) "9 Questions on Chrysotile and Health: Question 6," http://www.chrysotile.com/en/chrysotile/hltsfty/quest6.aspx

IEA (International Energy Association) (2020) "Global Energy Review 2020: Report Extract: Renewables," https://www.iea.org/reports/global-energy-review-2020/renewables

Ince, Muharrem and Ince, Olcay Kaplan (2019) "Heavy Metal Removal Techniques Using Response Surface Methodology: Water/Wastewater Treatment," in Muharrem Ince et al., eds. *Biochemical Toxicology: Heavy Metals and Nanomaterials* (Intech Open) https://www.intechopen.com/chapters/68822

ISRI (Institute of Scrap Recycling Industries) (2021) "ISRI Design for Recycling Award," https://www.isri.org/about-isri/awards/design-for-recycling

Maris, Elisabeth et al. (2014) "From Recycling to Eco-Design" in Ernst Worrell and Markus A. Reuter, eds., *Handbook for Recycling* (Waltham, MA: Elsevier), 429–437.

Maugh, Thomas H. II (1979), "Hazardous Wastes Technology Is Available," *Science* 204:4396, 930–933.

Miniwiz (2021) "7 Important Recycling Statistics from around the World," https://miniwiz.medium.com/7-important-recycling-statistics-from-around-the-world-53ac9d3b783c

National Cancer Institute (2021) *NIH* http://www.cancer.gov/about-cancer/causes-prevention/risk/substances/asbestos/asbestos-fact-sheet

Rapier, R (2020) Fossil Fuels Still Supply 84 Percent Of World Energy — And Other Eye Openers From BP's Annual Review, *Forbes* https://www.forbes.com/sites/rrapier/2020/06/20/bp-review-new-highs-in-global-energy-consumption-and-carbon-emissions-in-2019/?sh=39b9e18866a1

Reck, Barbara K. and Graedel, T.E. (2012) "Challenges in Metal Recycling," *Science* 337, 690–695 http://jupiter.chem.uoa.gr/thanost/papers/papers2/Science_337(2012)690.pdf

Revelle, Penelope and Revelle, Charles (1981) *The Environment* (Willard Grant).

Ritchie, Hannah and Roser, Max (2020) "Nuclear Energy," *Our World in Data* https://ourworldindata.org/nuclear-energy

Steger, Will and Bowermaster, John (1990) *Saving the Earth* (New York: Knopf).

UN (2015) "Transforming Our World: 2030 Agenda for Sustainable Development," https://sustainabledevelopment.un.org/content/documents/21252030%20Agenda%20for%20Sustainable%20Development%20web.pdf

UN Conference on Trade and Development (2020) "Developing Countries Pay Environmental Cost of Electric Car Batteries," *United Nations* https://unctad.org/news/developing-countries-pay-environmental-cost-electric-car-batteries

UN News (2021) "IPCC Report: 'Code Red' for Human Driven Global Heating, Warns NC Chief," *United Nations* https://news.un.org/en/story/2021/08/1097362

Union of Concerned Scientists (2013) "Environmental Impacts of Hydroelectric Power," https://www.ucsusa.org/resources/environmental-impacts-hydroelectric-power#:~:text=Environmental%20Impacts%20of%20Hydroelectric%20Power%201%20Land%20use.,impacts.%20…%203%20Life-cycle%20global%20warming%20emissions.%20

USGS (2019) "Hydraulic Fracturing," *United States Geological Survey* https://www.usgs.gov/mission-areas/water-resources/science/hydraulic-fracturing#:~:text=How%20is%20hydraulic%20fracturing%20related,injected%20into%20deep%20wastewater%20wells.

Watson, Claire (2021) "Researchers Find Evidence that Fracking Can Trigger an All-New Type of Earthquake," *ScienceAlert* https://www.sciencealert.com/scientists-find-evidence-that-oil-gas-extraction-triggers-new-type-of-slow-rupture-earthquake

Weber, James S. and Covey, James R. (1991) "Asbestos in Water," *Critical Reviews in Environmental Control* 21, 3–4 http://www.tandfonline.com/doi/abs/10.1080/10643389109388420

WHO (World Health Organization) (2016) "Ambient Air Pollution: A Global Assessment of Exposure and Burden of Disease," *World Health Organization* https://www.who.int/publications/i/item/9789241511353

World Bank (2018) "What a Waste 2.0: A Global Snapshot of Solid Waste Management to 2050," https://www.worldbank.org/en/news/infographic/2018/09/20/what-a-waste-20-a-global-snapshot-of-solid-waste-management-to-2050

World Commission on Environment and Development (1987) "Report of the World Commission on Environment and Development: 'Our Common Future'," *United Nations* https://www.un.org/ga/search/view_doc.asp?symbol=A/42/427&Lang=E

World Nuclear Organization (2021) "Nuclear Power in the World Today," https://world-nuclear.org/information-library/current-and-future-generation/nuclear-power-in-the-world-today.aspx#:~:text=Nuclear%20energy%20now%20provides%20about,of%20the%20total%20in%202018)

11

Appropriate Technology and Less Developed Regions (LDRs)

Special issues arise for engineering projects in parts of the world where technology is less widespread or less than state of the art. Large technological undertakings in such areas can have life or death consequences. The Aswan High Dam, for example, provided badly needed food and electric power but displaced 120,000 people and permitted saltwater to back up into groundwater. By trapping silt upriver, the Dam also increased use of chemical fertilizers, negatively affected the fishing industry, and created a rapid increase in the snail population, causing an outbreak of schistosomiasis.[1] It is important to understand that, depending on the particular case, decisions about such engineering projects need to take into account a broad and diverse range of factors, such as highly localized but very deeply rooted traditional ways of life; extreme poverty, famine, and disease; and cultural practices and expectations that may differ markedly from those familiar to outside engineers.

Even naming the issue is controversial. Disagreement abounds about which term is most appropriate. The term "Third Word" has been criticized for being overly judgmental, inaccurate, and outdated (the Soviet Union no longer exists). Some prefer to speak of "Globally South" or "Southern" nations, but Ulan Bator, Mongolia and Taskent, Uzbekistan are both further north than New York City, while Sydney, Australia is considerably further south than N'Djamena, Chad. The terms most widely used are "Less [or Lesser] Developed Nations [or Countries], abbreviated LDNs or LDCs.[2] Some prefer to avoid the term "developed" since there is disagreement, as Lori Keleher notes, about what development is and what the goal of development is (Keleher 2017). Whatever term is used, it is important not to overgeneralize. Technology levels vary within countries.

DOI: 10.1201/9781003242574-14

India includes large modern cities and a vibrant technology sector as well as remote rural regions where modern technologies may be spotty or unreliable. In addition, some cultural practices are quite different in, for example, Mumbai and Bishnoi Village. Thus, you may encounter some of the issues in this chapter if you are an engineer in New Delhi working on a project in Kibber or an engineer in Beijing working on a project in Dimen. Even within a particular location, one might find, for instance, a reliable electrical grid but sporadic satellite coverage. Thus, when discussing engineering projects (rather than politics or international law), it is more accurate to use the term "Less (or Lesser) Developed Regions" (LDRs), with the clarifications that "developed" refers to the state of technological development (and not, for example, cultural or artistic development) and that the term does not presuppose that technological development is necessarily or inherently better. Many engineering codes of ethics, such as ASCE's, express commitment not just to development as such, but to sustainable development. Sustainability includes not only environmental and economic goals, but social goals, such as a fair distribution of resources and cultural diversity. (See "Sustainability," Chapter 10.)

Seven general types of issues can be ethically problematic for engineering projects in LDRs, especially for engineers from more developed nations or regions participating in those projects:

(A) Substituting cash for subsistence crops; (B) products banned at home but legal abroad; (C) depleting natural resources (such as cutting down rainforests); (D) less stringent environmental regulation and unsustainable practices; (E) products safe at home but unsafe under local conditions; (F) cultural disruption; (G) doing as the Romans do; and (H) implementing regionally appropriate technology.

SUBSTITUTING CASH FOR SUBSISTENCE CROPS

One widespread factor setting the background for many engineering projects in LDRs is the substitution of cash crops for subsistence crops. Imagine four neighboring nations, A, B, C, and D. These four nations grow a subsistence crop, such as corn, and live by eating the crop. They are not, by and large, cash economies. For the most part, they grow corn and live off the corn they grow. Eventually, one of the industrialized nations approaches the leadership of country A and offers to sell tanks to country

A. Since A is a cash-poor nation, the industrialized nation offers to lend country A the money to purchase the tanks. When country A objects that it cannot repay the loan, much less the interest on the loan, the industrialized nation, perhaps in tandem with the World Bank or International Monetary Fund, offers to lend country A the money. China, for example, has lent approximately $140 billion to Latin America between 2005 and 2019 (Canuto 2019). At present, the corn grown by nation A is worth $1 an acre in world markets. If nation A stops growing corn and grows, instead, a cash crop, such as coffee, nation A can realize $1.10 per acre by selling the coffee in world markets. That would allow nation A to purchase the same amount of corn it would have grown, at $1.00 an acre, with ten cents an acre left over to repay the debt incurred from the purchase of the tanks. Thus, by substituting coffee for corn, nation A can obtain the tanks, in effect, for free. (Should nation A prove reluctant to accept this plan, various forms of pressure might be applied.) The result is that nation A substitutes coffee for corn and borrows money to pay for the tanks.[3] (Even without buying tanks, nation A would need to borrow money for the start-up costs of coffee farming.) At first, this transaction appears to work well. The industrialized nation benefits from the sale of the tanks as well as the interest on the loan, while nation A gains tanks without any loss in standard of living. Unfortunately, prices in world markets fluctuate. Earlier, the price of corn had little effect on country A: inhabitants did not sell or buy corn. They grew corn and ate the corn. Now, however, the inhabitants grow coffee, not corn. If the price of coffee goes down or the price of corn goes up, there is no longer enough cash, after the ten cents per acre are deducted for loan repayment, to purchase enough corn. For example, if coffee yields only $1.08 per acre while corn prices are stable, nation A has only $.98 per acre, after paying $.10 per acre toward its debt, with which to purchase corn. A significant number of people in nation A starve. The effect is the same if the amount (or quality) of coffee grown per acre diminishes, due to insect damage, drought, or other factors. (In addition, tariffs, agricultural subsidies in other nations, and other market factors affect cash crop income but have little or no effect on subsistence farming.) As a result, nation A possesses tanks and insufficient food. Meanwhile, neighboring nations B, C, and D would lack tanks but would have corn that nation A could appropriate to feed its starving population. B, C, and D cannot allow themselves to be in such a situation. Thus, when A switches from corn to cash crops and purchases tanks, B, C, and D have little choice but to follow suit. Since markets operate on supply and demand, when these four nations switch from corn to coffee, the price of corn will tend to increase (since demand increases), while the price of coffee will tend to decrease (since supply increases), thus virtually guaranteeing that starvation will occur in all four nations.

Several important consequences follow, raising troubling ethical issues.

PRODUCTS BANNED AT HOME BUT LEGAL ABROAD

When country A substitutes cash crops for subsistence crops, it becomes a matter of life and death that fields in country A produce the maximum yield at the minimal cost. One consequence that raises ethical issues for engineers concerns long-lived vs short-lived pesticides. All pesticides raise issues of sustainability, since their use and manufacture create environmental and health issues. But long-lived pesticides, like DDT, are more problematic than short-lived pesticides such as Malathion. Long-lived pesticides like DDT, since they do not degrade quickly, work their way up the ecosystem. They accumulate in lakes and aquifers, where they are imbibed by animals, including human beings. Livestock and human consumers of DDT-sprayed produce accumulate DDT in their bodies, where it is stored in fat instead of breaking down. In addition, as birds eats the poisoned pests, the DDT does not break down but accumulates in the birds. Increased DDT levels in birds result in thinner egg shells and hence reduction of the bird population. Meanwhile, a small percentage of insects with DDT immunity are able to repopulate quickly since they face less competition from other insects and less predation from the declining bird population. While a small number of birds might also be immune, birds and insects have different reproductive strategies. Insects lay large numbers of eggs, few of which, normally, reach reproductive age, while birds lay few eggs, a larger number of which, normally, reach reproductive age. Thus, (immune) insect populations recover from continued use of DDT much faster than (immune) bird populations do. After some years, the pest population, now immune to DDT, is greater than before, with fewer predators to keep them in control. For these and other reasons, DDT has been banned in Japan, Argentina, Finland, Ethiopia, the United States, and many other nations. India has a partial ban but uses DDT for malaria control. China has banned DDT for agricultural use. By 2018, 182 countries had agreed to the 2001 Stockholm Convention (which took effect in 2004) seeking, among other things, to reduce and ultimately eliminate the use of DDT. (The convention does recognize DDT may still play a role in vector disease control, such as controlling mosquitos that carry malaria, while alternatives are developed and implemented.) For nations like country A, however, long-lived pesticides such as DDT are attractive, while substituting short-lived pesticides such as malathion, which must be applied every two weeks, increases costs. Moreover, controlling insects that cause disease is an urgent matter in many LDRs, so DDT's use in disease vector control saves lives. Of course, there is much controversy about whether DDT is either the cheapest or most effective method of disease vector control. There are good reasons for striving to implement alternatives to DDT in disease vector control, and a World Health Organization special report in 2017 proposed strategies for eliminating the use of DDT (World Health Organization Regional Office for Africa 2017). Nonetheless, even if DDT ought to be replaced, implementing DDT alternatives on a large scale would take time. (For this reason, the WHO still recommends using DDT in certain circumstances.) Banning DDT, thus, can lead, in the short term, to many deaths by malnutrition or disease. If country A bans DDT, the short-term consequence is that people die. Country A, therefore, may decide to permit the use of DDT.

Note that when DDT is sprayed on a field, it doesn't stay in the field. Because DDT is long-lived and doesn't easily degrade, it works its way within and between ecosystems. Moreover, since A is selling its produce in world markets, DDT sprayed on fields in A may be consumed by someone half a world away. DDT sprayed in one location (whether for agriculture or disease control) can wind up in a distant city's drinking water.

Is it ethical for companies operating in the numerous countries that have banned DDT to import DDT to country A, where it is legal to import and sell DDT? Should engineers in China, Sweden, and Canada refuse to work on projects related to the legal importation of DDT?

DDT is just one example, of course. Even if DDT production is eventually halted worldwide, as long as nations have different laws, ethical questions will remain about engineers working on (or facilitating) the export of products and technologies that are banned in their home countries, from pharmaceuticals and cloning to manufacturing processes, to nations where those products and technologies are allowed.

DEPLETING NATURAL RESOURCES (SUCH AS CUTTING DOWN RAINFORESTS)

One way for country A to escape its dilemma is to develop sources of cash, which may be used as investment capital (e.g., in factories) or as a buffer against drops in coffee prices. If country A has virgin forests (such as rainforests), it may be tempted to cut down the forest, generating cash from the sale of the timber and creating more arable land (to produce more coffee, yielding more cash, and hence limiting starvation). For example, in 2019, worldwide concern was voiced about widespread fires in the Amazon started, in some cases, by individuals looking to employ the land cleared. Moreover, Brazilian President Jair Bolsonaro repeatedly turned down offers of international help in putting out the fires. (France's President Macron, among others, suggested that Brazil was dilatory about trying to extinguish the fires out of a desire to clear the land.) While the long-term ecological consequences of such actions are serious, refraining can mean more deaths here and now. (Country A is also likely to argue that, since developed nations like Europe and the United States cut down much of their native forests to promote development, it is hypocritical for developed nations to object to country A's doing the same.) Is it ethical for engineers to participate in such projects?

LESS STRINGENT ENVIRONMENTAL REGULATION AND UNSUSTAINABLE PRACTICES

To the extent that country A is able to invest (or attract foreign investment) in industry, remaining competitive internationally (often on a low budget) may call for or make attractive lax regulation. Moreover, if country A's regulations are stricter than country B's, market forces will induce or even force foreign manufacturers to shift their operations to country B. Thus, market forces make it difficult for A to strengthen its environmental protection laws. (The same logic applies to low wages for workers in country A.) While permitting industrial pollution has long-term costs, it may prevent immediate starvation. Similarly, heavy use of chemical fertilizers results in nitrogen run-off, polluting streams and watersheds. However, the immediate higher yield saves lives in the short run. Is it ethical for engineers to participate in projects and operations that create environmental damage not permitted at home?

PRODUCTS SAFE AT HOME BUT UNSAFE UNDER LOCAL CONDITIONS

Two examples illustrate the problem. A pesticide meant to be applied by airplane, flown by pilots wearing safety equipment, may be no worse, when properly applied, than other pesticides. A large container of that pesticide, replete with proper warnings and safety instructions, may be sold to a store in a village in rural India. Farmers in remote rural LDRs tend to lack aircraft and large farms. The norm, rather, is small family farms worked by hand and oxen. The store, therefore, breaks up the container and sells small batches of the pesticide in available containers (such as old Coca Cola bottles). Farmers then apply the pesticide by hand and get sick. Is it ethical to sell large containers of this pesticide to stores in rural India? Are engineers involved in sales or manufacture of the pesticide responsible? A second example was provided by Nestlé's Infant Formula. The powdered formula, like all formulas, lacks some virtues of breast milk, but otherwise presented no problem when sold in industrialized regions, where it is prepared by mixing with sanitary water under proper conditions. In rural India, however, the powder must be mixed with local water. (It is too expensive to ship either bottled water or fully constituted formula over rural roads to remote villages.) In the 1970s, Nestlé sent representatives, dressed as nurses, to give free samples of the formula to villagers. When families mixed the product with local water and fed it to babies, many babies died from pathogens and impurities in the local water. Adults, unlike infants, have developed immune systems and have developed immunity to those pathogens. Before Nestlé's arrival, babies nursed on breast milk filtered by the mother's body. However, as mothers used Nestlé's product instead of nursing, their own breast milk ceased to be produced. (Note: today, Nestlé says that it actively supports breastfeeding.) Is it ethical to sell or distribute baby formula that is a perfectly fine product but can be deadly when used under local conditions (and for engineers to be involved in any stage of the process)?

CULTURAL DISRUPTION

Traditional ways of life frequently center on indigenous technology. Village V traditionally wove rugs, which were traded for food and clothing to neighboring villages. Each family, for generations, carried out an assigned task: one family gathered flax, one family dyed the fabric, etc. The rugs were woven collectively by the village, after which a seasonal festival was held. During those festivals, marriages were arranged between families (for example, a gatherer-family groom always married a dyer-family bride) and rites of passage occurred. Individuals knew their lifelong trade from birth, depending upon the mother's family (e.g., the son of a dyer-family mother became a dyer). Similarly, the religious and social calendars of family farmers frequently center on harvest and planting cycles. The sense of belonging and purpose of a traditional community are often linked with traditional crafts and livelihoods. When a modern rug factory opens, village rugs are no longer competitive, so villagers abandon their traditional crafts for assembly line jobs in the rug factory, becoming anonymous staff members assigned repetitive routine tasks. When small family farms are conglomerated into modern agribusiness plantations, family members become rote (or migrant) workers. Economically, the family famers and villagers may be no worse off, although Ned Dobos (2019) points out that very low wages, even if they are above the local average, are exploitive and demeaning. However, traditional social structures may not be able to survive the change. While modern Western cultures, in which everything is a "life style" and disposability is widespread, easily assimilate changes and foreign influences, deep-rooted traditional cultures are built on a sense of historical inevitability. The result of large-scale farming and rug factories is often a cultural vacuum, resulting in widespread crime and drug abuse. Often, villagers leave for the city. Technological change has produced large-scale migration from the countryside to urban areas. "Between 1950 and 1960, urban areas in Africa grew by 69 percent, in Latin America by 67 percent, and in Asia by 51 percent, while rural areas grew by only 20 percent over the same period" (Fields 1975). The trend has only accelerated, resulting in vast increases in the number of unemployed and squatters, often in squatter cities with poor sanitation. In 2017, over 20% of the population of Metro Manila were "informal settlers" in communities where "one often finds the sight of listless children wanting food, care and support. The impoverished parents are often absent, their children used as fronts for begging and for charity....sanitation and cleanliness is wanting, sickness, crime and drugs are rampant" (Sicat 2017). In Lagos' 13 squatter settlements, many residents "live in makeshift homes made of wood and scraps and elevated on stilts as a precaution against flooding" (McDonnell 2017).

Is it ethical for engineers to participate in projects that create cultural disruption in LDRs?

DOING AS THE ROMANS DO

Ethical norms and practices vary, to some degree, from culture to culture. Engineers from one country (the home country) may work on projects in another (the site country). When the norms and practices of the home country and the site country conflict, which should the engineer follow? For example, in Europe and the United States, it is normally considered improper (or at least unusual) to bribe people to do their jobs. In the United States, the service industry counts as an exception. There, it is considered customary (even mandatory) to tip, for example, wait-staff in restaurants, beauticians, and certain hotel employees such as concierges, over and above the salaries they receive. In some apartment buildings in the US, the superintendent will fix a leaky sink eventually, but if tenants expect quick attention they may need to slip "the super" a little extra. However, in most developed countries, it is improper and/or illegal to pay a safety inspector extra for a prompt inspection or pay a professor extra to post grades promptly. In some nations, however, giving government officials such under-the-table or informal gratuities for prompt attention is expected and customary. "Pay to play" does sometimes occur in developed nations, though it results in scandal when revealed. Transparency International, in 2018, gave Italy a score of 52/100 (53rd out of 180 countries ranked), compared to Finland and Sweden (ranked third best) with a score of 85/100 each. In Israel, Elliott Broily admitted, in 2009, to bribing four Israeli officials to steer business his way. In the US, Illinois Governor Blagojevich (who maintains his innocence) was jailed after being convicted of demanding payment for appointing a replacement when Senator Obama became President Obama. Spiro Agnew was forced to resign as the Vice President as part of a plea bargain concerning his purported involvement, when Governor of Maryland, in a well-established system of contractor kickbacks to state officials in order to obtain contracts from the state. Contractors who did not participate did not get contracts. However, laws against such practices are generally enforced, as the fates of Blagojevich and Agnew attest.

While the 1997 OECD Convention on Combatting Bribery of Foreign Officials, signed by 44 nations, prohibits paying bribes overseas and many nations have laws criminalizing paying bribes abroad, the reality faced by many engaged in certain projects in certain LDRs requires a stark choice between paying and failure. Failure to "grease the palms" of safety inspectors or permit processors may result in exorbitant (and excessively costly) delays forcing the cancellation of a project or ensuring that the project cannot compete with rival firms or nations who don't indulge in such scruples. Should engineers (and the firms or enterprises for which they work) refuse to pay such "costs of doing business"?

IMPLEMENTING REGIONALLY APPROPRIATE TECHNOLOGY

What is "appropriate technology"? Broadly defined, technology is appropriate for a region when it fits the particular environmental and sociological needs and circumstances of that region. It is sometimes more narrowly defined as suited to specific sorts of regions: "An 'appropriate technology' is usually characterized as small scale, energy efficient, environmentally sound, labor-intensive, and controlled by the local community. It must be simple enough to be maintained by the people who use it. In short, it must match the user and the need in complexity and scale and must be designed to foster self-reliance, cooperation, and responsibility" (Amadei 2004, 28). Technology appropriate for a project or task may vary from region to region. A simple example is collecting the fare on a bus. In developed, industrial nations like Germany and the United States, labor is expensive and machines, especially simple machines, are comparatively cheap. For heavily used routes, having the bus driver collect and count fares creates too much delay. In Germany, thus, it is cost efficient for bus patrons to deposit their coins in a simple machine that displays the amount submitted to the driver. In Guatemala, labor is cheap and plentiful (there is high demand for unskilled jobs paying very little), while machines and their upkeep are comparatively expensive. Buses are typically crammed full. It is therefore generally preferable to have bus fare collected by an additional person (not the driver). Appropriate technology is, in Bonn and New York City, a bus coin box, while, in Guatemala, appropriate technology is a human collector (ayudante).

More ingenious examples of appropriate technologies in LDRs abound. Nokero's N100 solar-powered light bulb, at $15 per unit, stores energy from sunlight to provide up to four hours of LED light. Concrete Canvas offers a quickly deployable shelter, offering more protection than canvas and lasting up to 10 years, which is simply pumped up with air and then doused with water (salt or fresh). Mohammed Bah Abba's Pot-in-Pot refrigerator uses evaporation to create viable refrigeration without electrical power. Emily Cummins gave away the design for her version of a pot-in-pot evaporation refrigerator using water and sunlight. It

> consists of two metallic cylinders, one inside the other, between which a locally-sourced material such as sand or wool is packed tightly before being soaked with water. When the fridge is placed in a warm environment, the sun's energy causes the outer part of the fridge to "sweat." Water evaporates from the sand or wool and heat energy is transferred away from the inner cylinder, which therefore becomes cooler. The design is ideal for use in the developing world because it doesn't require electricity and can be built using barrels, spare car parts and ordinary household materials. Unlike previous pot-in-pot coolers, the contents are kept dry and hygienic because the water does not come into contact with the product.
>
> *Cummins 2018*

Emily Cummins also created a water carrier, designed to be manufactured in Africa, to replace carrying heavy water buckets on the head or with a yoke (which causes long-term back issues). Rocket stoves allow high efficiency cooking, directing heat to a small targeted area using fuel such as twigs. An Arberloo is a simple dry toilet, preventing run-off of waste from contaminating local water. It is easily moved when the pit is full, leaving nutrient-laden soil or fertilizer for agricultural use. Loband, which strips data-extravagant images from websites and converts them to plaintext, is useful in regions where bandwidth is very limited. The Leveraged Freedom Chair is a low-cost wheelchair designed for rugged terrain (such as an unpaved village).

Cultural appropriateness can be as important as physical or environmental appropriateness. Gender norms in Saudi Arabia often impact medical practice. For example, women's health decisions are often made by male guardians, who may refuse to allow treatment by a male nurse or physician. Eighty-one percent of Saudi women indicated, in a survey, that family problems should not be revealed to outsiders (Aldosari 2017). Engineers working on medical technology and medical information technology in Saudi Arabia need to take these factors into consideration. Similarly, a society may feel that no advantage gained from a dam justifies flooding a sacred, ancestral burial ground (Harris 1998).

ETHICAL FACTORS USEFUL IN DEALING WITH THESE CONCERNS

Engineers facing decisions raising any of these issues should employ the five-step process to make an ethically defensible choice. Any of the factors discussed in this volume might turn out to be relevant in a particular case but some are more likely to arise in decisions about appropriate technology and LDRs.

Most immediately useful, perhaps, is "Responsibility (Dual-Use)" in Chapter 6. As a reminder, engineers should ask themselves "what does my response to the issues raised above show about my values and the kind of person I am?" Promoting good consequences (thinking both in terms of policy and this particular case), a discounted but real obligation to future generations, safety (how widespread, severe, and likely the risks are, whether those at risk benefit from the risk and assume the risk knowingly and voluntarily), the duty to leave the world no worse, and the Weak Samaritan Principle (whether one can prevent harm or risk to others at low cost to oneself and without violating rights) are also likely to be relevant factors.

Engineers must find a delicate, often difficult balance between respecting the autonomy of those in LDRs who are affected by one's projects and being unwilling to be the agent of harm. For example, if nation N has decided that the short-term benefits of using DDT outweigh the long-term harms, it is somewhat arrogant and paternalistic for foreign engineers to say "we know better what is best for you." However, the right of nation N to permit DDT is mitigated by the fact that DDT use in N affects many others outside N, as DDT and its eco-effects migrate past N's borders. In addition, just as an ethical person may respect the right of another to choose to smoke cigarettes but refuse to be the one to supply the cigarette, an ethical engineer may respect the right of N to choose but refuse to be the one to provide the DDT. While nation N may have the right to set its own laws, it does not have the right to command the assistance of engineers abroad in projects those engineers find objectionable, and the engineer who "hands the gun" to nation N bears some responsibility for the consequences (see "Respect for Autonomy," Chapter 6). As discussed in Chapter 6, there are limits to the doctrine of intervening wills. The duty to set a moral precedent is relevant here.

One thing that engineers involved in projects in LDRs can do is make a genuine effort to understand the cultural and physical realities of the region, so as to better understand the implications of their designs and decisions. This is one reason diversity in engineering teams may be useful. An engineer who grew up in Uganda may be able to contribute perspective and understanding to a team working on a project in Uganda. In any case, standard engineering curricula do not spend vast amounts of time on the culture, social structures, and regional geographies of different parts of the world. Engineers working on such projects need to make special efforts to educate themselves about the region.

Case 61: Indonesia Textiles

The textile industry in Indonesia provides jobs for over 3.7 million people. In 2022, the average textile worker in the United States earned $23,290 a year. In 2016 in Indonesia, the minimum wage (set by regional governments) ranged, depending on the region, from US $80 to $200 a month (as opposed to $26 a month in Ethiopia), or $960 to $2,400, annually. (However, some factories receive exemptions for financial reasons.) By contrast, the median income in Indonesia is $788, so employment in the textile industry is competitive. Many factories lack proper safeguards against inhaling dust and fiber. Accidents, fires, sexual harassment, and abuse are not uncommon. Almost 60% of textile workers work more than 48 hours per week. Sixty-five percent of the lower-cost textiles produced in developing nations go to the United States and the European Union (Japan and China account for 8% and 5%, respectively). How should engineers asked to develop machinery for use in Indonesian textile mills balance: the low wages and harsh conditions of many factories against even lower wages available locally outside the industry; the profit to be made by selling the machinery; and respecting the autonomy of Indonesians. Note that if Indonesia substantially raised its minimum wage, textile manufacturers could simply relocate to a neighboring country with lower wages. (See Harris 1998 for a discussion of a similar case.)

Case 62: Sub-Saharan Plant Safety

Behemoth Manufacturing, headquartered in Paris, has developed a new medical device that uses enriched thorium. The radioactive thorium must be shaped and enclosed with miniature electronics in a small lead chamber. Behemoth asked Willingtonarial Consultants, an engineering firm, to design a plant to process the thorium and enclose it in the capsules. The plant will be located near a densely populated city in The Republic of Matumaini, an LDR in Sub-Saharan Africa, whose plant safety regulations are well below those of France. Stalwart, the Willingtonarial engineer heading the design project, suggests the plant be located in a more sparsely populated region and maintains that the plant needs multiple safety systems with built-in redundancy to ensure no radioactive materials escape. Behemoth, intent on lowering cost, insists that the plant need only meets Matumaini regulations and that the plant must be built near the city where there is an abundant supply of cheap labor. If Behemoth's demands are met, the new device will be cheaper and slightly superior to existing devices. If the cost of the new device is significantly higher than existing devices, says Behemoth, it will not sell. Behemoth plans to pay workers in the proposed plant wages that are 30% higher than the average wage in Matumaini. What should Stalwart do?

Case 63: Paying a Bribe

N corporation is building a plant for textile and clothing manufacture in country P, a lesser developed region constituting a promising market for the clothing N would manufacture. Establishing brand loyalty in P is important. Due to proprietary manufacturing techniques, N can manufacture clothing of equal quality for less, even while paying higher wages, than Q corporation, which is also developing plans for a clothing plant in P. Both plants require a safety inspection before they can become operational. N's plant is closer to completion than Q's, but P's Safety Inspector has been delaying the inspection, citing thinly veiled excuses and hinting that a "gratitude" payment under the table would greatly expedite the inspection. No date for the inspection has been given, and it is unclear, if N does not pay the bribe, when or even if the inspection will take place. The delay is costing a considerable sum every day. N is certain both that its plant would easily pass a fair inspection and that Q has no scruples about paying the bribe. Gratitude payments of this sort are common in P and the Safety Inspector is a relative of the Prime Minister. No one in P has ever been convicted of taking a gratitude payment. In N's country of origin, it is illegal to pay bribes to foreign officials. Should N pay the bribe, complete its plant, and get a jump on establishing brand loyalty, or should N refuse to pay, continue to lose money, and watch Q establish brand loyalty? Is there a third option N should pursue?

Discussion: Most ethicists would say N should refuse to pay the bribe. Many "realists" would say N should do what it needs to do. An act utilitarian would weigh the harm produced by paying the bribe against the good produced by paying it and opening the plant, such as higher wages for the workers. (See Appendix II and "Promoting Good Consequences" in Chapter 5.) Other relevant factors to consider, among others, are "When to Break the Rules" in Chapter 6 and "When to Fight a Battle" in Chapter 5. If the third option involves blowing the whistle, see "Whistleblowing" in Chapter 7.

NOTES

1. Much of the information in this chapter comes from Schlossberger (1997).
2. Such regions are sometimes defined in terms of income level, prevalence of disease and vulnerability, unemployment and illiteracy, and/or percentage of population engaged in agriculture. For the issues discussed in this chapter, the defining factor is the degree to which fully modern and up-to-date technology is widely available. Regions that are LDRs in this sense frequently also evidence those other characteristics.
3. For example, in Uganda, over half of farms 5 hectares or larger grow cash crops (Adjognon et al. 2017).

REFERENCES

Adjognon, S. G., Liverpool-Tasie, L. and Reardon, T. A. (2017) "Agricultural Input Credit in Sub-Saharan Africa: Telling Myth from Facts," *Food Policy* 67, 93–105. https://www.ncbi.nlm.nih.gov/pmc/articles/PMC5384443/#

Aldosari, Hala (2017) "The Effect of Gender Norms on Women's Health in Saudi Arabia," *The Arab Gulf States Institute in Washington* https://agsiw.org/wp-content/uploads/2017/05/Aldosari_Womens-Health_Online-1.pdf

Amadei, Bernard (2004) "Engineering for a Developing World," *The Bridge* (National Academy of Engineering) 34:2 https://www.nae.edu/7524/EngineeringfortheDevelopingWorld

Canuto, Otaviano (2019) "How Chinese Investment in Latin America Is Changing," *Americas Quarterly* https://www.americasquarterly.org/content/how-chinese-investment-latin-america-changing

Cummins, Emily (2018) "Inventions," https://www.emilycummins.co.uk/inventions

Dobos, Ned. (2019) "Exploitation, Working Poverty, and the Expressive Power of Low Wages," *Journal of Applied Philosophy* 36:2, 333–347.

Fields, Gary S. (1975) "Rural-Urban Migration, Urban Unemployment and Underemployment, and Job-Search Activity in LDCs," *Journal of Development Economics* 2, 165–187.

Harris, Charles H. (1998) "Engineering Responsibilities in Lesser-Developed Nations: The Welfare Requirement" *Science and Engineering Ethics* 4, 321–331.

Keleher, Lori (2017) "Toward an Integral Human Development Ethics," *Veritas* 37, 19–34.

McDonnell, Tim (2017) "Slum Dwellers in Africa's Biggest Megacity Are Now Living in Canoes," *National Public Radio (NPR)* May 15 2017 https://www.npr.org/sections/goatsandsoda/2017/05/15/528461093/slum-dwellers-in-africas-biggest-megacity-are-now-living-in-canoes

Schlossberger, Eugene (1997) "The Responsibility of Engineers: Appropriate Technology and Lesser Developed Nations," *Science and Engineering Ethics* 3, 317–326.

Sicat, Gerardo P. (2017) "Historical Roots of Urban Squatting," *The Philippine Star* Nov. 21 2017 https://www.philstar.com/business/2017/11/21/1761150/historical-roots-urban-squatting

World Health Organization Regional Office for Africa (2017) "Demonstrating Cost Effectiveness and Sustainability of Environmentally Sound and Locally Appropriate Alternatives to DDT for Malaria Vector Control in Africa," https://www.afro.who.int/sites/default/files/2019-03/AlternativeDDT-eng.pdf

12

Bioengineering and Medical Engineering

OVERVIEW

The future of bioengineering and medical engineering promises to bring enormous changes, from crops that produce their own insecticides and "designer babies" (parents using gene modification to select their baby's traits, such as eye color) to human cloning and drugs that prevent aging or increase our ability to remember. "Bio innovations could alleviate between 1% and 3% of the total global burden of disease in the next 10 to 20 years" (Evers and Chui 2020). This coming Brave New World raises many ethical challenges for bioengineers, the engineering profession as a whole, and society generally.

Bioengineering issues fall into at least *five subject areas of ethical concern*:

1. *Genetic engineering*, that is, modification, generally through gene splicing technologies such as CRISPR-Cas9 genome editing, either of somatic cell DNA that is not passed on to offspring or to germ cell DNA (sperm, ova, seeds, etc.) that is passed on to future generations. In addition, transgenic engineering (using genes from several species) differs from single species genetic modification.
   a. Genetic engineering of human beings, for the purpose of either:
      i. Fetal selection/creation ("designer babies")
      ii. Medical treatment (e.g., genetically modified embryonic stem cells to correct genetic defects or genetically modified skin transplants to cure epidermolysis bullosa);
   b. Genetically Modified Organisms (GMOs), such as GMO corn, for use as food, either:
      i. directly (such as drought resistant "golden rice") or
      ii. indirectly (e.g., as feed for livestock);

DOI: 10.1201/9781003242574-15

     c. Genetic engineering of other (non-food related, non-human) organisms, such as bacteria that break down oil spills (bioremediation), sequester carbon, or produce insulin (providing diabetic patients with a reliable, inexpensive, and organic source of insulin);

2. *New reproductive technologies* such as cloning;
3. *Biotechnology to enhance human beings or mitigate disability*, such as cochlear implants; exoskeletons; neural implants to operate robotic limbs or devices like computers; retinal implants; alertness drugs; and Google glasses;
4. *Brain research and psychological control*; and
5. *Genetic testing*, raising issues about the nature/nurture debate, insurance and fairness, the value of knowing v. not knowing, and privacy.

Medical and bioengineering raises a variety of *broad issues and concerns*. Some arise primarily within one of the above subject areas. For example, privacy concerns fall mostly within area 5, e.g., is privacy violated by performing, employing, or disseminating results of genetic testing for disease markers? Other concerns, such as safety, apply to several areas. Whether consuming GMO corn poses health risks is a safety concern in area 1b, while whether cloning poses health risks to offspring is a safety concern in area 2. Seven broad issues with wide application include *safety* (for example, some diabetics have experienced severe adverse reactions to GMO bacteria-produced insulin); effect on *the environment* (such as loss of biodiversity and risks from unexpected and unintended non-target gene expression of GMOs); *social implications* and consequences of bioengineering, for example, Michael Sandel's (2004, 2007) worry that designer babies will leave us indifferent to the "gifted' in life or Leon Kass' (2001) claim that human cloning will result in soulless automatons; disrespecting nature (many express a preference for what is considered "natural" and an aversion to what is deemed "unnatural" or "playing God"); *respect for humanity* and what it is to be human; *the difference between enhancement and treatment*; and concerns about *unequal access* to these technologies and its effect on social justice. Conversely, a particular technology may raise several of these seven issues. For example, Patuzzo et al. (2018) point out that 3D bioprinting raises issues of enhancement v. treatment, safety, and access.

BROAD ISSUES ARISING IN MULTIPLE CONTEXTS

Safety and the Environment

Safety and the environment apply to much more than bioengineering and are discussed throughout this book. Factors mentioned elsewhere in the book also apply to bioengineering. (Examples include Chapter 4's discussion of partnership with nature, evaluating how severe, likely, and widespread a risk is, and asking whether those taking a risk receive the benefit, as well as Chapter 10's question, "could we, instead, substitute another technology that does not cause this problem?") The use of GMOs has especially raised concerns about safety and the environment, but safety is also a major issue concerning any medical use of bioengineering technology (including biological enhancement and brain research). Engineers working on projects that raise concerns about safety or the environment must include those factors when engaging in the five-step process to make particular ethical decisions about their projects.

Social Implications, Playing God, Respecting Humanity, and the Slippery Slope

Some writers, both academic and popular, worry about the effects bioengineering will have on society and our conceptions of ourselves. Arthur Caplan nicely summarizes (before arguing against) these concerns: "First, biomedical research cannot continue on its present course without significantly altering human nature. Second, if, in the name of more cures, longer life, and improved quality of life, we continue on our present biomedical research course, we will commodify and objectify human life. Third, too much biomedical tinkering will produce a loss of authenticity and meaning in human experience." Finally, "biotechnology undermines or shifts our understanding of the nature of family, marriage, sexual relations, aging, and parenting" (Caplan 2004). Sandel (2004, 2007) argues that designer babies will undermine our sense of humility, our unconditional love of children, our charitable impulses and dedication to social justice, and our appreciation of "the gifted" in life (see below). Silver (2001, 659) points out "the real reason that people condemn" cloning and other biotechnologies "derives from religious beliefs," namely "that man is venturing into places he does not belong."

These concerns are usually followed by a slippery slope argument: once we start on that path, the claim goes, we will inevitably slide down the slope toward terrible consequences. Once we start allowing selective abortion after genetic testing, or genetic engineering to eliminate disease, the argument goes, where will it stop? We will quickly judge such lives are not worth living and begin a Nazi-style program of exterminating Down Syndrome children and those with disfavored eye color. James Watson, speaking of IVF in 1974, told a Congressional Committee that "a successful embryo transplant would lead to 'all sorts of bad scenarios.' Specifically, he predicted: 'All hell will break loose, politically and morally, all over the world'" (Garber 2012).

Most of these claims are highly overstated. "What it means to be human," in the deepest sense, has nothing to do with whether one was conceived in a test tube or has "unnaturally" perfect skin. The biotechnologies talked about here do not affect our ability to live according to ideals and moral commitments about which we are able to reason, our desire for a better future for ourselves and others, our exploration of meaning and the nature of reality (physical and personal), our need to rise above obstacles and strive toward excellence, the special quality of human intimacy, our capacity for empathy and concern, or any of the other things that make human existence both glorious and painful.

As for the claims that biotechnologies constitute a dangerous first step on a slippery slope leading to disaster, there is a reason that *logicians generally classify slippery slope arguments as fallacies*. After all, if I, from my present location in Munster Indiana, take a step north, I have taken the first step toward drowning in Lake Michigan. But that is no reason not to take a step toward my northern window, since, miles before I come to Lake Michigan's shores, I can just stop walking. Unless there is strong evidence that the process envisioned by these authors cannot be stopped (or will not stop, or is not likely to stop) short of those consequences, the fact that some biotechnology might be the first step toward an unwanted consequence is not a good reason to ban the technology. Everything is the first step toward something bad.

In fact, most of these technologies lie on a continuum along which we have been advancing for many years, without seeing the dire consequences predicted. Human beings have, for a very long time, been "playing God" by rescuing patients from death with medicines and organ transplants, extending life with better nutrition, improving human eyesight with eyeglasses and telescopes, making people smarter with education and pre-natal care (including vitamins and giving up alcohol during preg-nancy), and much more. IVF (In Vitro Fertilization) has been in use for over 40 years without creating a commodification of human life. (Video games are more plausible culprits for that.) The nature of families has been altered much more by changing social conditions (such as the prevalence of divorce) than by biotechnologies. Are the many families including a baby born using various reproductive technologies noticeably different from other families? It is significant that virtually none of the authors making these dire claims provide significant empirical evidence, while past experience casts doubt on these claims.

Engineering as a field is committed to clean, clear decision-making supported by careful study and evidence. Engineers should take seri-ously potential dangers, whether about safety or broad negative effects on society, when supported by reasonable evidence. Sweeping and dire predictions unsupported by any evidence should be taken with a heavy dose of salt.

The Enhancement/Treatment Distinction

In medical and bioengineering ethics, the difference between enhance-ment and treatment arises in several ways. The basic idea is that treatment restores or raises patient to some appropriate baseline of health, wellness, or functioning (such as being "normal" or "average"), while enhancement makes people better than that baseline, for example, by using "smart" drugs (cognitive enhancers or "nootropics"[1]) to increase the intelligence or creativity of someone whose abilities are average or above. "A therapy, roughly defined, is a treatment for a disorder or deficiency, which aims to bring an unhealthy person to health. An enhancement is an improvement or extension of some characteristic, capacity, or activity" (The President's Council on Bioethics 2002). The issue is further clouded if some treat-ments are "medically necessary" and others are not. Whether treatments can be separated into those that are medically necessary and those that are not depends on how "treatment" is defined. For example, if treatment is defined as raising a patient to a statistically average level of functioning, then some treatments may not be medically necessary. If Campos' ability to curl her tongue is well below average, a procedure that would give her an average ability to curl her tongue would, by that definition, be treatment. But, without the treatment, she can live a long, rewarding, and satisfactory life. She is not at a significant competitive disadvantage for anything of much importance in life. Hence, the treatment is not medically necessary. Also, there are two sorts of enhancement: (patient perceived) improvements that are not directly health or medicine related, such as enhanced eye color or cosmetic fat removal, and health-related improvement to a level well above average. "Anti-aging" procedures to remove wrinkles are examples of the former, while "anti-aging" procedures that prevent telomere shortening or keep arteries as pristine as a child's are examples of the latter.

The distinction between treatment and enhancement has been used in several ways.

1. Daniels (1985) has argued that treatment is a basic right respected by a just health care system, but enhancement is not. Thus, thinks Daniels, patients below the baseline ought, justly, to be given treatments, including treatments like cochlear implants, to reverse or mitigate deafness and modified gene therapy to treat disease.

2. Some have argued that the use of biotechnology in general, and, especially, genetic engineering to create "designer babies" (allowing parents to choose some part of their baby's genetic makeup), is morally permissible as a form of treatment (such as correcting or avoiding birth defects) but morally impermissible (or should be prohibited by law) for purposes of enhancement.

3. Insurers may use the distinction in deciding what to cover: an insurer might cover a procedure classified as "treatment" but not a procedure classified as "enhancement."

4. The distinction can be helpful when weighing the risks and benefits of a technology or treatment. A patient might decide against taking the risk of a procedure because the result is just an enhancement. An engineer might decide against developing or deploying a risky (or morally borderline) biotechnology because the potential benefits are enhancements, not treatment.

Two issues, therefore, must be addressed. First, how should treatment and enhancement be defined. Second, what moral difference does the distinction make. The two questions are closely connected. If treatment is defined as anything covered by US Medicare, then the line between treatment and enhancement is relatively easy to draw but the distinction is useless for ethical decision-making. For example, the fact that Medicare regulators chose not to cover a procedure doesn't make it unethical or not worth the risk. Treatment and enhancement must be distinguished appropriately for the way the distinction is used.

The line between treatment and enhancement has proven notoriously difficult to draw. Norman Daniels suggests that treatment is raising functioning, in those areas that characterize us as human, to the baseline of statistical normality. Daniel's definition is widely criticized and widely defended.[2] Suppose gene w, when condition z is present, produces a protein that causes the patient severe problems. When z is absent, w leads to well above average functioning. Daniel's definition seems to imply that giving the patient a regimen that prevents condition z is enhancement (since it produces well above average functioning), while not giving the regimen is denying treatment (since without the regiment the patient's functioning is well below statistical normality).

More importantly, it is impossible to define "health" without making value judgments. The World Health Organization's Constitution defines health as, "a state of complete physical, mental and social well-being" (World Health Organization 2006). The document continues: "the enjoyment of the highest attainable standard of health is one of the fundamental rights of every human being." Deciding what complete well-being means is a value judgment. (Moreover, since almost any enhancement might contribute to complete social or mental well-being, the WHO Constitution appears to imply that virtually any attainable biological enhancement is a fundamental human right.) Perhaps a healthy state is a statistically normal state, that is, a state within the appropriate region of the relevant bell curve. But then anyone whose eyesight is so good that it falls outside the "normal" region of the bell curve would be "unhealthy." Absurdly, making those people's eyesight worse would be "treatment" to restore a healthy state. The phrase "healthy states," clearly, should be replaced with the phrase "within the normal part of the bell curve *or better*" (thus modifying the claim to avoid the counterexample – see Chapter 2). But "better" cannot be defined in a morally neutral way. We might try saying that a "better" or "healthier" state is one that enables more choices. For example, being physically stronger allows us to choose more things than being physically weak. But, in fact, almost everyone has the same number of choices – an infinite (or indefinite) number. Even if all I can do is blink, I can blink while thinking of the number 1, or blink while thinking of the number 2, etc.[3] What matters is having *better* choices, and that is not a morally neutral criterion.

The fact is, enhancement and therapy form a muddy continuum. "The treatment-enhancement distinction and the somatic-germline distinction are not as clear-cut as they might initially appear. More importantly, they cannot be used to definitively differentiate right from wrong uses of the technologies in question" (Johnston 2020).

Still, "while some scholars may find the line between therapy and enhancement elusive or even illusory, the practice of medical care in the United States has created a *de facto* line that is very real" (Colleton 2008).[4] Some factors generally (but not necessarily always) count in favor of (but must be weighed against other factors) a procedure's being a treatment. One example is what I call the "state of nature" test. Medical intervention K eliminates a condition that would be a serious survival disadvantage for a human being in a "state of nature" (meaning the conditions under which human beings evolved). For example, K restores sight. Someone who cannot see and lives at the time when homo sapiens evolved has a much harder time surviving (avoiding saber tooth tigers, finding edible roots and berries, etc.).[5] This fact generally counts, I suggest, in favor of regarding K as a treatment. After all, if we evolved to need the ability to x in order to survive, a special reason is needed to think correcting the inability to x isn't treatment.

A final distinction was drawn by The President's Council on Bioethics (2002): there is "an important difference between enhancement mediated by biotechnology and enhancement advanced by social, cultural or educational means. The first sort tends to act on the body of the individual involved, while the second addresses itself to the individual's character, experience and psyche." The Council noted, however, that "the line between the two is not always clear."

The moral importance of the distinction between treatment and enhancement is also questionable. Patient N has severe cataracts in both eyes, making him almost blind. Current treatment calls for removing the clouded lenses and replacing them with artificial lenses, allowing the patient to see. Since current artificial lenses cannot be focused well, patients typically see distant objects 20/20 but need correction to see close up. Future technology may produce focusable lenses, allowing patients to have normal vision both close and far. It should be easy to further refine those lenses to produce somewhat better than average vision both near and far. It is truly immoral to use such a lens? External optical devices to sharpen or improve vision, like binoculars, already exist (and raise no moral objections). Are optical surgeons morally required to give patients worse lenses than are available in order to keep from bringing patients above average? Does morality demand ensuring that patients must use binoculars or other devices when needed instead of their own eyes? That seems like a perverse restriction. While the difference between treatment and enhancement may sometimes be a morally relevant factor, it cannot function as a sharp line between the permissible and the impermissible.[6]

Access and Social Justice

Because many biotechnologies are expensive, some have raised concerns about unequal access to these technologies and the effect of these technologies on issues of social justice. For example, if rich people are able to use expensive gene modification technologies to improve their offspring's intelligence, that confers a significant advantage unavailable to less wealthy parents, thus widening the growing gap between the poor and the wealthy. In addition, some object to vast sums of money going to research, development, and implementation of non-life saving biotechnologies when so many live in poverty.

These are, of course, legitimate concerns. They are not, however, unique or peculiar to biotechnology. Elite and expensive private education, for example, also confers a huge advantage on children of families able to afford it. Scholarships somewhat mitigate differences in access to top schools, but such scholarships are limited and do not address differences in academic preparation or support resources. Similarly, reproductive technologies for wealthy or middle-class couples may be less deserving of resource allocation than alleviation of global famine. However, the same is true of lipstick, and the total spent, worldwide, on cloning and other reproductive technologies is unlikely ever to approach the total spent, worldwide, by the cosmetics industry. In short, these concerns apply not only to more familiar medical care, but also to much of life far removed from biotechnology. While a case can be made for a drastic global shift of priorities across the entire spectrum of life, there is little reason to single out biotechnologies (leaving the rest of life the same). Moreover, historically, technologies become cheaper and more accessible over time (because of supply and demand, research costs being recouped, improvements and advances, and so forth). For example, in 1972 the HP-35 hand-held calculator sold for $395 (approximately $2,700 in today's dollars). In 1977 the Teal LC811 could be purchased for $24.95, while $5.95 could buy an EL-345 in 1985. Technological innovations have made ultrasound and dialysis widely available across the globe.

AREAS OF ETHICAL CONCERN

Bioengineering is a vast subject covering a wealth of technologies. Although the future is certain to raise new ethical concerns regarding new types of technologies, much of the current ethical debate focuses on five areas.

Designer Babies and Human Gene Engineering

CRISPR gene editing, either in human beings or in other species, allows insertion or deletion of genetic material, in human beings, other animals, plants, or microorganisms. The new genetic material may be implanted in somatic or germline cells and be from the same species, a different species, or even, potentially, created in a lab. Gene editing has engendered several kinds of ethical concerns. This section discusses what is perhaps the most sensitive aspect: editing of human DNA.

One distinction that some have thought important in these debates is between germline and somatic changes, that is, between gene alterations that are passed on to offspring and those that are not. I would suggest that this distinction is neither sharp nor ethically powerful. Johnston points out that some "germline" modifications may not be present in all germ cells, while in utero editing can result in a combination of somatic and germline changes, and "we can expect widespread public support for interventions that eliminate devastating genetic disorders like sickle cell anemia, even if those interventions involve germline modifications" (Johnston 2020). However, the distinction is often invoked in the literature. Eriksson et al. (2018) identify three levels of gene modification/editing: "(1) genome editing, where a few nucleotides are changed to variants already existing within the species (and then traditional breeding methods may be implemented); (2) making more extensive changes within a species, such as using vectors to transfer genetic code; and (3) introducing active genetic material from another species" (p. 5).

One area of controversy concerns living modeling. "Engineered entities are now being generated in laboratories from human stem cells to form biologically dynamic, living models of human biology. These models can be used to study various aspects of human development and to test new drugs and therapeutics. Prominent among these models are organoids (small stem cell-derived 3D structures that self-organize into functional cell types and recapitulate basic organ functions) and embryo models (stem cell-derived simulations of post-implantation embryos)" (Hyun 2020). Multi-cellular engineered living systems (M-CELS) derive from stem cells combined with non-biological components. "SHEEFS" stands for "synthetic human entities with embryo-like features." Aach et al. (2017) criticizes "the '14-day rule' for embryo research [that stipulates] that experiments with intact human embryos must not allow them to develop beyond 14 days." (At what stage of development killing functioning human embryos becomes morally problematic is, of course, a hotly debated topic. For philosophical discussions of abortion, see Cohen et al. (1974) and Gordon (2008).) Hyun (2020) points to the special challenges posed by "the potentially unpredictable nature of biologically autonomous, self-organizing human cells" and suggests that bioethicists and engineering ethicists routinely meet with scientists to discuss values and ethical concerns at the beginning and early design phase of projects, rather than reacting once an ethical concern is identified with a project already underway.

Perhaps the most sensitive issue in human gene modification focuses on "designer babies." The technology for altering human genes prior to or shortly after fertilization (or in utero) has begun to arrive. In 2018, He Jianqui announced he had implanted embryos with a CRISPR-edited CCR5 gene, resulting in the birth of human twins, for which he was sentenced to three years in jail (Mallapaty 2022). Will future parents be able to choose their baby's features, from intelligence to eye color, from a menu? Some, like Julian Savulescu insist creating designer babies is morally necessary, both because those babies will be happier (Savulescu 2009) and because, unless genetic enhancement creates morally superior human beings, we will not survive the ecological challenges facing us (Persson and Savulescu 2014). On the other side, Michael Sandel suggests that designer genes undermine our appreciation of the "giftedness" in life (natural endowment and chance) and enhances our sense of mastery over life. "That we care deeply about our children," Sandel (2004) writes, "and yet cannot choose the kind we want teaches parents to be open to the unbidden… The awareness that our talents and abilities are not wholly our own doing restrains our tendency toward hubris." Indeed, says Sandel (2007, 83), adopting "a stance of mastery and domination" toward the world may undermine the unconditional, accepting love of children that parents must balance against promoting excellence. Others worry that parents will indulge in sexism and choose boys over girls, thus making the technology a tool of sexual oppression, or that the costly tools of baby designing will give the rich a further advantage over the poor, further increasing the gap. Rich children, the argument goes, already have vast advantages over poor children. In a world in which rich people can design their babies, rich children will be born more intelligent, better looking, and healthier than poor children. (Of course, some object to designer babies on the grounds, discussed above, that the practice plays God, fails to respect humanity, etc.)

Designer babies and gene enhancement constitute just one step along a long, historical continuum, and many of the dire predictions about their use are unsubstantiated and overblown. Sexual selection is practiced by many animals, whether or not they know they are picking their mates to enhance the odds of desirable offspring. "Eugenics," in this sense, is a common evolutionary feature. Pre-natal care uses technological interventions, such as vitamins pills and Diclegis (doxylamine and pyridoxine, used to treat morning sickness during pregnancy), to enhance or optimize fetal development. Cochlear implants, athletic training regimens, sports medicine, the development of new techniques such as the split-fingered fastball in baseball, and artificial limbs are all technological (or technology-based) modifications of human beings or techniques for modifying human abilities. (See the discussion of slippery slope arguments above.)

Moreover, our genes, though important, are but one of the myriad factors that result in a human being, ranging from the womb to death. Even if parents chose every one of their child's genes, there will be plenty of "unbidden" in the life of a parent. Many factors affect gene expression and transcription, beginning with the uterine environment (Kristof 2010). "Recent studies indicate that environmental factors and diet can perturb the way genes are controlled by DNA methylation and covalent histone modifications. Unexpectedly, and not unlike genetic mutations, aberrant epigenetic alterations and their phenotypic effects can sometimes be passed on to the next generation" (Feil 2006, 46). Evolution itself shows that giftedness depends crucially on the environment, both social and physical, in which those gifts are evidenced. As Dan Brock notes, "there is no reason whatever to believe such a genetic determinism to be true" (Brock 2009, 255). Thus, designer babies will not eradicate the realm of the gifted. This is not to deny that designer babies would lessen the role of "giftedness" in life: genes certainly count for something. But virtually all of technology limits the role of giftedness. Cloud seeding and crossbreeding for drought resistance diminish the farmer's perception of rain as a gift by making them less dependent upon rain. Once again, designer babies do not constitute a radical discontinuity, but rather a large step in an ongoing process.

Again, Sandel suggests that changing ourselves instead of changing the world "deadens the impulse to social and political improvement." Why is building more crutches and wheelchairs preferable to fewer congenital leg problems? Are we really likely to care less about earthquakes or political torture because we are born with better vision? Have the multitude of changes in human beings over our history, from pre-natal nutrition (non-genetic) to natural selection (genetic), seriously diminished our impulse to social and political improvement?

Perhaps Sandel could grant all of this and even agree that designer babies are not radically discontinuous with a long history of technological advances. Critics like Sandel might, at heart, feel that it is not just designer babies but the entire history of achieving mastery that is ethically dubious.[7] If so, then designer babies are in the same boat as antibiotics, airplanes, heart transplants, audio recordings, etc. For a plethora of familiar reasons, most of us prefer being on that boat, rather than swimming unaided, when crossing the vast sea of life.

It remains uncertain whether parents will one day have the opportunity to choose their offspring's genetic makeup. Should that day come, it will bring changes and challenges, as all technological advances do. It is not clear that any powerful reason exists for engineers to refrain from working on every single technology related to choosing the genes of human progeny. Specific projects must be evaluated on a case-by-case basis.

GMOs in and outside of the Food Chain

GMOs, that is, (non-human) organisms whose DNA has been modified by gene splicing, have become highly controversial. Groups such as *Say No to GMO* and *GMO Free CT* are violently opposed to their use, particularly in agriculture, while others vociferously insist GMOs are safe and beneficial. Many foods advertise themselves to be GMO-free. The European Union requires labeling of foods containing more than 0.9% GMOs (but meat or products from animals fed with GMO feed need not be labeled). In the US, Congress passed a GMO labeling law in 2016 but implementation has been slow.

In one sense, there is a continuum between gene splicing and techniques like selective breeding and hybridization that have been in use for thousands of years. However, some worries are special to or more potent in

GMOs. Some GMO opponents voice an aversion to "unnatural" foodstuffs and technologies, but the largest concern about GMO is safety and environmental disruption.

GMOs fill a broad spectrum. Some genetic modifications are relatively minor and/or relatively predictable in their effects, comparable to the genetic variation and exchange occurring in nature or through cross-breeding. Other modifications are more radical departures from what evolution has produced, and so their long-term effects, when widely disseminated in different climates and ecologies, are much harder to predict, particularly when field mutations (from background radiation, gene copying errors, etc.) are taken into effect. Not all GMOs are employed in food production. Hydrocarbon degrading bacteria, such as Alcanivorax borkumensis, are useful in bioremediation of oil spills. Genetic modification could be used to produce "supereaters" for a variety of bioremediations. *E. coli* have been modified by recombinant DNA techniques to produce a generally safe,[8] reliable, organic, relatively inexpensive supply of insulin without killing thousands of pigs. Human growth hormone, which used to be obtained from the pituitary glands of cadavers (posing a risk of neurological disease), is now produced by genetically modified yeast and bacteria. The bacterium Avastin bevacizumab, a product of genetic engineering, is used to treat several kinds of cancers. GMO techniques are employed in the production of a variety of vaccines as well. However, it is the use of GMOs in agriculture, either directly for human consumption or as livestock feed, that produces the most controversy. Genetically modified AquAdvantage Salmon was approved for human consumption by the FDA in 2015. In cattle, efforts have largely focused on dehorning and udder improvement (Eriksson et al. 2018).

Some agricultural GMOs bring clear benefits. "Genetic engineering is used in agriculture to make plants with increased yield, disease resistance, and pest resistance like Bt genes to kill selectively pests that eat crops. There have also been fruits and vegetables modified for long term storage or delayed ripening that remain fresh for a long time, which is useful also during transportation to the market," as well as GMOs with "high nutritional content and improved food quality like golden rice, plants that can tolerate high salt levels in the land or are modified so that they can grow in harsh conditions like drought" (Macer 2004). Benefits of GMOs include "Enhanced taste and quality, reduced maturation time, increased nutrients, yields, and stress tolerance, improved resistance to disease, pests, and herbicides, new products and growing techniques, increased resistance, productivity, hardiness, and feed efficiency, better yields of meat, eggs, and milk, improved animal health and diagnostic methods, 'friendly' bioherbicides and bioinsecticides, and conservation of soil, water, and energy" (Anon. nd).

Some GMOs raise genuine concerns. First, GMOs that are benign may not remain that way in the field. Natural mutations occur, for example as a result of earth's background radiation. In addition, bacteria and viruses share DNA. Thus, an organism designed to be safe may trade DNA with other organisms and wind up being unsafe, much as bacteria may become immune to antibiotics through gene trading. Transfer of antibiotic resistance markers is a concern. In addition, "studies have shown that viruses, lacking the gene needed for movement, can easily gain it from neighboring genes" (Maghari and Ardekani 2011, 112). Second, absorbing modified DNA into the body, some think, might pose health risks such as unexpected allergic reactions. "Rats exposed to transgenic potatoes or soya had abnormal young sperm; cows, goats, buffalo, pigs and other livestock grazing on Bt-maize, GM cottonseed and certain biotech corn showed complications including early deliveries, abortions, infertility and also many died. However, this is a controversial subject as studies conducted by [the] company producing the biotech crops did not show any negative effects of GM crops on mice" (Maghari and Ardekani 2011, 112). Third, GMOs may have deleterious effects on the environment.

Bt-maize, which produces crystal proteins that kill the larvae of western corn rootworm, exemplifies the complexity of the issue. On the one hand, by replacing applied pesticides, Bt-maize avoids the environmental harms produced by spraying. Bt-cotton in India, for example, has halved the use of pesticides (Jones 2011). On the other hand, Bt-maize raises potential concerns over consumers ingesting toxins derived from Bacillus Thuringiensis. (It has also been suggested and disputed that BT is harmful to Monarch Butterflies.) In addition, the use of Bt-maize exacerbates the concern that pests will develop immunity, since farmers could alternate the use of different pesticides but Bt-maize remains present during the full growing season. For similar reasons, vast fields of GMOs raise the specter of creating "superweeds."

Of course, plants that produce their own pesticides are nothing new in nature. Tobacco leaves contain nicotine, a natural insecticide. Rhubarb leaves contain oxalic acid. Brachen fern contains the carcinogen ptaquiloside. The young fronds of the plant also release hydrogen cyanide when ingested. In nature, however, such plants are found within a field containing many other species. Insects and other animals can usually simply bypass toxic plants. Agricultural practice poses the danger that pests will become immune: in a vast field of rhubarb or Bt-maize, only immune insects or pests survive. Similarly, while concerns that planting vast fields of a single variety of GMO crop will reduce biodiversity are valid, such concerns are equally pertinent to vast fields of a single variety of crop produced by cross-breeding. Sustainable agricultural practices must be followed whatever the source of the seed employed.

Where there is evidence that a particular crop may pose a danger, prudence dictates restraint. Some GMO plants contain antibiotic resistance genes (used as markers). As ACRE (The Advisory Committee on Releases to the Environment) suggests, genes that confer resistance to medical or veterinary antibiotics should not be used. Further investigation is called for in such cases.

My own view (about which intelligent people may disagree) is that it makes no more sense to be "for" or "against" GMOs than it does to be for or against drinking liquids. Drinking water is essential while drinking cyanide is deadly. One cannot generalize from one transgenic crop to all transgenic crops, much less all GMOs, any more than one can conclude that all grains are dangerous because some people are allergic to wheat. The point is that one must not paint with too broad a brush. Because some people are allergic to peanuts, it is useful to label foods that contain peanut products. Both safety and autonomy (allowing consumers to make informed choices) call for it. However, the usefulness of peanut labeling is lost if, instead, foods containing any kind of plant product simply bear the label "contains plant products." Making labels too broad defeats both autonomy and safety, because it fails to convey the information consumers need. However, legitimate concern over particular GMOs suggests that some form of screening is needed before a new GMO is released for general use. Indiscriminate and unregulated dissemination of GMOs is as unwise and potentially dangerous as automatic opposition to every GMO is foolish. Some form of labeling may become a part of the regulatory process, but "GMO" (by itself) is much too broad a label to convey much useful information.

Further problems arise because GMO seeds or organisms can migrate to other farms without the consent or even knowledge of the farm's owner. As a result, farmers of neighboring farms may lose control over whether their own harvests contain GMOs, and labeling becomes problematic. Moreover, corporate patentholders of GMO seed, to protect their patents, have sued farmers whose crop contains patented GMOs as a result of natural migration without the defendant's knowledge or control. Producers of GMO seed may use legal and economic pressure to force farmers to buy new patented seed each year instead of planting seeds reserved from last year's crop. The increased cost causes economic hardship, particularly in Lesser Developed Regions, where the increased cost can become a life-or-death issue. (See Chapter 11.) Global use of GMOs may lead to "domination of world food production by a few companies, increasing dependence on industrialized nations by developing countries, [and] biopiracy, or foreign exploitation of natural resources" (Anon. nd). These genuine concerns call for political, economic, and legal scrutiny.

Ethical engineers involved in GMO technologies should follow the debate, play a leadership role, and help inform the public, regulators, legislators, and public action groups about the engineering and scientific dimensions of these issues.

New Reproductive Technologies

New reproductive technologies, from artificial insemination using partner sperm (performed by John Hunter in the 1770s) to human cloning (not yet successfully performed as of this date), have engendered a host of objections. Some have insisted they amount to playing God, change the meaning of being human, or are an abomination against nature (sees Issues C. D. and E. above). Artificial insemination using donor sperm was claimed to constitute adultery, since sperm from someone other than the husband was placed in the vagina. Cloning was said to violate the offspring's right to "be a surprise" (Putnam 1999) or be genetically unique, or that incalculable psychological harm would result from the offspring knowing it was not genetically unique. Presumably, Putnam must think identical twins suffer incalculable harm and rights violations. (In fact, identical twins are examples of human cloning occurring naturally.) Tiefel (1982) argues that using reproductive technologies is unethical since the risks are not definitively known and the offspring cannot give consent. Edwards (1974) notes that the same is true of giving pre-natal vitamins to pregnant women. In addition, if a parent can give proxy consent for a neonate, why not for prospective offspring? Leon Kass insists, without a hint of evidence, that parents engaging in cloning would be unable to accept any differences from themselves in the offspring, and so would "inevitably" brainwash them to the point where it was questionable to call them moral agents (Kass 2001). Reproductive human cloning has been banned in many countries, including France, Germany, Canada, several states in the US, Japan, and the UK. The General Assembly of the UN adopted a non-binding resolution urging the banning of all forms of human cloning. Some of the antipathy toward cloning comes from a misunderstanding of what (reproductive) cloning is. DNA is removed from an egg cell. Donor DNA (e.g., from a skin cell) is inserted into the egg (or an electric current is used to fuse the egg and the donor DNA). The egg is then transplanted (immediately or after developing into an early stage embryo in vitro) into the uterus of an adult female, where it grows to term like a normal baby. The result is a perfectly normal baby, indistinguishable from any other, naturally produced baby. Clones do not have the memories of the DNA donor and they are not robotic automatons. Reproductive cloning should be distinguished from therapeutic cloning, which does not produce a baby but rather creates stem cells with the same DNA as the donor. Objections to therapeutic cloning tend either to focus on the destruction of an embryo or to mistakenly conflate therapeutic and reproductive cloning. Note that therapeutic cloning used to produce a baby for the purpose of harvesting its organs raises different quite issues about the legitimate treatment of babies, whether cloned or born naturally. That is, the issues involved are the same whether the baby whose organs are harvested were produced by cloning or natural childbirth.

While reproductive technologies remain controversial, it is hard to find arguments against them that stand up to critical scrutiny. Historically, each new development was greeted with alarm and moral handwringing, including techniques now largely accepted, such as artificial insemination with husband donor sperm.[9] Of course, safety and other concerns apply here as much as they do in any other form of medical technology. Like all medical technologies, from pills to organ transplants, human reproductive cloning should not be widely practiced until the technology is properly developed and shown to be safe. In general, it is hard to find sound reasons for singling out reproductive technologies as fundamentally different from other medical technologies.[10] For a detailed discussion of this issue, see Schlossberger (2008).

Biotechnology to Mitigate Disability or Enhance Human Beings, Brain Research, and Psychological Control

Dealing with disabilities is a complex matter. Even the term "disability" is controversial. For example, some members of the deaf community on

Martha's Island, where there is a vibrant community and a rich way of life in which sign language is the norm, regard their way of life as simply different from hearing communities and in no way deficient. For them, "correcting" hearing with cochlear implants or gene modification is insulting and discriminatory, akin to offering to "fix" an African American's skin color by making it white. [11] Lee (2016) notes that some in the deaf community have argued that "[cochlear implant] technology represented an attack on the Deaf culture since it aimed to ensure that deaf children grow up to use spoken languages instead of the signed languages of the Deaf. This inevitably decreases the population of the community which communicates using signing." As Ashley Shew points out, "not every person who is disabled needs your help. The way in which people go about turning disabled people into service projects can be really dehumanizing and patronizing" (Crane 2020). Some advocate for some forms of barrier removal, such as ramps, mechanical limbs, crutches, etc., but object to gene modification to regrow limbs or fetal genetic engineering to prevent missing or underdeveloped limbs. Some find exoskeletons objectionable. There are a wide variety of positions on these topics and much debate. (See, for example, Barnes 2016; Glover 2006; and Shew 2022.) Engineers may ethically work on technologies offsetting differences in ability from the norm for those who do want such technologies, but sensitivity and understanding of those who do not is imperative.

Biochemical and electrical control of the brain is still in its infancy. Controversy and uncertainty surround the extent to which biotechnologies, whether chemical or electrical, can control addiction, depression, violence, memory retention, and a host of other psychological features. Because so much remains unknown, detailed discussion of the ethics of such technologies is premature. Possible unethical applications are unknown and distant, while the quest for knowledge and understanding is an important and worthwhile human enterprise. Thus, at present, it would seem that engineers may ethically work in the area of brain research and technologies, as long as limitations required by other ethical concerns (such as safety and respect for autonomy) are respected. As the field develops, however, more and more ethical questions will emerge that engineers must address.

Genetic Testing

Another area in which ethical issues arise is the use and dissemination of individuals' genetic information. A person's genetic makeup may be used to predict a higher risk of a variety of diseases. That information, in turn, might be used for preventive care (such as Angelina Jolie's pre-emptive radical mastectomy because the BRCA gene was present), innovative treatments for cancer patients, raising or lowering insurance rates, or even in making hiring decisions. Genetic information might be used in lawsuits, for example, when deciding whether a plaintiff's condition was genetically or work induced. Most controversially, genetic makeup has been thought to be a predictor of a higher likelihood of mental illness and/or anti-social behavior, including rape and murder. On that basis, individuals could be targeted for surveillance, treatment, or even pre-emptive arrest.

Some argue that differential treatment on the basis of a person's genetic makeup is unjustified genetic discrimination. "At root,' says Carol Isaacson Barach (2008, 47), "is the ethical question of when (if ever) biology should provide a basis for differential treatment." Opponents object to differential treatment on the basis of genetic information for one or more of three reasons:

a. Genes do not decree destiny (nature v. nurture). Is it fair, they ask, to base differential treatment on merely statistical information? In answering this question, one might consider racial profiling in police work, which is, at present, considered generally unethical by a majority of courts and ethicists. (For discussions of race and police discretion, see Kleinig 1996.)

b. People are not responsible for their genetic makeup. Even when a genetic marker is a 100% reliable predictor, what constitutes a legitimate basis for differential treatment? For example, should people with congenital diseases that are expensive to treat pay more for health insurance? Is it legitimate for employers not to hire them on that basis? (For more on this question, see Schlossberger 2006.)

c. Genetic testing is unduly invasive and/or violates privacy. Two questions are at issue. The first question concerns the invasiveness of the testing itself – is it too invasive to require genetic testing? For example, does an employer overstep the bounds of what may reasonably be demanded by requiring job applicants to undergo genetic testing? The second question concerns what is done with genetic information once it is obtained. Is the information private, even when others could legitimately use that information if they had it? For example, there is robust debate about how scarce medical resources should be appropriated, but one factor some see as relevant is the likelihood the recipient will survive or be able to benefit. For example, if the need for liver transplants greatly outstrips the number of available livers, it is at least arguable that a scarce liver should not go to a patient unlikely to survive the transplant for more than a short time. Genetic test results might be helpful in evaluating the odds that a particular patient will survive the treatment for a given length of time (e.g., five years). Similarly, it is at least arguable that, other things being equal, preference in allocating scarce treatments should go patients who are more likely to benefit from the treatment. Genetic test results might be helpful in determining that. Would privacy be violated if applications for scarce medical resources required genetic test information? (See Chapter 13 for more on privacy.)

These are issues that engineers working on projects related to genetic testing must confront. Again, individual projects must be evaluated on a case-by-case basis.

FURTHER ISSUES

Many other areas of medical and bioengineering have or will raise concerns. Bowen (2014) mentions "the move to reduce the number of patient visits to clinics, and to reduce the number and duration of patient stays in hospital, by increasing the availability of monitoring and treatment in the home" (Bowen 2014, 54). A "device may use a computer, with monitoring and feedback, to offer encouragement to the patient to carry out the repetitive exercises that physical rehabilitation requires. Thus, the device may be designed to fulfil some of the motivational role previously carried out by the healthcare staff" (Bowen 2014, 55). Computer-assisted surgery can help avoid mistakes but carries high costs in obtaining and maintaining the equipment. Telehealthcare and monitoring, such as GPS monitoring of dementia patients, provides access to the homebound or remotely located, increases patients' and providers' ability to monitor, and can give patients a more active role. However, these technologies raise concerns about altering patient-provider relationships as well as about those with little training correctly using sophisticated equipment. In Japan, robot assistants used by disabled and/or socially isolated individuals can reduce loneliness. Robot pets can replace some of the functions of a pet for those unable to provide proper pet care. Are these deceptive, since their value is in part replacing real interactions and making patients feel that the robots are "real" persons or pets? (Bowen 2014).

While fMRI (functioning magnetic resonance imaging) of the brain has proven useful in brain research, the possibility exists that, with significant advances in imaging and our understanding of brain physiology, fMRI might be able to reveal personal and potentially embarrassing information about the subject. To what extent should this possibility concern and constrain current work on fMRI and related technology?

SUMMARY

Bioengineering raises a wide spectrum of complex ethical issues for engineers and society. Ethical considerations range from difficult and sophisticated philosophical questions (such as the difference between treatment and enhancement) to factors, such as safety and the environment, that apply to a wide range of ethical issues in engineering (and beyond). Engineers working on projects with bioengineering application must recognize which ethical issues apply to their work and use the five-step process to make particular decisions.

Case 64: Genetic Engineering of Mosquitos

The Aedes aegypti mosquito is a vector for several serious diseases, including Dengue, Zika, and Yellow Fever. Males detect females via wing beat frequency vibrations. Mating typically occurs near a blood host (such as a human being). Males perform a dance, during which the male and female modulate their frequencies in response to each other, sometimes reaching harmonic convergence, which is believed to enhance the male's chances of reproductive success. Males attempt to catch a female for mating. Females sometimes evade capture by flight maneuvers or by kicking the male away. Company K is contemplating a plan to use CRISPR to modify the genes of Aedes males, making them sterile but also more sensitive to female frequencies and more adept at achieving harmonic convergence, resulting in increased numbers of females copulating with sterile males. It is anticipated that the program will vastly reduce or eliminate the local population of

Aedes. Incidence of disease is expected to decrease, but the effects on the ecosystem are not well understood. Aedes predators include tadpoles, fish, other mosquitos (who eat the larvae), and the carnivorous plant Utricularia macrorhiza. There are other possible methods of mosquito control, all of which have either limitations or undesirable consequences, including pesticide spraying, introduction of predators, and sterilization by radiation. Should Company K proceed with the plan, and, if so, what precautions should they take and what should they monitor?

Case 65: Vaginal Device for Urinary Incontinence Exercises

Urinary incontinence is a common problem in women of all ages. One meta-study found that in the vicinity of 40% of women over age 70 worldwide are afflicted (Batmani et al. 2021). Pelvic floor exercises, when consistently and correctly performed, help treat or prevent female urinary incontinence. Company I is developing an unobtrusive vaginal device to be surgically implanted that monitors muscle activity continuously and sends the results via smartphone to a database that can be accessed by the patient's medical team and correlated with a patient and provider-maintained diary of leakage episodes. This allows the team to monitor progress and suggest new exercises. The information can also be incorporated in a database for research purposes. Because the monitoring is continuous, it also reveals readily inferred personal information about the patient, such as when and for how long the patient is having sex. Should Company I proceed with developing and manufacturing the device, and, if so, what precautions or adjustments should it make? (For instance, an information sheet can be printed and given to the patient, before the patient agrees to implantation, detailing what kind of knowledge the medical team will be able to infer from the information it receives.)

NOTES

1. Compounds currently touted as nootropics, such as Aniracetam, have not (yet) been proven either effective or safe.
2. For example, Glover (2006, 36) writes, "making some enhancements may add to flourishing as much as eliminating some disabilities....if what we care about is really

not disability but flourishing, the medical boundary may be impossible to defend." Harris (2010), Bostrom and Roache (2007), and Miah (2010) argue convincingly that the distinction between therapy and enhancement cannot be non-arbitrarily drawn in a way that bears significant moral weight. Kamm (2010) criticizes the attempt made in Schwartz 2005 to define dysfunction as falling too far below the mean.

3. Georg Cantor proved that an infinitely big set of n members has the same cardinality as a set of n times 2 members.

4. Cf. also Schermer (2007): "although it is difficult to make a clear-cut distinction between treatment (of disease) on the one hand, and enhancement (of normal functioning) on the other from a philosophical point of view, this counter-positioning of therapy to enhancement is clearly at work in medical and social practice and in policy debates."

5. This idea is somewhat complicated by the fact that *homo sapiens*, from the beginning, existed in social groups. The state of nature test sets aside purely social disadvantages due to transient preferences. The ability to speak and thus establish relationships with others counts, even though it is social, while having a socially preferred body shape does not (since preferences for body shape can readily change).

6. This is, of course, my own view, with which some philosophers and ethicists will disagree. For a general philosophical discussion of human enhancement, see, in addition to the sources cited above, Juengst and Moseley (2016).

7. Sandel does try to draw various distinctions between permissible and impermissible improvements on nature, including the distinction between enhancement and treatment discussed earlier. But, if these fail (and it must be said he does not think they do), perhaps he could fall back on the claim that what is morally objectionable is something of which designer babies is merely an instance, namely, the human disposition to master (the hubris involved in trying to control and mold nature). After all, Sandel does speak (on page 93) of the sanctity of nature (which, he takes, can be either secularly or religiously based). Perhaps, at heart (although he does not say this), Sandel's criticism is simply another form of anti-technologism.

8. Some users have reported adverse reactions from bacteria-created insulin but insulin from pigs also posed the danger of contamination and was less effective.

9. "The moral and social implications of artificial insemination were debated in both the medical and popular press in the United States since 1909, [while] in Europe the debate started in the 1940s. The Catholic Church objected to all forms of artificial insemination, saying that it promoted the vice of onanism and ignored the religious importance of coitus. The main criticism was that artificial insemination with donor semen was a form of adultery promoting the vice of masturbation. Other critics were concerned that AID could encourage eugenic government policies" (Ombelet and Van Robays 2015).

10. For example, as noted above, legitimate concerns about access and social justice apply equally to education and cosmetics. Concerns about safety apply to all medical interventions, from surgery to pharmaceuticals, as well as nonmedical or biological technologies, such as fracking.

11. Against those who claim that it is rude or insulting to suggest that deaf people may be missing something of value, musicians might claim that it is rude or insulting to suggest that someone who can't hear music isn't missing anything of value.

REFERENCES

Aach, J., Lunshof, J., Iyer, E. and Church, G. (2017) "Addressing the Ethical Issues Raised by Synthetic Human Entities with Embryo-Like Features," *eLife* 6, e20674 https://elifesciences.org/articles/20674

Anon. (nd) "GM Products: Benefits and Controversies," (information from the Human Genome Project) available at http://osarahjayneo.tripod.com/id7.html

Barach, Carol Isaacson (2008) *Just Genes: The Ethics of Genetic Technologies* (Westport, CT; London: Praeger).

Barnes, Elizabeth (2016) *The Minority Body: A Theory of Disability* (Oxford University Press)

Batmani, Sedighe, Rostam, Jalai, Mohammadi, Masoud and Bokaee, Shadi (2021) "Prevalence and Factors Related to Urinary Incontinence in Older Adults Women Worldwide: A Comprehensive Systematic Review and Meta-Analysis of Observational Studies," *BMC Geriatric* 21:1, 212 https://pubmed.ncbi.nlm.nih.gov/33781236/

Bostrom, Nick and Roache, Rebecca (2007) "Ethical Issues in Human Enhancement," in Jesper Ryberg et al., eds., *New Waves in Applied Ethics* (Palgrave Macmillan),120–152.

Bowen, W. Richard (2014) *Engineering Ethics: Challenges and Opportunities* (Springer).

Brock, D.W. (2009) "Is Selection of Children Wrong?" in Julian Savulescu and Nick Brostrum, eds., *Human Enhancement* (Oxford: Oxford University Press), 251–276.

Caplan, Arthur L. (2004) "Is Biomedical Research Too Dangerous to Pursue?" *Science Magazine* 303:5661, 1142

Cohen, Marshall, Nagel, Thomas, and Scanlon, Thomas eds. (1974) *The Rights and Wrongs of Abortion* (Princeton: Princeton University Press).

Colleton, Laura (2008) "The Elusive Line between Enhancement and Therapy and Its Effects on Health Care in the U.S." *Journal of Evolution and Technology.* 18:1, 70–78 http://jetpress.org/v18/colleton.htm.

Crane, Brent (2020) "How Big Tech Gets Disability Wrong," https://folks.pillpack.com/how-big-tech-gets-disability-wrong/

Daniels, Norman (1985) *Just Health Care* (Cambridge, UK: Cambridge University Press).

Edwards, R.G. (1974) "Fertilization of Human Eggs in Vitro: Morals, Ethics, and the Law," *Quarterly Review of Biology* 40:3, 3–26.

Eriksson, S., Jonas, E., Rydhmer, L. and Rocklinsberg, H. (2018) "Breeding and Ethical Perspectives on Genetically Modified and Genome Edited Cattle," *Journal of Dairy Science* 101:1, 1–17.

Evers, Matthias and Chui, Michael (2020) "The Bio Revolution Is Changing Business and Society," *McKinsey Global Institute* https://www.mckinsey.com/mgi/overview/in-the-news/the-bio-revolution-is-changing-business-and-society

Feil, Robert (2006) "Environmental and Nutritional Effects on the Epigenetic Regulation of Genes," *Mutation Research* 600:1–2, 46–57.

Garber, Megan (2012) "The IVF Panic: 'All Hell Will Break Loose, Politically and Morally, All Over the World'," *The Atlantic* June 25, 2012 https://www.theatlantic.com/technology/archive/2012/06/the-ivf-panic-all-hell-will-break-loose-politically-and-morally-all-over-the-world/258954/

Glover, Jonathan (2006) *Choosing Children: Genes, Disability, and Design* (Oxford: Oxford University Press).

Gordon, John-Stewart (2008) "Abortion" *Internet Encyclopedia of Philosophy* https://iep.utm.edu/abortion/

Harris, John (2010) "Enhancements Are a Moral Obligation," in Julian Savulescu and Nick Bostrom, eds., *Human Enhancement* (Oxford University Press), 131–154.

Hyun, Insoo (2020) "Towards a New Bioengineering Ethics," *Harvard Medical School Center for Bioethics* https://bioethics.hms.harvard.edu/journal/stem-cell-ethics

Johnston, Josephine (2020) "Shaping the CRISPR Gene-Editing Debate: Questions about Enhancement and Germline Modification," *Perspectives in Biology and Medicine* 63:1, 141–154.

Jones, J.D.G. (2011) "Why Genetically Modified Crops" *Philosophical Transactions of the Royal Society A* 369:1942, 1807–1816.

Juengst, Eric and Moseley, Daniel (2016) "Human Enhancement," *The Stanford Encyclopedia of Philosophy* https://plato.stanford.edu/archives/spr2016/entries/enhancement/

Kamm, Frances (2010) "What Is and Is Not Wrong with Enhancement," in Julian Savulescu and Nick Bostrom, eds., *Human Enhancement* (Oxford University Press), 91–130.

Kass, Leon. (2001) "The Wisdom of Repugnance: Why We Should Ban the Cloning of Humans," in Michael C. Brannigan, ed., *Ethical Issues in Human Cloning* (New York; London: Seven Bridges Books), 43–68.

Kleinig, John, ed. (1996) *Handled with Discretion: Ethical Issues in Police Decision Making* (Lanham, MD: Rowman and Littlefield).)

Kristof, Nicholas (2010) "At Risk from the Womb," *The New York Times* October 2, 2010 http://www.nytimes.com/2010/10/03/opinion/03kristof.html?_r=2&scp=1&sq=%2b%22American+Medical+Association%22&st=nyt

Lee, Joseph (2016) "Cochlear Implantation, Enhancements, Transhumanism and Posthumanism: Some Human Questions," *Science and Engineering Ethics* 22, 67–92.

Macer, D.R.J. (2004) *Bioethics Education for Informed Citizens across Cultures* (Eubios Ethics Institute) www.eubios.info/BetCD/Bet10.doc

Maghari, B. M. and Ardekani, A. M. (2011) "Genetically Modified Food and Social Concerns," *Avicenna Journal of Medical Biotechnology*, 3:3, 109–117.

Mallapaty, Smiriti (2022) "How to Protect the First 'CRISPR Babies' Prompts Ethical Debate," *Nature* https://www.nature.com/articles/d41586-022-00512-w

Miah, Andy (2010) "Towards the Transhuman Athlete: Therapy, Nontherapy and Enhancement," *Sport in Society* 13:2, 221–233.

Ombelet, W. and Van Robays, J. (2015) "Artificial Insemination History: Hurdles and Milestones," *Facts, Views, and Visions in ObGyn* 7:2, 137–143 https://www.ncbi.nlm.nih.gov/pmc/articles/PMC4498171/

Patuzzo, S., Goracci, G., Gasperini, L. and Ciliberti, R. (2018) "3D Bioprinting Technology: Scientific Aspects and Ethical Issues," *Science and Engineering Ethics* 24, 335–348.

Persson, Ingmar and Savulescu, Julian (2014) *Unfit for the Future: The Need for Moral Enhancement* (Oxford: Oxford University Press).

Putnam, Hilary (1999) "Cloning People," in Justine Burley, ed., *The Genetic Revolution and Human Rights* (Oxford: Oxford University Press), 1–13.

Sandel, Michael J. (2004) "The Case Against Perfection," *The Atlantic* April, 2004. http://www.theatlantic.com/magazine/archive/2004/04/the-case-against-perfection/302927/

Sandel, Michael J. (2007) *The Case Against Perfection* (Cambridge, MA: Harvard University Press).

Savulescu, Julian (2009) "The Moral Obligation to Create Children with the Best Chance of the Best Life," *Bioethics* 23:5, 274–290.

Schermer, Maartje (2007) "The Dynamics of the Treatment-Enhancement Distinction: ADHD as a Case Study," *Philosophica* 79, 25–37.

Schlossberger, Eugene (2006) "Setting Premiums Ethically: Seven Models for Distributing Risk Costs," *International Journal of Applied Philosophy* 20:2, 331–337.

Schlossberger, Eugene (2008) *A Holistic Approach to Rights: Affirmative Action, Reproductive Rights, Censorship, and Future Generations* (Lanford, MD: University Press of America).

Schwartz, Peter (2005) "Defending the Distinction between Treatment and Enhancement," *American Journal of Bioethics* 5:3, 17–19.

Shew, Ashley (2022) "The Minded Body in Technology and Disability," in Shannon Vallor, ed., *The Oxford Handbook of Philosophy of Technology* (Oxford University Press).

Silver, Lee. M. (2001) "Cloning, Ethics, and Religion," in Wanda Teays, and Laura Purdy, eds., *Bioethics, Justice, and Health Care* (Belmont, CA: Wadsworth), 656–660.

The President's Council on Bioethics (2002) "Staff Working Paper: Distinguishing Therapy and Enhancement," https://bioethicsarchive.georgetown.edu/pcbe/background/workpaper7.html

Tiefel, H.O. (1982) "Human in Vitro Fertilization: A Conservative View," *Journal of the American Medical Association* 247, 3235–3242.

World Health Organization (2006) *Constitution of the World Health Organization*, 45th ed., Supplement www.who.int/governance/eb/who_constitution_en.pdf

13

Information Technology, Artificial Intelligence, and Software Engineering

OVERVIEW

Information technology is a broad field encompassing many aspects of engineering, including software design and testing, system design and maintenance, data management, telecommunication design, implementation, and maintenance, artificial intelligence (AI), and much more. "Software engineering is an occupation that builds practical skills based on the theoretical foundations of computer science together with practical expertise in the engineering methodologies aimed at the development of software systems" (Lurie & Mark 2016, 423). AI and machine learning raise many sorts of issues. Some particular applications, such as facial recognition technology (FRT) and autonomous vehicles, are discussed below. Replacement of human workers by intelligent machines is another concern, as are issues of social justice: bias, inequality, and the digital divide (see below). Other concerns voiced about AI, but not directly addressed here, include lack of human compassion and understanding in decision-making and increasing control by and dependence on intelligent networks. Some worry about a point of "singularity," when AI is self-aware, more intelligent than human beings, and irreversibly powerful. Although current technology has not yet reached this point, philosophers and scientists ponder the appropriate status of self-aware robots and AI systems. Do they, for instance, merit rights?

While projects in information technology and related fields involve most of the ethical concerns and factors found in other areas of engineering, there are some characteristics of information technology that either exacerbate those ethical problems or create new ones.

DOI: 10.1201/9781003242574-16

Characteristics of Information Technology
Producing Special Ethical Concerns

Five distinctive features of information technology deserve special notice: (a) processing speed and computational power; (b) data capacity; (c) connectivity of materials and locations/devices, as well as widespread access on both sides (user and provider); (d) reproducibility; and (e) shortened lifecycle and rapidity of technological change. Together, they create four characteristics that engender special problems:

- *Malleability and Capacity for fabrication*: Fabricating a painting or physical document leaves traces and forgers often need special access or materials. A convincing forgery of a Rembrandt requires exceptional painting skill, historical knowledge, and, ideally, 17th-century materials (paint, canvas). By way of contrast, program or technology modification, photoshopping, site imitation, hacking, etc., can be done remotely, can be difficult to detect or trace, and require only widely available tools.
- *Virality* (fast proliferation and decisions and processes rapidly spin out of control of originators): A single post by an ordinary individual (even a six-year-old child) can be reproduced millions of times across the globe in a very short time. A program can be copied, modified, or made malicious and employed internationally: once the program is available, the original creators often have little control. Once the genie is out of the bottle, it cannot be put back.
- *Nonlocality*: Because the internet seamlessly connects users and providers across the globe, regulation becomes problematic. Collecting tax and banning harmful or fraudulent practices, for example, raise problems when traffic jumps across legal or political boundaries such as nations or states. How should an international platform or provider like Facebook respond if, for example, Poland were to make it a crime to display images of breastfeeding? What is unproblematic in one cultural or environmental setting may be problematic in another, yet the same content and services go everywhere.
- *Widespread integration*: Information technologies are both widely connected with each other and integrated into a wide variety of other systems, such as power grids, medical care systems, financial systems, factory operating systems, and smart homes. As a result, a small problem in an information technology system might have widespread, unforeseeable, and disastrous results.

List of Issues in Information Technology

Ethical issues in information technology fall into three groups:

- *Issues about software and technology development*
  Issues about rushing to market and cutting costs; Issues about results of failure and unintended consequences; Issues about Dual Use; and Issues about intellectual property/copyright/patents/trade secrets.
- *Issues about content (data)*
  Bias; Issues about privacy and security (e.g., FRT); Hosts' responsibility for user/client code and content.
- *Information technology and society*
  AI dominance; Autonomous technologies; and Job displacement and the digital divide.

Three General Reminders (Value-Sensitive Design, Broader Context, Anti-Patterns)

Engineers should, in general, embrace value-sensitive design (VSD), which "is a theoretically grounded approach to the design of technology that accounts for human values in a principled and comprehensive manner throughout the design process" (Friedman et al., 2006). As Nissenbaum (2005, 62) states, "Ethical and political values ought to be added to traditional considerations and constraints guiding the design and regulation of [technical] systems, such as functionality, efficiency, elegance, safety, and, more recently, usability." VSD "also seeks procedures to incorporate and balance the values of different stakeholders in the design process" (Brey and Søraker 2009). It involves, says Friedman (2008), philosophical, empirical, and technical questions. When considering philosophical analysis of relevant concepts, engineers should ask:

How are values supported or diminished by particular technological designs? Who is affected? How should we engage in trade-offs among competing values in the design, implementation, and use of information systems….Empirical investigations involve both social-scientific research on the understandings, contexts, and experiences of the people affected by the technological designs as well as the development of relevant laws, policies, and regulations. Technical investigations involve analyzing current technical mechanisms and designs to assess how well they support particular values, and, conversely, identifying values, and then identifying and/or developing technical mechanisms and designs that can support those values.

Friedman 2008

Lurie and Mark (2016) point out that program designers and developers must also consider ethical questions about the broader context of use. For example, when designing software packages for health care providers, engineers should ask "to what information about the availability and costs of treatments should the health provider have access when designing a treatment protocol for a patent?" (p. 425). Even if this is a decision made by others, the engineer can raise the issue with decision-makers. "Software engineers [should] attempt to understand and match the needs of the client with the best technical response" (Lurie and Mark 2016, 428). The role of ethics continues through the design, development, and maintenance stages. "There are three types of maintenance: corrective, perfective and adaptive. At each phase, there are fundamental ethical considerations that must be taken into account" (Lurie and Mark 2016, 429).

Another thing engineers should watch out for are "anti-patterns," that is, common bad practices that should be avoided: Shannon Vallor and Arvind Narayanan have identified several examples relevant to software engineering:

• Analysis Paralysis (a risk-averse pattern of unending analysis/discussion that never moves forward to the actual decision phase) • Blind Coding/ Blind Faith (implementing a bug fix or subroutine without ever actually testing it) • Boat Anchor (a totally useless piece of software or hardware that you nevertheless keep in your design, often due to its initial cost) • Bystander Apathy (everyone can see impending disaster, but no one is motivated to do anything about it) • Cut and Paste Programming (largely self-explanatory; reusing/cloning code) • Death March (a project everyone knows is doomed to fail but is ordered to keep working on anyway) • Design by Committee (no intelligent vision guiding and unifying the project) • Error Hiding (overriding the display of error messages with exception handling, so neither the user nor tech support can actually see them) • Escalation of Commitment (adopting a misguided goal or poor strategy, then completely refusing to rethink it despite clear evidence of its failure) • Improbability Factor (refusing to expend resources on a fix on the assumption that the problem is unlikely to actually occur in use) • Input Kludge (failure of a software program to anticipate and handle invalid or incorrect user input) • Gold Plating ('gilding the lily'; continuing to develop a product or adding features that offer insufficient value to justify the effort)

• Golden Hammer (using the same favored solution at every opportunity, whether it's the appropriate tool for the job or not) • Lava Flow (old code, often undocumented and with its function poorly understood, therefore left in) • Moral Hazard (insulating decision makers from the risks/negative consequences of their decisions) • Mushroom Management (keeping non-management employees in the dark, information-starved) • Software Bloat (successive iterations of software using more and more memory/power/ other resources with little or no added functionality) • Spaghetti Code/Big Ball of Mud (unstructured, messy code not easily modified or extended.

Vallor and Narayanan 2015, 58

ISSUES ABOUT SOFTWARE AND TECHNOLOGY DEVELOPMENT

Rushing to Market and Cutting Costs

Berenbach and Broy (2009) identify nine specific problematic situations that might arise under this heading,[1] including agreeing to schedules that are impossible to meet, making agreements whose terms are not clearly understood, delivering incomplete product with known defects or missing functionality, promising "fictionware" (some of the features are infeasible) or "vaporware" (a product that does not exist), shoddy workmanship or missing documentation, and ignoring best practices. For example, Berenbach and Broy (2009) write, "error codes might be hard-coded in the software rather than placed in a table," making them easy to overlook when making revisions or updates. Such problems may occur because of inadequate review or failure to listen to team members' advice or objections, because problems are swept under the rug, and because a team member or manager shifts the burden onto others. For example, a project manager might lie about progress during meetings with clients or management ("red lies"), putting the burden on project teams to make up the difference.

These problems are driven and made worse by the rapid pace of technological change. "The shortened lifecycle has weakened and in some cases obliterated software review by management and legal teams….The cost-cutting imperative often leaves little room for user studies or consultations with experts that would allow software development firms to acquire this familiarity" (Vallor and Narayanan 2015).

Results of Failure and Unintended Consequences

Software failure is both costly and widespread. Software Fail Watch (2017) reported over 600 software fail stories in 2017, mentioning 274 companies, affecting 3.7 billion individuals, and resulting in $1.7 billion in losses and 268 years of person hours lost.

A software failure nearly resulted in the end of civilization. In 1983, a bug in the Soviet Union's early warning software caused the system to announce (at the highest level of reliability) that the United States had launched five ballistic missiles. Fortunately, the duty officer, Lt. Colonel Stanislav Petrov, reasoned that a real attack would involve more than five missiles. Petrov dismissed the warning as a false alarm, chose to disregard his instructions, and did not relay the warning to his superiors, thus averting World War III. *The more catastrophic the result of failure, the more important it is to have a human monitor who is able to override.* In 2010, an international incident occurred when Nicaraguan troops accidentally invaded Costa Rica. The soldiers were sent to an island (Isla Calera in the San Juan River, constituting part of the border with Costa Rica) because Google Maps, mistakenly, marked the island within Nicaragua. Even after the error was exposed, Nicaragua refused to remove the troops because it decided the island should be part of Nicaragua (Mackay 2010). Other software failures resulted in actual deaths. In 2001, software developed by Multidata Systems International Corp. delivered a radiation overexposure (20% to 100% over the prescribed amount) to 28 patients at the Panama National Institute of Oncology. Doctors at the Institute found digitizing each radiation block time consuming, so they treated up to five blocks as a single continuous block. "The Multidata system did not alert the user that an improper data sequence had been entered" (Borras 2006, 177). At least five deaths resulted, according to the FDA. Canada's Therac-25 was a computer-controlled medical linear accelerator designed to give cancer patients radiation therapy. Hardware interlocks to prevent unsafe errors in previous versions were eliminated from the Therac-25 in favor of software control. Poorly designed error messaging and a race condition (the failure of the turntable to reposition correctly when the technician switched from x-ray to electron beam within 8 seconds) permitted technicians, in at least six incidents between 1985 and 1987, accidentally to fire a high-power electron beam without proper patient shielding or unknowingly to subject patients to multiple exposures. At least three of the patients died. Significantly, the race condition was missed during testing because operators require practice before they can operate the equipment quickly enough to trigger the issue (Leveson 2017; Leveson and Turner 1993). *When feasible, "on the ground" testing should supplement lab testing.* In 2003, an Ohio power line brushed against trees and shut down. A bug in the XA/21 alarm and monitoring system prevented an alarm from registering in FirstEnergy Corporation's control room, so that operators, unaware they were looking at outdated data, failed to take the appropriate steps to contain the problem: when three other lines switched off, over-taxing other lines, system operators did not realize what was happening. A cascading series of issues, including the fact that failure to employ dynamic mapping and data sharing meant that operators lacked knowledge of systems outside their control, produced the 2003 Northeast US Blackout, affecting 55 million people. In New York City, 800,000 people were trapped in the subway. "The event contributed to at least 11 deaths and cost an estimated $6 billion" (Minkel 2008). *In complex systems, it is beneficial to have operators be aware of what is happening in other parts of the system.* In August 2012, Knight Capital's trading algorithm lost $440 million dollars because the automated software market maker, which is supposed to buy at bid and sell at offer, instead bought at offer and sold at bid, 2,400 times a minute for 45 minutes. Knight rushed the software to activation in roughly a month and a half, in order to be ready for the debut of the newly approved Retail Liquidity Program. Knight updated the program (SMARS) that receives orders and breaks them into smaller packages for execution. The update re-utilized a flag that, originally, initiated code (Power Peg) that had not been used in eight years. According to a 2013 SEC filing (Release # 70694), "During the deployment of the new code, however, one of Knight's technicians did not copy the new code to one of the eight SMARS computer servers. Knight did not have a second technician review this deployment and no one at Knight realized that the Power Peg code had not been removed from the eighth server, nor the new RLP code added. Knight had no written procedures that required such a review" (Seven 2014). While the first seven servers functioned properly, "Orders sent to the eighth server triggered the supposable [sic] repurposed flag and brought back from the dead the old Power Peg code" (Seven 2014). The 1990 AT&T phone service outage, which prevented 60,000 customers from making long distance calls, occurred when a new software installation malfunctioned. All 114 4ESS switches kept rebooting when ATT sped up the communication process (see Neumann 1990 and 1994). *Reusing outdated code is problematic. When unavoidable, written policies requiring careful review are essential.* The explosion of the first Ariane 5 rocket in 1996 occurred because the inertial reference system converted 64-bit floating point data into 16-bit signed integer data, resulting in an arithmetic overflow. Software with the ability to deal with the problem had been disabled (see Lions 1996). In 1998, the $193 million-dollar Mars Climate Orbiter was lost because the team working on the thrusters used the English unit pound-force-seconds, while the others used the metric unit Newton-seconds (see Sawyer 1999). *Review and co-ordination of all elements of a project and all teams working on the project are required.*

Unintended consequences of fully functional software employed more or less as intended can be more significant than one might think. A Google Doodle in celebration of Les Paul's 96th birthday led Google.com's 740 million daily unique visitors to spend (extrapolating from RescueTime's time management software) an average of 26 seconds more on site, which, over two days, comes to 10.7 million hours wasted (RescueTime 2011).

In general, Stewart Baird identifies six major causes of project failure: project objectives not fully specified, bad planning and estimating, technology new to the organization, inadequate or no project management methodology, insufficient senior staff on the team, and poor performance by suppliers of hardware and/or software (Garrett and Lewis ND). Ammanath (2022) encourages red teaming – intentionally trying to make a system in development fail, to explore its weaknesses and shortcomings.

In sum:

- Consider and anticipate potential failures and unintended consequences before development and review periodically during development, before deployment, and after release.
- The more catastrophic the result of failure, the more important it is to have a human monitor who is able to override.
- In complex systems, it is beneficial to have operators be aware of what is happening in other parts of the system.
- When planning and devising systems, use red teams and fault-tolerant programming to minimize the chance of failure and implement strategies and procedures to minimize harm and facilitate recovery when failures do occur. Include senior staff in the team and make sure project objectives, constraints, management methodology, cost estimates, and timetables are clear and well-thought out.
- Oversee suppliers and sub-contractors thoroughly.

- Reusing outdated code is problematic. When unavoidable, written policies requiring careful review are essential.
- Review and co-ordination of all elements of a project and all teams working on the project are required.
- When feasible, "on the ground" testing should supplement lab testing. Separate Failure Mode Effects Analysis (FMEA) teams or units should be employed.
- Conduct retrospective or after-action-reviews after development but before release.

Dual Use

Informational technology presents particularly troubling examples of dual-use (technologies that may be harmfully used in ways not intended by the creator). For example, the term "dual-use" is sometimes used to mean technology that has both a civilian and a military application. Microsoft employees objected when Microsoft planned to provide its HoloLens technology (an augmented reality device) to the army's Project Maven, for use in analyzing drone imagery. In some cases, whether a program is good or evil depends upon who is using it. Cyber weapons can be beneficial when used to constrain a terrorist activity or harmful when used to disrupt or steal from legitimate activities such as commerce or research. In other cases, the question is how the technology is used. A clever algorithm, for instance, can be useful in vastly different programs that do quite different things. While issues of dual-use arise in virtually all engineering fields, digital technologies in particular are instantly available worldwide and originators of the technology have significantly limited ability to restrict or regulate access. In addition, software code is highly malleable, so that even squeaky-clean technologies and programs can be modified into dubious ones. Furthermore, other features of the digital world invite dubious usage. Cryptocurrencies are widely used in illegal financial transactions (such as child pornography and the drug trade) because cryptocurrencies are decentralized and cryptocurrency transactions are hard to trace. Engineers must weigh several factors in dealing with possible dual-use. See "Responsibility (Dual-Use)" in Chapter 6.

Intellectual Property/Copyright/Patents/Trade Secrets

Software can assist in violating copyrights, such as software that abets pirating music, or can itself skirt the margins of copyright, e.g., software only marginally different from copyrighted software. Thus, software engineers face ethical issues about copyright in at least two ways. First, they must ensure that their own code does not infringe on another's copyright. Two famous examples of companies claiming that another company infringed their copyrights are Oracle v. Google and Lotus v. Borland. Second, engineers need to think about how the software they create can be used by others to infringe on copyrights.

Many of the issues relevant here are discussed in Chapter 8 ("Confidentiality and Trade Secrets" and "Patents and Copyrights"), such as reverse engineering software so as to evade copyright laws by making small changes and incorporating portions of someone else's code into one's own; "using open-source code in their own code without properly crediting the source;" and "using illegal software to perform their tasks" (Garrett and Lewis ND, 4). Others are discussed in "Responsibility (Dual-Use)" in Chapter 6. Software that has a legitimate use may also or even primarily be used to pirate copyrighted or patented content, and thus constitutes a dual-use issue. Balancing legitimate use against illegitimate use may prove difficult. "What U Hear" was eliminated from Windows and "Stereo Mix" disabled by default. Those functions were often employed by some users to record copyright-protected audio being streamed but by others to record their own compositions produced on an analog synthesizer.

ISSUES ABOUT CONTENT (DATA)

Bias

As noted in Chapter 9, software can evidence bias in a number of ways. For example, from 1982 to 1986, initial software screening of applicants to St. George Hospital Medical School, based on analysis of past admission decisions, resulted in "women and those from racial minorities [having] a reduced chance of being interviewed independent of academic considerations" (Lowry and Macpherson 1988). (St. George cooperated fully with the subsequent investigation and the screening software was removed.) "Pulse oximetry overestimated arterial oxygen saturation among Asian, Black, and Hispanic patients compared with White patients," leading to "a systematic failure to identify Black and Hispanic patients who were qualified to receive COVID-19 therapy and a statistically significant delay in recognizing the guideline-recommended threshold for initiation of therapy" (Fawzy et al. 2022). Bias can also occur when program analysis conflates correlation with causation. For example, a medical triage program that assigned treatment priority on the basis of mortality statistics could assign a higher priority to Welsh speakers in a region after a chemical disaster in a Welsh-speaking community. Friedman and Nissenbaum (1996) identify three types of bias computer systems may display: pre-existing (social biases prior to the system), technical, and emergent (biases that arise from changes after the system is designed). Designers should (a) identify social biases, such as race and gender, and proactively look for ways the system might reflect those biases, (b) think about how technical limitations of the system might cause bias problems as the system is used, and (c) think about any likely design changes or changes in use or circumstances and make sure to incorporate avenues for modifying the system in response to unforeseen changes. They recommend rapid prototyping, formative evaluation, and field tests including members of populations likely to be overlooked or disadvantaged. "The time to begin thinking about bias is in the earliest stages of the design process, when negotiating the system's specifications with the client" (343). (See also Brey and Søraker 2009.)

Other remedies include ensuring diversity among programmers and data providers; adding explainability to neural networks; making sure training data is not restricted to a distinctive population subset (e.g., gender, income, and ethnicity) to avoid sample bias; keeping alert for possible sources of measurement bias; and establishing procedures to monitor for bias the software's performance in the real world, including AI audits, which employ machine learning to detect biased outputs, code, or data.

Privacy, Security, and FRT

Privacy is a concern in many engineering decisions, both in information technology and in other areas (such as bioengineering). Yet, although people argue heatedly about privacy in ethics, law, and personal life, it is far from clear what exactly privacy is. The issue is particularly troublesome because conceptions of privacy vary widely across the globe. Chinese engineers, for example, operate in a very different legal and political environment, with respect to privacy, than French and Canadian engineers.

In general, the right to privacy has been claimed to include decisions, such as the choice to abort or use contraception; information, such as one's location and whether a patient is HIV positive; keeping others from seeing one's genitals or other body parts (or even one's face or license plate number in Google Map views); keeping others from recording a conversation one has with them or viewing one's emails; and much more. Body parts considered "private" in one culture may not be in another. What is public information in one society may be considered extremely private information in another. Many different kinds of things might be called "private," but, in ethics, what matters is the morally important notion of privacy, that is, the kind of privacy to which one has a right and which an ethical person should try to respect.

What philosophy can provide is a general framework for thinking about this. A good starting point in understanding privacy (in this sense) is the fact that human beings are social individuals. As philosophers since Aristotle have noted, being part of a thriving social world is important for human flourishing. But, unlike ants, human beings participate in society as individuals. Important elements of our lives are separate from the social life of the community. They belong to ourselves as individuals: they are private rather than public. Human beings thrive when both elements are strong.

> Our individuality depends in deep ways on our shared social goals, concepts, understandings and interactions....Much of what is significant in a human life depends upon conceiving of oneself as part of a community, as belonging to a social order. Conversely, belonging to a good human community means bringing to the social order one's individual values and perceptions. Human communities are not like ant farms; they depend essentially upon differences of aims, views, outlooks and so forth.
>
> *Schlossberger 2008, 149*

What particular things count as privacy, in this sense, will always be relative to a time and a social setting, because which particular things we need to keep private (to keep separate from our community or societal life) varies from culture to culture. In the Philippines, where it is not uncommon for large families to live together in small quarters, privacy will have a different meaning than it does in areas where the norm is each individual having a separate bedroom and each nuclear family having a separate domicile. Many who grow up in a religiously conservative culture may have a different notion of how much of the body others may see without invading privacy: necklines and hemlines in Paris and Dhaka may be markedly different. In the US today, "the features of one's genitalia are an essential element of the private, i.e., that which pertains to life as an individual and not to life as part of the community. For most Americans, control over who views their genitals is central to feeling that one's body is one's own" (Schlossberger 2008). In another (more rational) society, a requirement to disrobe (e.g., for health examinations) might not be an invasion of privacy. Public nudity is legal and not especially considered remarkable in several European nations. In Finland, family saunas are traditional. Community-oriented cultures place more value on social connections and see more of human life as community-based than do cultures that value and emphasize individuality. So, while the need to respect privacy is universal, what violates privacy very much depends on where and when one lives. This is particularly important to keep in mind for technologies, like the internet, that span the globe.

The right to privacy, when understood in this way, consists in adequate provision for keeping a reasonable part of our lives apart from the social realm – enough that people can flourish and lead satisfying human lives. Schlossberger (2008) argues that what counts as adequate depends on the culture and society and is a feature of the whole picture. The law may need to draw lines in the sand about what violates privacy and engineers must make decisions about particular practices, products, and processes, and projects. But what ultimately counts, ethically, is how robust the overall domain of the private is for each individual. Whether requiring that everyone's fingerprints be on file with the government violates privacy can't really be decided in isolation, because it depends on how that file is used and how much overall privacy people in that society have.

It must be noted that there are other conceptions of privacy and privacy is an area of philosophy that remains very much debated. Some have viewed privacy as just an application of other rights, such as property and bodily security or autonomy. Some construe the right of privacy as the right to control access to us. Some feminists argue that privacy rights are an invidious shield for the control and domination of women. (See, for example, DeCew 2018; Nissim and Wood 2018; Tavani 2007, 2013; and Thomson 1975.)

In addition to protecting the sphere of the private from the community, *issues concerning privacy frequently involve two additional factors: autonomy and respect (avoiding embarrassment).* Often things a person would withhold from the community (the private) are also things the person would choose not to let others know or sense (that is, see, hear, smell, taste, or touch). Thus, often, violating a person's privacy is also violating that person's autonomy. Both factors (respecting privacy and respecting autonomy) need to be considered. In addition, in making ethical decisions about privacy, engineers should wish to avoid causing people embarrassment or humiliation and making them feel demeaned. Sometimes issues about privacy are really more about avoiding humiliation or embarrassment, something an ethical engineer takes seriously.

Privacy of location illustrates some of this complexity. In most countries, including India and the US, it is not generally illegal (and it is not, in general, considered unethical) to hire a private investigator to shadow an individual and report that individual's whereabouts, provided the investigator follows legally permitted techniques (e.g., does not impersonate an officer) and the surveillance is not for an illegal or immoral purpose. The subject of surveillance is likely to raise personal and, perhaps, moral objections if the surveillance violates a relationship of trust. Stalking is a crime in many nations. In Japan, stalking consists in interfering with the tranquility of another's life by, e.g., loitering around a person, making repeated unwanted phone calls or leaving unwanted social media messages, or making e-mail threats. The Netherlands prohibits intentional, systematic intrusion into another person's personal space intending to frighten that person or force or pressure the victim to do something. In the US, stalking, which is a crime in all 50 states, generally involves more than merely following, although the definition of stalking varies from state to state. California State Penal Code § 646.9 2008, for example, defines a stalker as one who "willfully, maliciously, and repeatedly follows" another and "makes a credible threat with the intent to place that person in reasonable fear for his or her safety." Illinois Statutes Chapter 720. Criminal Offenses §.3 5/12-7.3 specifies "Knowingly and without lawful justification following or surveiling another on at least 2 separate occasions and threatening or placing the victim in reasonable apprehension." Generally, stalking is not a crime without some element of threat – surveillance without threat is permissible. Why, one might ask, should electronic forms of location surveillance, like GPS data, fall into a different ethical category? One consideration, as the Supreme Court emphasized in Katz v. the United States 389 US 347 (1967), is the importance of the reasonable (predictive) expectation of privacy. [2] People are aware that others can see them when they are visible to people who are on the street or (through unshaded windows) in adjacent buildings, but they expect their GPS data is not visible to others. Given prevailing policies and general knowledge, a relevant moral factor is people's reasonable expectations of privacy. However, those expectations change as policy changes. When it becomes permissible for others to scan or log one's GPS data, the reasonable expectation that one's GPS data is private disappears. Thus, our reasonable expectations about privacy can't be the moral basis for setting policy because policy is partly what creates those expectations.

Sometimes embarrassment alone is enough moral reason to keep from revealing location information, whether the information was obtained by personal observation or electronic observation. For example, Herb Smith walked down a street in a red-light district of town. Since he was walking in plain sight in a public place, the fact that he was walking there cannot plausibly be said to be protected from the sphere of the public – it was a public, not a private, act. Whether the information was obtained by personal surveillance or through collecting GPS data, Smith should have been aware that his presence could be noted. It is not obvious why publicizing the fact of Smith's presence would violate Smith's autonomy. Smith chose to walk there in plain sight. Nonetheless, posting online that Smith was walking down that street might cause Smith embarrassment, and so, in the absence of a strong reason to post, publicly posting the information might impinge on Smith's privacy.

Privacy issues are usually about balance. Avoiding demeaning, embarrassing, or humiliating people is generally preferable, as is giving people autonomy and room for the private in their lives. Most often, those who design and implement processes that might conflict with those goals do so for a reason. Their reasons may be legitimate or dubious, or anywhere in between. Reporters might publish information that embarrasses a political candidate but which the public has a right to know. A government might wish to keep track of every inhabitant's location in order to deter crime or respond quickly to an emergency or improve traffic flow, all of which are worthwhile goals. But a government might wish to intimidate dissent, which is anything but a worthwhile goal. Companies want to track users' activities in order to better target advertisements that might both increase sales and inform users of products or services of use to them. They may also use that information for malign purposes, such as tricking or deceiving users. News programs might publish embarrassing information about a politician's family member in order to boost viewership. Impinging on privacy for a malign reason is almost never justifiable. But even when the purpose or goal of an activity or technology is a good or legitimate one, it must be balanced against the extent to which it impacts privacy. Different nations and societies will weigh these competing goals differently. The

Chinese government's view on these matters is very different from that of the Swedish government.

In the case of privacy, dual-use applies not just to the technology or system itself but to theft of or manipulation of the product of the technology (the data). Data compiled for sound reasons may be misused by others. Sometimes aggregate data is sold from which identifying information has been hidden or stripped. Unfortunately, it can be distressingly easy to unmask such data. Indirect identifiers are surprisingly revealing. "63% of the population can be uniquely identified by the combination of their gender, date of birth, and zip code alone" (Lubarsky 2017, 203). NDR reporter Svea Eckert was able to obtain a colleague's browsing history from her free trial of an "anonymous" database by using her company's login page URL (Oberhaus 2017). Data fusion techniques (data mining) also permit relatively harmless data from different sources to be combined in a way that constitutes serious breaches of security or privacy. Two databases may contain information about x that, separately, are not unduly problematic, but become problematic when combined. Transactional information (or transactional data), such as the amount withdrawn from an ATM machine, the time and location of a motorist using EZ Pass, and website behavior recorded by cookies, must be considered as well as traditional information such as age, weight, phone number, and so on. Data mining can merge transactional and traditional information to be used in ways that are deleterious to the subject of the information.

The simplest form is when separate pieces of data can be combined to unlock restrictions on further data or activities, e.g., taking out a credit card or generating a "lost" password to an account. NORA (non-obvious relationship awareness) software probes databases in multiple languages seeking obscure matches of pertinent information. ANNA (anonymized data) software searches encrypted data.

In addition, there is always a significant risk that compiled data may be hacked or stolen. Epstein (2007) suggests that issues of security involve confidentiality, integrity, and accessibility. Nissenbaum (2005) divides security concerns into six categories. Technical security issues include making systems unavailable to users; destroying or compromising information; and providing unauthorized access; while cyber-security issues include "the use of networked computers as a medium or staging ground for antisocial, disruptive, or dangerous organizations and communications…threats of attack on critical societal infrastructures … [and] threats to the networked information system itself (p. 64)." Tavani (2013, 176–177) identifies three kinds of vulnerabilities: "I. Unauthorized access to data, which are either resident in or exchanged between computer systems. II. Attacks on system resources (such as computer hardware, operating system software, and application software) by malicious computer programs. III. Attacks on computer networks, including the infrastructure of privately owned networks and the Internet itself." Companies whose customer credit card information has been hacked include Target, Delta Airlines, K-Mart, Lord & Taylor, Orbitz, and many more. In March 2016, data from the Philippine's Commission on Elections regarding 55 million Philippine voters was leaked online. Leaked data included almost 16 million fingerprint records and the passport numbers of 1.5 million overseas voters. Before compiling data, data collectors must ask whether the benefits of compiling the data are worth the risks of its being stolen, whether less sensitive data would suffice, whether separating the data is feasible, and how best to safeguard the data.

Many of these issues arise about FRT. Software can match a facial image with a database, identify the individual, and search for any accessible information about the person. I could point my phone at a stranger in a restaurant and learn his name, address, phone number, employer, age, marital status, social media posts, etc. Smartphones can be unlocked with the owner's face instead of a password or fingerprint. Travelers could use their faces as their passports. Mounted cameras linked to databases allow police to scan for potential terrorists at public events. FRT can tell advertisers who is seeing their ads, provide highly detailed demographic data, and deliver fine-tuned personal messages. ("Hi Tim! For that Hawaii vacation next week, you'll want Brand X sunscreen, especially formulated for delicate skin like yours.") By August 2017, New York's driver license FRP program had generated 4,000 arrests and 16,000 non-criminal administrative actions. (For example, the program identified residents with suspended licenses who took out a second license.) VIP-identifying software for use by retail employees has been developed. Sophisticated FRT can identify people despite sunglasses, beards, and surgical masks. FRT is still in its infancy. The London Metropolitan Police's use of FRT at two Notting Hill Carnivals and the 2017 Remembrance Sunday events at the Cenotaph, as well as South Wales Police's recent employment of the technology, produced a fair number of false positives. However, Facebook's FRT was able to determine correctly whether two images are of the same face more than 97% of the time (human beings did only a quarter of a percent better). FRT programs seem to make more mistakes in identifying non-white faces, raising worries about racial bias.

"Critics of face-recognition technologies in particular, and biometrics in general, point to at least three problems: error, abuse, and privacy" (Brey 2005, 215). FRT data could generate a whole new level of identity theft. Passwords can be changed after a data breach. Faces can't be. False positives present a danger. Residents can lose their driver's licenses because FRT mistakenly identified them with someone else whose license was suspended. Yet worse, innocent citizens can be arrested or forcibly apprehended because an FRT system mistakes them for terrorists or dangerous suspects. In 2019, Nijeer Parks spent 11 days in jail because an FRFT system matched a photo of him with a fake license left behind at the scene of a crime. Fortunately, Mr. Parks was eventually able to prove that he was 30 miles away at the time of the crime. In addition, government surveillance and crushing of dissidents are vastly improved by FRT. A hidden camera across the street from a secret meeting of dissidents would yield the name of every attendee. Moscow has added FRT to 5,000 units of its citywide closed-circuit TV (CCTV) monitoring system.

Many find particularly disturbing the possibility that anyone at all can discover their personal information with a glance from across the street. At present, we have the option of refusing to give strangers our names, addresses, and phone numbers. That ability helps protect us from stalkers, unwelcome phone calls, unexpected strangers at the door, or worse. If FRT is widely available and unregulated, we all lose that choice. The stranger behind you in line in the grocery store can, without exchanging a word with you, decide to call you or show up at your doorstep or workplace. This poses an obvious threat to safety. Is it also a violation of our autonomy? At present, we can decide not to tell someone our age. FRT removes that choice, since it can search birth or other records and produce the age of anyone at whom the camera is pointed. Is any part of the wealth of information available to anyone who can see you something that should be kept in the sphere of the private?

Engineers working in any aspect of the developing field of FRT need to think about such questions and the many others that will arise as the technology and its uses grow. What safeguards or limitations can be built into the technology to best balance these concerns with the benefits? In addition, engineers have a duty of public outreach (see "Models of the Profession" in Chapter 3 and "Codes of Ethics" in Chapter 6). Engineers with special knowledge about FRT should lend their voices and expertise to the public debate and discussion about these questions.

Peterson (2020) mentions several other design areas where privacy concerns arose: designing traffic cameras to photograph license plates from the rear so as not to reveal car occupants, using Automatic Target Recognition in security scanners that show generic body outline instead of a fine-grained image of the skin, and developing VPN and WhatsApp to hide user data

In general, engineers' need to keep current about the status of privacy issues in their societies cannot be overstressed. Engineers in general have a duty to obey the law, and the laws in different countries vary greatly and will continue to change, often rapidly because circumstances (such as available technology) relevant to privacy are in rapid transition. Engineers making privacy decisions must also take into account both the local culture affected and the realities with which they must deal. Chinese engineers work in a very different environment, with respect to privacy, than do Norwegian or Canadian engineers. Pakistani engineers (and engineers from anywhere working on projects that largely affect Pakistanis) must be sensitive to Pakistani understandings of what is private. Many regions are culturally diverse and may contain groups with sharply contrasting feelings about privacy and what is embarrassing. As a result, engineers need to be sensitive to and recognize privacy issues when they arise and engineer responses that are sensitive to relevant local laws, customs, and outlooks.

Hosts' Responsibility for User/Client Code and Content

To what extent are online hosts responsible for monitoring and overseeing the code and content that clients place on their platforms and venues? This question has received much attention lately because of a variety of publicized abuses. Cambridge Analytica harvested the personal data of millions of Facebook users without their knowledge or consent. Facebook users were invited to take a quiz, which collected not only the personal data of those who took the quiz but also of all those in their social network. Approximately 270,000 users took the quiz, resulting in the harvesting of data from, according to Facebook's Mark Schroepfer, as many as 87 million users (Confessore 2018). The data was allegedly sold to US political campaigns, who used it to target ads and spread false rumors about opponents. In any case, Cambridge Analytica was paid for services by the campaigns of Donald Trump and Ted Cruz, among others, as well as by a PAC headed by John Bolton (Pearle 2018). Individuals and associations affiliated with or working for the Russian government bought ads and posted content through individual accounts on Facebook in 2016 to spread fake news in an effort to help elect Donald Trump (Parks 2019). Facebook's WhatsApp, widely used in India, is end-to-end encrypted, thus making it impossible to monitor communications. Users need be identified only by cellphone number, thus remaining anonymous, and were able to forward unlimited numbers of messages to an unlimited number of groups before forwarding limits were introduced. In 2018, mobs enraged by false WhatsApp messages about child-trafficking rings and organ harvesters killed more than twelve innocent individuals, prompting India's Ministry of Electronics and Information Technology to call upon WhatsApp to undertake "immediate action to end this menace" (Gowen 2018). WhatsApp responded by, in 2018, offering funding for technologies to inhibit the spread of false news, advertising the danger of fake news in newspapers, and imposing forwarding restrictions in 2018, 2020, and 2022. In 2020, WhatsApp said the spread of viral messages had decreased by 70%. Despite these efforts, however, another WhatsApp rumor-fueled lynching occurred in 2020. Fake news on social media is often compared to a Hydra: chop off one head and two others take its place.

In addition to false or misleading claims and misinformation, "several types of speech have been proposed as candidates for censorship. These include pornography and other obscene forms of speech, hate speech such as websites of fascist and racist organizations, speech that can cause harm or undermine the state, such as information as to how to build bombs, speech that violates privacy or confidentiality, and libelous and defamatory speech" (Brey and Søraker, 2009, 1394). For example, various platforms now delete false information about vaccines and COVID-19. Capurro (2005) suggests these issues can be addressed in five ways: through self-control by individuals and the internet community, by campaigns (he cites the Blue Ribbon Campaign against censorship), through codes of ethics, via legal regulation, and by employing technical regulations such as software filters.

The international nature of the internet raises questions of jurisdiction. Because of nonlocality, an Indian or Sudanese resident's use of Gmail is in critical ways regulated by the US law, not Indian or Sudanese law, because Google is governed by the US law. (However, in 2022, Russia made it illegal to refer to its military action in Ukraine as a "war," forcing Google, out of concern for the safety of its local employees, to instruct its Russian translators to comply.) Many of the largest platforms are US-based, and the US law in this area is in a state of flux. The US Code Title 47 § 230

(Title 47, Chapter 5, Subchapter II, Part I, § 230) of the Communications Decency Act 1996 protects hosts of platforms against civil and criminal liability for what users or other information content providers post, provided they meet certain conditions. For example, they must inform clients of available parental control protections. At the same time, it protects hosts who perform good faith efforts to restrict access to content that is obscene, harassing, or objectionable. The Digital Millenium Copyright Act (DCMA) of 1998 17 USC. § 512 designated information on platforms provided by other users a "safe harbor" from copyright infringement liability. In 2018, US Public Law No: 115-164 (04/11/2018) criminalized operating a facility of interstate commerce to promote or facilitate prostitution, with up to 25 years imprisonment for acts with reckless disregard concerning the possibility of conducing to sex trafficking. It also permitted victims to recover damages.

The clash between the nonlocality of the internet and the diversity of local situations across the globe creates a variety of problems. Local liability laws may permit someone harmed in, for example, Indonesia to sue operators of foreign platforms for damages. Local censorship laws in one nation may conflict with freedom of speech law in another. Thus, a situation could arise where a host's permitting some particular content is illegal in one country while restricting it is illegal in another.

Morally, those in charge of platforms must keep in mind the potential for harm posed by user posts and other actions taken through their facilities. In addition to scrupulously adhering to applicable law, platforms should respect the balance between respecting the autonomy and freedom of expression of users, on the one hand, and, on the other, public welfare, truth and accuracy, and the safety of content viewers. There is a range of acceptable balances between respecting the freedom and autonomy of users and taking responsibility for the way the platform is used. There is some value in the existence of different platforms with different policies, some of which are more geared to the integrity of information (trustable sources) while others are more open to a plurality of voices (the marketplace of ideas). However, it is clearly wrong for a platform to take no steps when a user's post contains the home address of a particular physician performing abortions, a call for the firebombing of their home, and complete instructions for building firebombs. Platforms should make it clear to both providers and consumers of content just where they stand on this, so everyone knows whether they are visiting a free and open but *caveat emptor* environment or a well-monitored environment.

The rules of usage for platforms and facilities, as well as the review process itself, are usually not under the control of engineers, although engineers can certainly weigh in on these matters within the organization. Nonetheless, there are some things that engineers and programmers can do, when developing the software and system architecture and hardware for such platforms and facilities.

1. Include mechanisms for monitoring third party code and content and flagging harmful and objectionable content, such as defamation and hoaxes, and malicious code, such as trojans and viruses. (A) Incorporate a process (*viewer-flagged*) whereby viewers can flag such codes and content. (B) Implement automatic (*system-based*) flagging as feasible and consistent with privacy requirements. (C) Provide a mechanism for *review* of flagged cases and removal of content deemed inappropriate for the facility.
2. Ensure that the system provides users with clear guidelines and detailed information about rules of usage, which types of code and content are inappropriate, how to flag inappropriate code and content, how their own code and content may be flagged, how the review process works, and the types of steps taken as a result of flagging and review.

INFORMATION TECHNOLOGY, AI, AND SOCIETY

Autonomous Technologies

Information technology systems, especially AI systems, increasingly make autonomous technologies possible, from self-regulating houses to self-driving vehicles and autonomous weapons. Autonomous systems promise many advantages. They can save lives by decreasing the number of automobile accidents or replacing people in dangerous occupations. They are never tired, do not take lazy shortcuts, never indulge in road rage, and are more precise. They are not only more efficient soldiers but can potentially lessen "collateral damage" to non-combatants. But autonomous technologies also raise some ethical issues. (For more about these issues, see Schlossberger 2020.)

One issue receiving a lot of attention is the "trolley problem," which has leapt from the pages of philosophy journals to the popular press. If a self-driving car must choose between hitting a pedestrian or swerving and injuring a passenger, what should it choose? What factors should the system be programmed to weigh? Should the ages of the passengers and the pedestrians be considered? Should car passengers' safety be prioritized, or should all persons be weighed equally? Some ethicists think that killing is worse than allowing to die. For example, we shouldn't kill a living patient, say by cutting out his liver, in order to keep another patient from dying (by offering a transplant). Other ethicists argue that sometimes there is little or no moral difference. Rachels (1975) famously argued that allowing someone to die is sometimes much worse for him than killing him, which is why we "put down" a beloved pet. Should autonomous vehicles be programmed to prioritize not killing over allowing to die? How much of a difference should numbers make? Most people would think it wrong to allow all of humanity to die to avoid killing a single person. Should the autonomous vehicle choose killing one person over allowing two persons to die? If not, what is the cut-off – is allowing ten people to die better than killing one? Risk assessment also plays a role. If swerving poses an extremely small risk of serious or fatal injury to passengers and not swerving is almost certain to kill the pedestrian, most of us would agree the car should swerve. How big must the risk to passengers be to change the result? How well equipped to estimate the risk is the system? Programmers must also keep in mind that the more complex and sophisticated the program's moral algorithm, the greater the chance of failure or malfunction. Moreover, if the system is designed for consumer purchase in a free market, cost and marketability are also factors. Since any of the scenarios described could arise, programmers must make these decisions. After all, we don't want the autonomous vehicle to "freeze" because its programming doesn't cover the situation. In any case, ignoring the issue is making a decision by default.

Currently, industry consensus is tending toward protecting the vehicle's passengers above all else. There is some justification for this, in my view. Human drivers have the ability to prioritize the safety of themselves and their passengers – they can choose, for example, to risk hitting a pedestrian in the road rather than swerving off a mountain pass, especially when the driver is not at fault (for example, the driver is not speeding). They are not legally required to swerve off the road, and it is not incontrovertibly clear that morality requires it. It seems reasonable, then, to think programmers' fiduciary responsibility to the customer buying the car would include not taking that option away. Most people would not buy or use an autonomous vehicle knowing the car is designed to sacrifice them in this way. (See also

Bonnefon et al. 2016) However, a driver who chose to kill a pedestrian rather than incur a small risk of minor injury would be subject to legal and moral sanctions. The priority programmers give to vehicle passengers should not be absolute.

An interesting question that has not received much attention is whether self-driving vehicles should be programmed to reflect local mores. Views about self-sacrifice and duties to others vary across the globe. Should vehicles sold to cultures that emphasize independence and self-reliance (individualist cultures) give greater priority to passengers than vehicles sold to cultures that emphasize community and commitment to group needs (collectivist cultures)?

Engineers working on projects involving autonomous weapons need to give thought to several ethical issues.

> Arkin (2010) suggests that under battle conditions, robotic systems may ethically outperform human agents for six reasons: robots can act conservatively, as they do not need to protect themselves; are free from judgment-clouding emotions such as anger and fear; can monitor and report ethical violations by others; are in an epistemologically superior position because their sensors are more accurate; can more rapidly integrate information; and are not subject to "scenario fulfillment" (filtering out information and options that do not fit prematurely fixed beliefs). These advantages must be weighed against concerns about ceding responsibility for killing to machines, e.g., creating an accountability gap because robots cannot be held responsible (Sparrow 2007), and making it easier to kill by distancing ourselves from the act.
>
> *Schlossberger 2020, 563*

Job Displacement and the Digital Divide

The "digital divide" refers to "the gap between those who can effectively benefit from information and computing technologies...and those who cannot" (Ryder 2005), that is, socioeconomic inequality in access to, use of, sophistication concerning, and skills pertinent to using computer technology in general and the internet in particular, both within a nation and internationally. When the term was first coined in the 1990s, it applied mostly to having access to a computer and, subsequently, to people's ability to get on the internet. According to Joseph Kizza, while 74% of individuals in the developed world use the internet, only 33% globally and 26% in developing nations do so (Kizza 2013). However, the difference is gradually shrinking. Smartphones, internet cafes, and public access points have made internet access more widespread across the globe. Moja, for example, is a free, public wi-fi network in Nairobi and other locations in Kenya. As a result, quality of use (and not just access) is becoming increasingly important.

The digital divide creates a vicious cycle. Greater wealth and education create (or at least correlate) with higher quality use of information technology. At the same time, two dominating trends, worldwide, are globalization and an increasing reliance on information technology rather than manual labor or conventional manufacturing. As a result, "the wealth of people and national economies is increasingly decided by the quality with which the internet can be used" (Bartikowski et al. 2018, 373). In addition, as government increasingly goes online, effective citizenship increasingly requires digital access and sophistication. Many nations have regions or social groups for which the nature and quality of access is much greater than it is for others. In such countries, the digital divide effectively disenfranchises large segments of the population. In countries in which mobile technology is less widespread, smartphone internet access (as opposed to laptop or desktop) has become a symbol of status and accomplishment.

Closing the gap does more than benefit those on the "wrong" side of the chasm. Everyone benefits when broader markets are created. The communal and cooperative nature of engineering, the cumulative nature of progress, and the value of diverse perspectives means that bringing more talent into the pool and having more gifted minds contribute to progress enhances everyone's prospects. After all, today's jobs exist because of yesterday's innovation across the globe. The compass, printing, gunpowder, rocketry, and bristle toothbrush came from China. The USB and flash drive came, respectively, from India and Israel. Many chemical advances stem from the work of the Persian alchemist Jabir ibn Hayyam. Immigrants like Louis Joseph Chevrolet, Ernst Stuhlinger, and Werner von Braun helped make possible the careers of today's automotive and space engineers at Toyota and GM, ISRO and NASA. At present, technological progress is fueled by innovation occurring in every continent with the possible exception of Antarctica, such as Vodafone's M-Pesa (a mobile monetary transfer and finance system) in Africa and Renew Power's wind farms in India. In a very real sense, in the information world, we are all in this together.

Closing the digital divide is a long-term, large-scale project. It requires diverse efforts from a wide variety of individuals and institutions. Initiatives that can contribute to this project include creating more community access/technology centers (and supporting those that exist); continuing and expanding attempts to achieve computer and information technology literacy, beyond the bare minimal, as a core element of basic education; establishing government and non-governmental organization programs to provide/support/foster education, tools, and social opportunities; taking account of less affluent users in creating programs and technologies (e.g., options to assist those with more limited internet speed); and providing IT job training for disadvantaged workers.

While engineers in information technologies cannot do all this alone, they can take a leadership role, both as engineers and as individuals, in the worldwide effort to bring everyone into the 21st century. It can be as simple as volunteering to assist a community technology center, making accommodations in the program one is developing for slower technology for some users in poorer nations, contributing to non-governmental programs, or lending one's voice. Closing the gap is an effort to which we can all contribute and from which everyone benefits.

ADDITIONAL RESOURCES AND ILLUSTRATING CASES

Joint ACM/IEEE-CS Software Engineering Code of Ethics and Professional Practice (1997) https://ethics.acm.org/code-of-ethics/software-engineering-code/.

Maner, Walther (1996) "Unique Ethical Problems in Information Technology," *Science and Engineering Ethics* 2:2, 137–154. http://www.skypig.info/pdf/unique/5.pdf

Quinn, Michael J. (2016) *Ethics for the Information Age* 7th Edition (Pearson).

Reynolds, George (2019) *Ethics in Information Technology* 6th Edition. (Boston: Cengage).

Vallor, Shannon and Narayanan, Arvind (2015) "An Introduction to Software Engineering Ethics." *Markkula Center for Applied Ethics.* https://www.scu.edu/media/ethics-center/technology-ethics/Students.pdf

Case 66: Inadequate Security System

Abeni Abimbola is designing a computer system for Alpata Eniyan, a cyber retailer. The system will store customer names and credit card numbers as well as sensitive employee data. The CEO of Alpata Eniyan wants to use the cheapest security available. Engineer Abimbola feels strongly that this security system is inadequate, since Alpata Eniyan is a tempting target for hackers. She explains her concerns both verbally and in writing. The CEO, however, concerned about a possible proxy battle, insists on the cheaper system to make the company's near-term bottom line more attractive. What should Abimbola do?

Case 67: Expanding a Faceprint Database

The Ubiquitous4U Company is compiling a Faceprint database, which, for a fee, can be accessed, for purposes of FRT, by law enforcement, corporate advertisers, and others. Ubiquitous4U does not monitor its customers or the use to which its database is put. Doormonitor Inc. produced a widely installed doorbell and monitoring program that stores home surveillance video on a company server. To expand its database, Ubiquitous4U offers to pay Doormonitor for access to its server, so that Ubiquitous4U's computers can pick out, for each address, faces of presumed residents (based on an algorithm using timing and frequency of entrance as well as key use). Should Doormonitor proceed with this plan, and if so, what further steps should it take?

NOTES

1. The eight dilemmas are identified as "mission impossible," "mea culpa," "rush job," "fictionware versus vaporware," "not my problem," "nondiligence," "canceled vacation," and "red lies."
2. "Reasonable expectation" has two senses, a normative and a predictive one. Jones may be reasonably expected to tell the truth, in the sense that it is reasonable to require Jones to tell the truth, but, if Jones is known to be a compulsive liar, it may be unreasonable to expect him to tell the truth, in the sense that it is foolish to predict that Jones will tell the truth. Thus, there is no contradiction in saying "I expect you will fail to do what I expect of you," because the first "expect" is predictive, while the second is normative. The court's ruling involves predictive expectations (otherwise, the court's reason would be circular).

REFERENCES

Ammanath, Beena (2022) *Trustworthy AI: A Business Guide for Navigating Trust and Ethics in AI* (Wiley).

Arkin, Ronald (2010) "The Case for Ethical Autonomy in Unmanned Systems," *Journal of Military Ethics* 9:4, 332–341.

Bartikowski, Boris, Raroche, Michel, Jamal, Ahmad and Yang, Zhiyong (2018) "The Type-of-Internet-Access Digital Divide and the Well-Being of Ethnic Minority and Majority Consumers: A Multi-Country Investigation" *Journal of Business Research* 82, 373–380.

Berenbach, Brian and Broy, Manfred (2009) "Professional and Ethical Dilemmas in Software Engineering," *Computer* 42:1, 74–80.

Bonnefon, Jean-Francois, Shariff, Azim and Rahwan, Iyad (2016) "The Social Dilemma of Autonomous Vehicles," *Science* 352:6293, 1573–1576.

Borras, Cari (2006) "Overexposure of Radiation Therapy Patients in Panama: Problem Recognition and Follow-Up Measures," *Revista Panamericana de Salud Pública* 20: 2/3,173–187.

Brey, P. (2005) "Freedom and Privacy in Ambient Intelligence," *Ethics and Information Technology* 7:4, 157–166.

Brey, P. and Søraker, J. (2009). "Philosophy of Computing and Information Technology," in D. Gabbay, P. Thagard and J. Woods, eds., *Philosophy of Technology and Engineering Sciences*, 1341–1408, Vol. 9 of A. Meijers, ed., *The Handbook for Philosophy of Science* (Amsterdam: Elsevier).

Capurro, Rafael (2005) "Information Ethics," *CSI Communications* 28:12, 7–10.

Confessore, Nicholas (2018) "Cambridge Analytica and Facebook: The Scandal and the Fallout So Far," *NY Times* April 4, 2018 https://www.nytimes.com/2018/04/04/us/politics/cambridge-analytica-scandal-fallout.html

DeCew, Judith (2018) "Privacy," *Stanford Encyclopedia of Philosophy* https://plato.stanford.edu/entries/privacy/

Epstein, Richard G. (2007) "The Impact of Computer Security Concerns on Software Development," in K. Himma, ed., *Internet Security: Hacking, Counter-Hacking, and Society* (Sudbury MA: Jones and Bartlett), 171–202.

Fawzy, Ashraf, Wu, Tianshi David, Wang, Kunbo, Robinson, Matthew, Farha, Jad, Bradke, Amanda, Golden, Sherita, Xu, Yanxun and Garibaldi, Brian (2022) "Racial and Ethnic Discrepancy in Pulse Oximetry and Delayed Identification of Treatment Eligibility among Patients With COVID-19," *JAMA Internal Medicine* https://jamanetwork.com/journals/jamainternalmedicine/fullarticle/2792653

Friedman, Batya (2008) "Value Sensitive Design," https://www.publicsphereproject.org/content/value-sensitive-design

Friedman, Batya, Kahn, Peter H. and Borning, Alan (2006) "Value Sensitive Design and Information Systems," in P. Zhang and D. Galletta, eds., *Human-Computer Interaction in Management Information Systems: Foundations* (New York: M.E. Sharpe, Inc.). file:///C:/Users/philo/Downloads/Value_Sensitive_Design_and_Information_Systems.pdf

Friedman, Batya and Nissenbaum, Helen (1996) "Bias in Computer Systems," *ACM Transactions on Information Systems* 14:3, 330–347 https://doi.org/10.1145/230538.230561

Garrett, Ron and Lewis, Jennifer (ND) "Ethical Issues in Software Development," https://www.scribd.com/doc/10880744/Ethical-Issues-in-Software-Development

Gowen, Annie (2018) "As Mob Lynchings Fueled by WhatsApp Messages Sweep India, Authorities Struggle to Combat Fake News," *Washington Post* July 2, 2018 https://www.washingtonpost.com/world/asia_pacific/as-mob-lynchings-fueled-by-whatsapp-sweep-india-authorities-struggle-to-combat-fake-news/2018/07/02/683a1578-7bba-11e8-ac4e-421ef7165923_story.html

Kizza, Joseph M (2013) *Ethical and Social Issues in the Information Age*, 5th ed. (Springer).

Leveson, N.G. (2017) "The Therac-25: 30 Years Later," *Computer* 50:11, 8–11 https://www.computer.org/csdl/magazine/co/2017/11/mco2017110008/13rRUxAStVR

Leveson, N.G. and Turner, C.S. (1993) "An Investigation of the Therac-25 Accidents," *Computer* 26:7, 18–41 https://ieeexplore.ieee.org/document/274940

Lions, J.L. (1996) "Ariane 501 Inquiry Board Report," *European Space Agency* https://esa-multimedia.esa.int/docs/esa-x-1819eng.pdf

Lowry, Stella and Macpherson, Gordon (1988) "A Blot on the Profession," *British Medical Journal* 296, 657–658 March 5 1988 https://europepmc.org/backend/ptpmcrender.fcgi?accid=PMC2545288&blobtype=pdf

Lubarsky, Boris (2017) "Re-identification of "Anonymized" Data," *Georgetown Law Technology Review* 202–213 https://www.georgetownlawtechreview.org/wp-content/uploads/2017/04/Lubarsky-1-GEO.-L.-TECH.-REV.-202.pdf

Lurie, Yotam and Mark, Shlomo (2016) "Professional Ethics of Software Engineers: An Ethical Framework," *Science and Engineering Ethics* 22, 417–434.

Mackay, Robert (2010) "The Google Maps War that Wasn't," *New York Times* https://the-lede.blogs.nytimes.com/2010/11/19/the-google-maps-war-that-wasnt/

Minkel, J. R. (2008) "The 2003 Northeast Blackout—5 years later," *Scientific American* 13 https://www.scientificamerican.com/article/2003-blackout-five-years-later/

Neumann, Peter G. (1994) *Computer Related Risks* (Pearson).

Neumann, Peter G. (1990) "Some Reflections on a Telephone Switching Problem," *Communications of the ACM* 33:7.

Nissenbaum, Helen (2005) "Where Computer Security Meets National Security," *Ethics and Information Technology* 7, 61–73.

Nissim, Kobbi and Wood, Alexandra (2018) "Is Privacy Privacy?" *Philosophical Transactions of the Royal Society A* 376:2128 https://royalsocietypublishing.org/doi/pdf/10.1098/rsta.2017.0358

Oberhaus, Daniel (2017) "Your 'Anonymous' Browsing Data Isn't Actually Anonymous," *Motherboard* August 3, 2017 https://motherboard.vice.com/en_us/article/gygx7y/your-anonymous-browsing-data-isnt-actually-anonymous

Parks, Miles (2019) "Fact Check: Russian Interference Went Far Beyond 'Facebook Ads' Kushner Described," *NPR* April 24, 2019 https://www.npr.org/2019/04/24/716374421/fact-check-russian-interference-went-far-beyond-facebook-ads-kushner-described

Pearle, Lauren (2018) "Cambridge Analytica Accused of Violating US Election Laws in New Legal Action," *ABC News* March 26, 2018 https://abcnews.go.com/Politics/exclusive-cambridge-analytica-accused-violating-us-election-laws/story?id=54010145

Peterson, Martin (2020) *Ethics for Engineers* (Oxford).

Rachels, James (1975) "Active and Passive Euthanasia," *New England Journal of Medicine* 292:2, 78–80.

RescueTime (2011) "Google Doodle Strikes Again! 5.35 million hours strummed," *Rescuetime* June 9, 2011 https://blog.rescuetime.com/google-doodle-strikes-again/

Ryder, Martin (2005) "Digital Divide," in Carl Mitcham, ed., *Encyclopedia of Science, Technology, and Ethics, Volume 2* (Detroit, MI: Macmillan Reference USA),

525–527. http://go.galegroup.com.pnw.idm.oclc.org/ps/i.do?p=GVRL&u=hamm1 1355&id=GALE%7CCX3434900199&v=2.1&it=r&sid=exlibris

Sawyer, Kathy (1999) "Mystery of Orbiter Crash Solved," *Washington Post* Oct. 1, A1.

Schlossberger, Eugene (2008) *A Holistic Approach to Rights: Affirmative Action, Reproductive Rights, Censorship, and Future Generations* (Lanham. MD: University Press of America).

Schlossberger, Eugene (2020) "Autonomy in Engineering," in Diane P. Michelfelder and Neelke Doorn, eds., *The Routledge Handbook of the Philosophy of Engineering* (Routledge), 558–568.

Seven, Doug (2014) "Knightmare: A DevOps Cautionary Tale," https://dougseven. com/2014/04/17/knightmare-a-devops-cautionary-tale/

Software Fail Watch 5th Edition (2017) https://badr.blog/wp-content/uploads/2018/09/20180207_software-fails-watch.pdf

Sparrow, R. (2007) "Killer Robots," *Journal of Applied Philosophy* 24:1, 62–77.

Tavani, Herman (2007) "Philosophical Theories of Privacy: Implications for an Adequate Online Privacy Policy," *Metaphilosophy* 38:1, 1–22.

Tavani, Herman (2013) *Ethics and Technology: Controversies, Questions, and Strategies for Ethical Computing*, 4th ed. (Wiley).

Thomson, Judith Jarvis (1975) "The Right to Privacy," *Philosophy and Public Affairs* 4:4, 295–314.

Vallor, Shannon and Narayanan, Arvind. (2015) "An Introduction to Software Engineering Ethics," *Markkula Center for Applied Ethics* https://www.scu.edu/media/ethics-center/technology-ethics/Students.pdf

14

Consulting Engineering

Consulting engineers face special ethical problems. An engineer who works for government, Samsung, or Exxon does not have to compete for clients or advertise their services. Most of the time, their organization has control over most aspects of the project on which they are working, and so they are primarily answerable to one employer. Consulting engineers, however, usually collaborate with other firms, companies, or agencies. For a particular project, the X corporation may employ firm Y, which, in turn, subcontracts work to Z consultants, which employs mechanical engineer Jones. Jones must address the needs of X, Y, and Z, whose interests may conflict. This chapter helps you untangle some of the knotty problems that confront consulting engineers.

ADVERTISING

Before 1976, many professional codes considered it unethical for engineers to advertise their services. Others, such as the NSPE, proscribed self-laudatory advertising, misleading or untruthful advertising, and advertising that diminished the dignity and honor of the profession. The US Supreme Court ruled in 1976 that such bans on professional advertising were unconstitutional because they restrained trade. As a result, professional societies no longer ban advertising. Indeed, "engineering practice has become much more commercial, competitive, and market-driven, with marketing, sales, and advertising playing an increasingly more important part" (NSPE 2015). Instead, they seek to establish guidelines for improper forms of advertising. The Board of the NSPE "recognizes its continuing role in carefully evaluating situations involving non-truthful

DOI: 10.1201/9781003242574-17

or misleading and deceptive advertising claims and notes that these judgments will need to be made on a case-by-case basis" (NSPE 2015). Moreover, societies are free to impose their own rules for advertisements in their own publications. IEEE, for instance, requires of advertisers in IEEE publications that "the advertising intent should be to serve 'common interests' of the user/reader of the print or electronic product. It should contain technical content or be of informational usefulness and appropriate to the member demographic" (IEEE 2017).

Ethical concerns about advertising fall into four categories: advertisements of questionable honesty, advertisements that demean the profession of engineering, advertisements with morally questionable messages, and new forms of advertising.

Dishonest or Misleading Advertising

Because engineering is built upon trust, consulting engineers have an institutional duty to be scrupulously honest in attracting clients. This duty goes beyond not telling lies. The consulting engineer must avoid, as much as possible, creating a false impression of the capacity, expertise, experience, personnel, or facilities of the firm.

The criteria for honesty are three-fold: *advertisements must be accurate, relevant, and straightforward*. First, is the information given strictly true? Consulting engineers should scrupulously avoid making any statements that are not strictly true. Second, does the information presented provide a legitimate reason for clients to select the firm? Since the point of engineering advertising is to give clients a reasonable basis for making a choice, consulting engineers should do their best to present clients with information that gives legitimate reasons for considering their firm. Third, would the information presented give a false or misleading impression to a reasonable client? Any statement could be misconstrued. However, even the most misleading statement, if it is not strictly false, might be understood correctly. The engineer must ask herself, "how would a reasonable client understand this statement?"

It may be useful to look at a few examples of dishonest advertising. It is dishonest to use the names, experience, or achievements of engineers who are no longer employed by the firm. Of course, care must be used in distinguishing between the personal achievements of prior employees or partners, which may not be used, and the projects handled by prior employees or partners while they were with the firm, which may be mentioned. Thus, it is improper to say that "our engineers have won the X award," when the engineer who won the X award is no longer with the firm. However, it is not improper for the firm to list Y project among its achievements, even though the engineer who worked on Y project for the firm is no longer with the firm. Another form of dishonesty consists in exaggerating, explicitly or by insinuation, the role played by the firm in a particular project. If the firm played a minor role in project Y, it is misleading simply to list project Y among the firm's achievements, as this falsely suggests that the firm had primary responsibility for project Y. (A reasonable client might have this false impression.) However, it is not improper to say that the firm "participated in" or "assisted in" project Y, since the "vote of confidence" given to the firm when their assistance was requested is a legitimate reason for considering the firm. Finally, it is dishonest to list an area or function the firm is not fully qualified to handle. For example, it is improper to take out an ad stating, "Z Firm: Site and Risk Assessments; Electrical Power Systems; Hazardous Waste Hydrographic Surveys," when all hydrographic surveys are subcontracted out, and Z firm has neither the personnel nor facilities to carry out all but the simplest site and risk assessment assignments.

Unseemly or Demeaning Advertising

Advertising by consulting engineers should maintain the dignity and high social values of the engineering profession. Engineers should remember that their advertisements represent engineering to the public and should avoid advertisements that undercut the values of the engineering profession. For example, it is inappropriate for consulting engineers to advertise by cost alone. Although price is a legitimate factor in selecting a consulting engineer, advertising that suggests to clients that price is the only factor undermines crucial values of the engineering profession, such as the values of safety, excellence, and thoroughness. Providing irrelevant information in an advertisement is also inappropriate. It is important that clients choose consulting firms on the basis of relevant and defensible values, such as expertise, safety, excellence, and experience. The profession as a whole is harmed when other factors become the basis of choice. Thus, for example, it is improper to advertise that the firm's engineers were born locally, as this is not a relevant criterion, given the values of the engineering profession. Finally, advertisements whose tone demeans the profession are unseemly. Advertisements represent the profession to the public.

Advertisements that are brash, tasteless, or undignified demean the profession in the public's eyes and should be avoided.

Advertising with Morally Questionable Messages

Advertisements can convey moral messages explicitly or implicitly. This claim may surprise some readers, so it is worth unpacking. Advertisements may convey association values, decision values, situational messages, and direct moral messages. Virtually, all advertisements convey an association value – they attempt to associate the product (a firm, process, material, or technology) with something the viewer deems good. This can be accomplished directly, by, for example, claiming to be competent, or indirectly. For example, a soft-drink commercial showing people having fun while intoning the name of the drink is attempting to associate the drink with having fun. An advertisement can convey a moral message by how it grabs one's attention. Consider an advertisement that says, in bold letters, "wouldn't you love to get your hands on this baby" over a large picture of a scantily clad woman or man standing suggestively next to a CAD station. The smaller print then extolls the virtues of the CAD system, with the name of the vendor in largish print at the bottom of the ad. This ad makes no false or misleading statements. The "baby" in the ad presumably refers to the CAD system. But it attracts attention by conveying approval of a certain kind of objectifying attitude toward, depending on who is pictured, women or men. (After all, a reader who is repulsed by the idea of thinking of a person as a thing to "get your hands on" might avoid rather than linger upon the ad.) That ad is an example of an advertisement that conveys an objectionable moral message (namely, one should regard and treat women/men as consumable objects). Advertisements can also convey decision values (suggesting, for example, that one should decide on the basis of emotion instead of facts) and situational values, that is, values implied by the situation depicted in the advertisement, such as a story enacted in the ad. An ad in which a couple deceives their friends by making espresso machine noises in the kitchen while really serving them instant espresso not only suggests that the instant espresso is so good that it is indistinguishable from freshly brewed espresso but also conveys the message that we should use deception to impress our friends. Finally, advertisements may convey moral messages directly. Thailand's TrueMove H service provider's commercial emphasizes the importance of small acts of kindness and concludes with the line "Giving is the best communication."

The power of such messages is magnified both by the fact that they are repeated by many advertisements for a vast array of products and by the fact that they are generally implicit, so that readers do not think about the values and messages, good or bad, they are internalizing.

An ethical engineer will strive to ensure that the advertisements with which they are associated (e.g., those of their company or firm) convey morally acceptable and not morally dubious association, decision, direct, or situational messages.

New Forms of Advertising

In addition to traditional advertisements in printed publications, engineering firms may now avail themselves of several other methods of reaching potential clients: blogging, mail marketing, sponsored content, social media, and creating a website, perhaps with a chatbot. Each of these raises new ethical challenges: new forms of the ethical issues raised above (dishonesty, demeaning the profession, and morally objectionable content) as well as new issues. *Clickbait, sponsored content, and dangling* are three questionable practices found in some internet-based advertising. Is it ethical for a firm to employ a misleading headline in order to entice viewers to click on their link? An egregious example might be a link labeled "Is Obama in Love with Melania Trump?" Clicking the link leads to a site saying, in effect, "no, he is not, but you will love our work." The link did not make a false or misleading statement – it simply asked a question – but it is deceptive in that the reader expects to find some basis, even if ultimately refuted, for the supposition that a romantic relationship exists between the two. Sponsored content is cast in the form of an informative article that, in fact, pushes the sponsor's product or service. Another questionable practice is dangling, that is, creating a web item enticing readers to learn, basically, the answer to a single question, but requiring them to click through screen after teasing screen, each full of advertisements, to learn the answer. (I made up a name for this common practice of "dangling a carrot" since I could not find a generally used term for it.)

The landscape and ethical challenges will keep shifting as technology and the social milieu change. Engineers dealing with these challenges should keep in mind the proclamative principle: *would engaging in that form of advertising be a good model for how an engineer should behave* – does it proclaim what it is to be a good example of an engineer? Clickbait and dangling, generally, do not.

COMPETING WITH OTHER FIRMS

Competitive Bidding

Before 1978, competitive bidding based on price was banned by most professional codes and many regulatory boards. Although the Supreme Court ruled in 1978 that such bans were unconstitutional, there are some loopholes remaining. Engineers should check local laws and regulations before entering such bids.

Ethically, the key issue is whether such bids undermine professional values. It is improper for consulting engineers to sacrifice quality, safety, care, and excellence in order to put together a price-competitive bid.

Contingency Fees

The key element here is making sure that your professional judgment is not compromised. Avoid situations in which you personally stand to benefit from one result rather than another. (See also "Conflict of Interest" in Chapter 8.) For example, suppose you are asked to make a safety or feasibility assessment of a project, knowing that you will probably get the job if the client decides to go ahead with the project. It is hard to turn in a negative report, knowing that this will cost you the project. Similarly, if you are hired to review a project, knowing that you will be paid only if you save the company money, there is a built-in incentive to sacrifice safety to cost.

Bribes and Kickbacks

Kickbacks or bribes to or from state officials or individual corporate officers are forbidden. (See, however, Case 63 in Chapter 11 for a more problematic example.)

Some firms enter into mutual referral arrangements (you recommend me and I'll recommend you) or offer discounts or commissions to firms that recommend or refer them. Mutual referral arrangements are to be avoided, unless the situation does not compromise professional judgment and the arrangement is known to the client. Discounts should be passed on to the client.

Derogatory Remarks about Other Engineers

In general, it is both unethical and bad practice to make derogatory remarks about another engineer or engineering firm. However, one should not cover up unethical or incompetent practice. If you know a particular engineer or firm to be incompetent, or not qualified for a particular kind of task in which it seeks to engage, taking formal action may be appropriate, or even morally required (but only after speaking to the individual or firm in question). "Informal" derogatory remarks are unfair, however, as they give the victim no opportunity to explain or defend himself. If, when reviewing the work of another engineer, you find flaws, inaccuracies, safety problems, or can devise a more efficient solution, you should (as feasible) make your recommendation in a way that does not unnecessarily reflect badly on the other engineer, without compromising honesty.

REVIEWING THE WORK OF OTHERS

Reviewing, Checking, and Stamping the Work of Unlicensed Individuals

Is it ethical for an engineer to review, check, and stamp plans or designs of an unlicensed individual, not prepared under the engineer's supervision? In many cases, this would violate state or local laws or regulations. *Make sure that no applicable local laws are being violated, in letter or in spirit.* If it is legal to do so in your area, there are nonetheless some purely ethical constraints on reviewing, checking, and stamping such work.

There may be some special situations in which, if lawful, an unlicensed individual may legitimately seek the stamp of a licensed engineer. For example, a small church may need a new facility but be unable to build the facility without substantial volunteer work. One of its members, though unlicensed, is competent to design the facility. The church group wishes to employ a licensed engineer to review, check, and stamp the member's plans. However, it is definitely not ethical to assist an unlicensed individual to carry on an engineering practice in violation of the spirit of the licensing laws. If the individual who drew up the plans is not the original client, the engineer should consult the original client before stamping, to make sure that (a) the client understands and approves of the arrangement, and (b) the final plans continue to suit the client's needs. The complexity and potential risk involved in the plans is an important factor. The greater the complexity or potential risk involved, the more an engineer ought to hesitate before reviewing, checking, and stamping work not prepared under their supervision. In any case, before stamping, the engineer ought to check the work as thoroughly as if it were their own, as they are assuming responsibility for the work.

Reviewing the Work of Other Engineers

In many circumstances, it is both ethical and sensible for a client to ask another consulting engineer to review the work of the original engineer, for example, in the event of a physical failure, or when plans are substantially complete, but the client believes that the plans are deficient or uneconomical.

The reviewing engineer should not exaggerate minor objections and should not hesitate to commend the original engineer's work, when warranted. However, serious drawbacks should be clearly documented. When feasible, the original engineer should be notified of the review, and, if the client consents, be given a copy of the reviewing engineer's report.

SAFETY AND LIABILITY

Consulting firms face special problems of legal liability for jobs with which they are associated. For example, in Krieger v. J.E. Greiner & Co., Inc. et al., Maryland Court of Appeals 1978, the court ruled that because the engineering consultant had previously inspected the work site for safety, this could be taken as conduct indicating that the consultant is responsible for supervising the safety of the work. As a result of this ruling, consultants are sometimes reluctant to do any extracontractual inspection, even though such inspections sometimes prevent disasters. To limit this problem, make sure the contract is clear about responsibilities.

Case 68: Reviewing by an Unregistered Subordinate

The city of Urbanium, population 4 million, assigned its employee, Hector Cordoba, P.E., to design a five-story municipal office building. Cordoba assigned a junior engineer, Singh, to create the design. When Cordoba received the completed design from Singh, he assigned his subordinate, Solossa, a non-registered engineer recently hired by Urbanium, to check the plans. Are Cordoba's actions problematic? What should Solossa do?

REFERENCES

IEEE (2017) "IEEE Guidelines on Advertising, Accessibility, and Data Privacy," https://journals.ieeeauthorcenter.ieee.org/become-an-ieee-journal-author/publishing-ethics/guidelines-and-policies/ieee-guidelines-on-advertising-accessibility-data-privacy/#:~:text=Advertising%20in%20IEEE%20Publications&text=It%20should%20contain%20technical%20content,what%20the%20IEEE%20brand%20represents

NSPE (2015) "Advertising—Use of Technical Information by Contractor," https://www.nspe.org/resources/ethics/ethics-resources/board-ethical-review-cases/advertising-use-technical-information

Appendix I: Links to Codes of Ethics from across the World

[Note: Internet Links Are Subject to Change. www.ethicscodescollection. org is a good place to search for additional Codes]

AUSTRALIA

Institution of Engineers, Australia

http://www.ethicscodescollection.org/detail/05490c69-782e-4b84-8b91-42108a786ef1

CANADA

Engineers Canada

https://engineerscanada.ca/sites/default/files/Yukon/APEY_Code_Of_Ethics.pdf

CHILE

Colegio de Ingenieros de Chile A.G.

https://www.ingenieros.cl/codigo-de-etica/

CHINA (PEOPLE'S REPUBLIC OF)

[See PAN ASIAN below]

CHINA (TAIWAN)

Code of Ethics of the Chinese Institute of Engineers (1976)

http://ethicscodescollection.org/detail/da88d773-dbaa-4b06-ab33-b65171a24177

ETHIOPIA

Ethiopian Society of Chemical Engineers (ESChE)

https://www.opcw.org/sites/default/files/documents/SAB/en/2015_Compilation_of_Chemistry_Codes.pdf

[Code is on pp. 113-114]

EUROPEAN UNION

European Council of Engineers Chamber

https://www.ecec.net/fileadmin/pdf/ECEC-Code-of-Conduct.pdf

European Council of Civil Engineers

http://www.ecceengineers.eu/about/code_of_conduct.php

GREECE

Professional Code of Greek Licensed Engineers (Technical Chamber of Greece)

http://www.ethicscodescollection.org/detail/9e9e9771-e392-4094-ba4f-4eaf1034a9c

INDIA

The Institution of Engineers (India)

https://www.ieindia.org/webui/IEI-Publication.aspx#downloads

JAPAN

The Japan Society of Civil Engineers

http://www.jsce-int.org/system/files/Code_of_ethics.pdf

Institution of Professional Engineers, Japan

https://www.engineer.or.jp/c_topics/003/003856.html

KOREA

Korean Society for Precision Engineering

http://jkspe.kspe.or.kr/_common/do.php?a=html&b=12

National Academy of Engineering of Korea

http://www.ethicscodescollection.org/detail/f3f331ed-be2e-44fb-9297-a3135774d77b

NEW ZEALAND

Code of Ethical Conduct Engineering New Zealand:

https://www.engineeringnz.org/resources/code-ethical-conduct/

MALAYSIA

Board of Engineers Malaysia

http://apec-emf.org/wp-content/uploads/2013/12/BEM-code-of-professional-conduct.pdf

PAN AMERICAN

Pan American Academy of Engineering

http://www.ethicscodescollection.org/detail/3b342e80-8afe-4d77-9fb6-add79b4e2454

PAN ASIAN

Chinese Academy of Engineering, Engineering Academy of Japan, and National Academy of Engineering Korea (Declaration on Engineering Ethics)

http://www.ethicscodescollection.org/detail/f3f331ed-be2e-44fb-9297-a3135774d77b

PERU

Peruvian Engineers Association (APEC)

http://apecengineerperu.cip.org.pe/code-of-ethics/

PHILLIPINE ISLANDS

Philippine Society of Mechanical Engineers

https://psme.org.ph/page/code_me_ethics_ph

RUSSIAN FEDERATION

Russian Center of Certification and Registration of APEC Professional Engineers

https://portal.tpu.ru/apec_eng/certification/Requirements%20to%20be%20registered/Ethics

SAUDI ARABIA

Saudi Council of Engineers (Engineer Agreement)

https://www.saudieng.sa/English/EngineerCorner/Pages/CharterEngineer.aspx

SOUTH AFRICA

Engineering Council of South Africa

https://www.ecsa.co.za/regulation/RegulationDocs/Code_of_Conduct.pdf

TANZANIA

Engineers Registration Board

https://erb.go.tz/index.php/code-of-ethics

UGANDA

Uganda Institution of Professional Engineers

http://www.uipe.co.ug/files/downloads/Code%20of%20Professional%20Ethics%20and%20Disciplinary%20Procedures%202011.pdf

UNITED STATES

ASCE

https://www.asce.org/-/media/asce-images-and-files/career-and-growth/ethics/documents/asce-code-ethics.pdf

IEEE

https://www.ieee.org/content/dam/ieee-org/ieee/web/org/about/corporate/ieee-code-of-ethics.pdf

NSPE

https://www.nspe.org/resources/ethics/code-ethics

WORLD

World Federation of Engineering Organizations

http://www.wfeo.org/code-of-ethics/

International Federation of Consulting Engineers

http://www.ethicscodescollection.org/detail/64e0b482-495e-4dc2-a1d6-c5d0523312ec

Appendix II: Summary of Major Western Ethical Theories

Traditionally, Western ethicists since the 1700s viewed ethical decision-making as (a) selecting a general moral theory or principle, then (b) applying that theory or principle to the case at hand and (c) acting in accordance. For example, if I used this method of principle and application to decide how to start my day, I might (a) select the rule "always do the hardest task first," then (b) apply the rule by deciding which of the tasks on my agenda is the hardest, such as washing last night's dishes, and then (c) begin my day by washing the dishes. Earlier ethicists (such as Aristotle and Confucius) tended to focus on attaining virtues, such as wisdom and courage, realized in living one's life. While neither (alone) is the procedure recommended in this volume, it is helpful for engineers to have at least a tincture of acquaintance with ethical theories. (I like to say that ethical theories are meant to be wrong in a helpful way, by which I mean that each points to frequently relevant factors to be balanced against other relevant considerations. Of course, there are contemporary defenders of each of these theories who would disagree with me.) Accordingly, here is a quick run-through of six categories of Western ethical theory. Some non-Western approaches to ethics are discussed in "World Religious Traditions," Chapter 5.

The first three of our theories try to answer the question, "What should I do?" Virtue theory has a different approach – it asks, "What kind of person should I be?" I should note that in the long history of these ideas and vast literature about them, there are many complex and subtle variations. What follows is a very brief and overly simplified overview to get you started. Rachels and Rachels (2012) is a good starting point for further reading. Slightly more advanced is Timmons (2022). *The Stanford Encyclopedia of Philosophy* (2022) is always a useful tool. (Useful articles include "Egoism," "Virtue Ethics," "Consequentialism," "Kant's Moral Philosophy," "Feminist Ethics," and "Moral Particularism.")

1. *Consequentialism* holds that we should select the course of action with the best consequences. There are several different forms of consequentialism. *Egoism* mostly holds that one should do whatever produces the best consequences for oneself.[1] In its simplest form, egoism is largely dismissed by ethicists (with a few notable exceptions).[2] Arguably, Plato was a kind of egoist, since he held that what is best for us is to be just and that the only way to really hurt people is to make them less just. Such views are often referred to as "enlightened egoism." *Act utilitarianism* advocates choosing the option that produces the most overall happiness, now and in the future, for everyone involved. For example, if I grab that hot tamale off your plate and eat it, I will get 3 units of pleasure from the taste but you will experience 4 units of negative pleasure from having your lunch stolen, plus I will experience 6 units of negative pleasure tonight when my ulcer objects to the spices in the tamale. If I just eat my own lunch of dry toast, I will experience only 1 unit of pleasure but you will experience 3 units of pleasure from eating your tamale and my quiet ulcer will be neutral (0 units). Let us assume that no one else is affected. Then, grabbing the tamale results in a total of negative 7 units of pleasure, while eating my toast instead results in positive 4 units of pleasure. Since $4 > -7$, I ought to keep my hands off your tamale and eat my toast. Jeremy Bentham is a famous example of an act utilitarian. The units are just a heuristic device – the theory needn't assume that happiness comes in discrete measurable units, although it does assume that we can at least roughly estimate the amount of happiness produced. Mill distinguished between higher and lower pleasures. Happiness need not be defined as pleasure and utility need not be defined as happiness. There are well-known problems with act utilitarianism, including several famous counterexamples. What if framing an innocent person or enslaving a small group of unfortunate individuals would produce the most overall happiness? Should we carve up an innocent person to save the lives of six people in need of organ transplants? *Rule utilitarianism* is, in part, a response to these worries. It says that we should follow the rules that would produce the most happiness if people generally followed those rules. For example, it would work out best if everyone didn't steal and left their homes unlocked (and valuables unguarded): since no one would steal, we wouldn't need locks. Unfortunately, not everyone does follow those rules, so it would not be a good idea for me to leave my valuables unguarded on a busy city sidewalk. So, rule utilitarians need to fine-tune their principle. There are many versions of rule utilitarianism trying to make this idea more precise, and each faces difficulties. A major objection

to rule utilitarianism is that if the idea is to maximize happiness, and I happen to know that in this one case breaking the rule would produce more happiness, it seems perverse to follow the rule anyway (rather like watering a dead plant). (See also Sinnott-Armstrong 2019.) There are many sophisticated versions of both act and rule utilitarianism responding to these and several other objections.[3] Sufficing utilitarianism, for example, does not insist on maximizing utility but only on promoting sufficient utility (and so actions need only be "good enough"). Consequentialism is discussed more in Chapter 5, "Promoting Good Consequences."

2. *Kantian*[4] *Deontology*. Deontologists suggest that in making ethical decisions, we should look at the character of the action rather than its consequences. An act utilitarian would tell a lie if telling the lie resulted in greater overall happiness, while a deontologist might insist that lying is simply wrong, whatever the consequences. The philosopher Immanuel Kant held that one should always act on the universally correct set of maxims, where a maxim is a rule of conduct such as "never lie" or "never steal." He gave several versions of the rule to generate these maxims (the "categorical imperative"), which he thought were equivalent. First, always show respect for persons. Second, always treat people as ends, not just means. For example, a soda machine is just a means for me to get my soda. I needn't worry about its rights or feelings. As noted in Chapter 5, when I buy a soda from a store clerk, the clerk is also a means of getting my soda but the clerk is also important in themselves – I mayn't kick the clerk if they are being too slow, for example, because the clerk's rights and needs and feelings are important too. Third, we should only act on maxims we can will as a universal law (universality). By that he means morality doesn't play favorites. As Example 10 in Chapter 5 explains in more detail, if I lie when it's convenient, I have to be prepared to say that everyone should lie when convenient. But that is impossible – if we all understand that no one is required or expected to tell the truth, then whatever you say isn't lying. For example, it is not lying to say "I am Henry" in an improvisation game. Everyone understands that you are picking a name at random. Universality and respect for persons are discussed more in Chapter 5.

3. *Social Contract Theory*. We all agree, explicitly or implicitly, to follow certain rules because it benefits us all if most people do. For example, if people drove just anywhere on the road, there would be constant accidents. If we all agree to drive on the same side of the road – on the right in the US, Brazil, and Nigeria, but on the left in Japan and the UK – we

all benefit. Social contract theory says that we should follow the rules to which we as a community implicitly or explicitly agree. *Hypothetical social contract* views say that we should follow the rules it would be reasonable to agree to under certain specified conditions, for example, if everyone were rational and knew all the facts. Perhaps the most influential modern example of social contract theory is John Rawls' *A Theory of Justice* (1971). Rawls held that just social arrangements are those that a rational person would agree to under a "veil of ignorance," that is, they did not know their own individual circumstances or place in society.

4. *Virtue Theory.* Virtue theory identifies character traits, such as courage and wisdom, for which we should strive in leading our lives. Aristotle identified the virtues as the "golden mean" between extremes. For example, too little regard for danger makes one rash and foolhardy, too much makes one a coward, while just the right balance between facing and avoiding danger makes one courageous. What is the right balance? We discover, through practice and reflection, which balance leads to eudaimonia (flourishing or happiness), to leading a good human life. There are other types of virtue theory besides Aristotle's. For example, Zagzebski (2004) suggests that wrong actions are those that virtuous persons would not normally do and would feel guilty about doing.[5] Prominent modern virtue theorists include Philippa Foot, Rosalind Hursthouse, and Alasdair MacIntrye. Those wishing to learn more about virtue ethics might wish to read Hursthouse and Pettigrove (2018) and van Zyl (2018).

5. In addition, there are ethical views that deny that there exists a single, universal ethical answer. These may be called "heterogenous" or pluralistic views. *Individual Relativism*, for example, says that morality is relative to the individual: we should do whatever we honestly believe is right, and so what is right varies from person to person. For example, if you sincerely believe that abortion is wrong, then it is wrong for you to have one but not wrong for me if I believe abortion is permissible. *Social Relativism* says that morality is relative to a society: we should each follow our society's ethical demands. There are well-known problems with both forms of relativism, including the argument that relativism contradicts itself. (See the discussion of relativism in Chapter 2. For a defense of relativism, see Wong 2006.) *Intuitionism* says that we can directly intuit what is right. W.D. Ross, for example, says that we can intuit general moral demands, such as "keep your promises," and that when general moral demands (which he calls prima facie duties) conflict, as when saving a child from a burning building would make us

break our promise to attend a friend's party, we can just see which of the two demands (keep your promises and save a life) has priority in that case. *Piecemeal intuitionists* think that we can simply see what is right in any given case or what the morally salient reasons are. Situational ethics and Dancy's (2004) sophisticated particularism are examples. This volume adheres to *reason-guided particularism*, which holds that the role of ethical theory is to articulate a wide variety of sometimes competing ethical factors and that reasons (not just intuitions) can be given for balancing relevant factors in particular cases. *Amoralism* says that there simply is no right and wrong. (For an argument against amoralism, see Schlossberger 1992.)[6]

6. Some ethicists propose a critical revision of ethics to eliminate oppression and marginalization. *Feminist Ethics* strives to expose and correct the oppressive role gender has played in societal and academic moral beliefs, institutions, and practices.

More specifically, feminist ethicists aim to understand, criticize, and correct: (1) the binary view of gender, (2) the privilege historically available to men, and/or (3) the ways that views about gender maintain oppressive social orders or practices that harm others, especially girls and women who historically have been subordinated, along gendered dimensions, including sexuality and gender-identity (Norlock 2019).

Critical Ethical Theory, broadly speaking, seeks to correct traditional ethics by examining the reality of social contexts and structures that, it holds, present a flawed understanding of how the world is and create marginalization and oppression. Critical ethical theory is sometimes more narrowly construed as the claim that all morality is merely a system for maintaining the power of a privileged group over an oppressed group, but there are ethicists who do not hold this view who might properly be described, in the broader sense, as critical ethical theorists. There is no contradiction in holding both that, as a sociological fact, the prevailing morality in a society has served to bolster existing power relationships (and thus needs to be critically corrected) and that there are some legitimate ethical truths. In current widespread (and often heated) debates about "critical theory," there is often little clarity about what the term actually means (and therefore which views it includes).

Finally, it should be stressed that not all ethical views will fit neatly into one of these six categories. Once again, it must be emphasized both that these are just rough descriptions of sophisticated views and that there exists a plethora of nuanced views.

NOTES

1. This view, called "normative" or "ethical" egoism, should be distinguished from "descriptive" or "psychological" egoism, which holds that people always act, ultimately, to maximize their own self-interest (act from selfish motives) or always do, ultimately, what they think will make them happy.
2. For example, Eric Mack is a self-described egoist.
3. For a description of a famous and oft-discussed argument against utilitarianism by Robert Nozick (the "experience machine" argument), see Buscicchi (2022).
4. Kant is a complex and difficult philosopher and Kant scholars often disagree about the finer points of interpreting Kant. What I have presented is what might be called "Intro Textbook Kant," simplified to a point that would make many Kantians cringe.
5. "A wrong act = an act that the *phronimos* [roughly, a person with practical wisdom] characteristically would not do, and he would feel guilty if he did = an act such that it is not the case that he might do it = an act that expresses a vice = an act that is against a requirement of virtue (the virtuous self)" (Zagzebski 2004, 160).
6. More sophisticated views sometimes classified as "amoralist" include "error theory" (which holds that while people who make moral statements generally intend them to make a claim that is true, they are mistaken because moral statements actually lack truth values) (see Mackie 1977) and "quasi-realism" (which holds that moral statements project sentiments as if they were statements of fact: see Blackburn 1993). Blackburn himself calls his view "projectionism," not "amoralism." Roughly, Blackburn is a realist about human needs and holds that only a few systems of moral discourse serve human needs.

REFERENCES

Blackburn, Simon (1993) *Essays in Quasi-Realism* (New York, Oxford: Oxford University Press).

Buscicchi, Lorenzo (2022) "The Experience Machine," *Internet Encyclopedia of Philosophy* https://iep.utm.edu/experience-machine/

Dancy, Jonathan (2004) *Ethics without Principles* (Oxford: Oxford University Press).

Hursthouse, Rosalind and Pettigrove, Glen (2018) "Virtue Ethics," *The Stanford Encyclopedia of Philosophy* https://plato.stanford.edu/archives/win2018/entries/ethics-virtue/

Mackie, J.L. (1977) *Ethics: Inventing Right and Wrong* (Harmondsworth, NY: Pelican Books/Penguin).

Norlock, Kathryn (2019) "Feminist Ethics," *Stanford Encyclopedia of Philosophy* https://plato.stanford.edu/entries/feminism-ethics/

Rachels, James and Rachels, Stuart (2012) *The Elements of Moral Philosophy*, 7th ed. (New York: McGraw Hill).

Rawls, John (1971) *A Theory of Justice* (Cambridge, MA: Harvard University Press).

Schlossberger, Eugene (1992) *Moral Responsibility and Persons* (Philadelphia, PA: Temple University Press).

Sinnott-Armstrong, Walter (2019) "Consequentialism," *Stanford Encyclopedia of Philosophy* https://plato.stanford.edu/entries/consequentialism/

Stanford Encyclopedia of Philosophy (2022) Zalta, Edward N., principal ed., Metaphysics Research Lab (Stanford University) https://plato.stanford.edu/

Timmons, Mark (2022) *Moral Theory: An Introduction*, 3rd ed. (Lanham, MD: Rowman and Littlefield).

Van Zyl, Liezl (2018) *Virtue Ethics: A Contemporary Introduction* (New York: Routledge).

Wong, David (2006) *Natural Moralities: A Defense of Moral Relativism* (New York: Oxford University Press).

Zagzebski, Linda (2004) *Divine Motivation Theory* (New York: Cambridge University Press)

Appendix III: Additional Cases

Case 69: 1889 South Fork Dam Failure, Johnstown PA, USA

The original earth and stone dam over the Conemaugh River, completed in 1853, often leaked and was hastily patched. Iron dispatch pipes were sold for scrap. In 1879 the dam was sold to a developer who planned to build a resort for the wealthy. The dam crest was lowered two feet to permit carriages to cross and damaged portions were repaired with hay and clay. Meshed screens were placed in the spillways to keep fish from escaping. In 1889, A heavy storm overtopped the lowered dam and debris accumulating in the mesh decreased spillway capacity. 20 million tons of water descended on Johnstown, killing 2200.

NPS 2022; Blakemore 2020

Case 70: 1940 Tacoma Narrows Bridge Collapse, Washington, USA

At the time the bridge over the Tacoma Narrows was constructed, aerodynamic factors were not generally considered. Instead of the lattice beams used in early suspension bridges, in which wind passed through the lattice with relatively little effect on the beams, the Tacoma Narrows Bridge used solid I-beams. The substitution of plate girders for web trusses, coupled with the flexibility of the two-lane roadway, only 39 feet wide but spanning 5959 feet, proved unwise. During construction, the bridge was known as "Galloping Gertie" because it moved during even mildly windy weather. On November 7, 40 mph winds created tortional oscillation. There is some dispute about whether this was primarily due to mechanical resonance, vortex shredding as wind passed around the girders, or aeroelastic flutter. In any event, the bridge swayed wildly (the differential between the height of the sidewalk at its ends reached 28 feet) before collapsing. Fortunately, authorities were able to close the bridge before the collapse occurred, and thus no lives were lost. At the time, theory and experience regarding aerodynamics were inadequate. Expanding knowledge via aerodynamic modelling may have prevented the disaster.

WSDOT 2004

Case 71: 1970 West Gate Bridge Collapse, Melbourne, Victoria, Australia

To rectify a 4.5 inch camber difference between two half spans, 80 tons of concrete block were placed on the north half span. This created a buckle. In attempting to flatten the buckle, bolts were removed, resulting in the fall of a 2000 ton, 367 foot span between piers 10 and 11. The span fell 164 feet onto worker huts and the Yarra River, followed by gas explosions and the ignition of spilled diesel fuel. 35 workers died. While the immediate cause was "the application of kentledge [concrete ballast] in an attempt to overcome difficulties caused by errors in camber," the Royal Commission cited two additional fundamental factors: inadequate margins of safety on the part of the designers and an unusual erection method requiring "more than usual care on the part of the contractor"

Royal Commission 1971

Case 72: 1979 Three Mile Island Nuclear Plant Meltdown, Pennsylvania, USA

The Three Mile Island nuclear facility, consisting of two reactors (one of which was running at the time of the accident), was located near Harrisburg, PA. Due to a mechanical or electrical failure, the main feedwater pumps did not supply the steam generators responsible for removing heat from the core. The turbine generator and the reactor automatically shut down, increasing pressure in the primary system. In response, the pilot-operated relief valve opened. Although the valve was stuck in the open position and failed to shut when the pressure normalized, readings in the control room indicated the valve was shut. Thus, plant operators were unaware that coolant was being lost through the open valve. In addition, plant workers mistakenly believed that when instruments indicated pressurizer water level was sufficient, the core was covered in water. When reduced pressure due to the open valve caused the reactor coolant pumps to begin to vibrate, plant operators turned off the coolant pumps. To avoid emergency cooling water from overfilling the pressurizer, the water flow was reduced. As a result, the core overheated (USNRC 2018). The zirconium fuel cladding was exposed and reacted with superheated steam to create large amounts of radiated hydrogen gas, some of which was released into the environment (Brittanica, Editors of Encyclopedia 2016). Almost two million individuals were exposed to low levels of radioactivity. In sum, the highly complicated system was prone to multiple failures, instrumentation was inadequate and confusing, and staff were not properly trained to handle the situation.

Case 73: 1981 Hyatt Regency Walkway Collapse, Kansas City, USA

Two walkways collapsed during a tea and dance competition in the lobby, killing 114 and injuring 216. At the time of the collapse, 40 people were on the second-floor walkway and 20 people were on the fourth-floor walkway. Unfortunately, the fourth-floor walkway was held up by rods connected to the roof, while the second-floor walkway was connected to the fourth-floor, meaning the fourth-floor supported double the weight. Thus, the rods connecting the fourth-floor walkway to the ceiling held two walkways and 60 people. This was too much weight. The fourth-floor walkway broke free from its support rods and collapsed onto the second. Both crashed into the lobby. The story behind the design flaws is complicated and controverted, but a simplified version is provided by Newson and Delatte (2011): "The original connection for the walkway showed a continuous rod from the lower walkway, up through the upper walkway to the roof, with a nut halfway up the rod supporting the upper walkway" (1023). To simplify assembly and avoid the risk of damage during hoisting, the steel fabricator "suggested a change to two rods, one from the roof to the upper walkway, and another from the upper walkway to the lower walkway. The structural engineer approved the change without calculating the revised forces, but the change doubled the bearing force between the nut and the cross beam of the upper-level walkway. This connection failed, with the nut punching through the beam" (Newson and Delatte 2011, 1023). It should be noted, however, that the nature of the communications about the rod change between the fabricator and the consulting structural engineer is disputed. "The fabricator, in sworn testimony before the administrative judicial hearings after the accident, claimed that his company (Havens) telephoned the engineering firm (G.C.E.) for change approval. G.C.E. denied ever receiving such a call from Havens"

Staff 2006 (for additional details, see also NBS 1982,
Texas A & M 2009, and Morin and Fischer 2006)

Case 74: 1984 Union Carbide Toxic Gas Release, Bhopal, India

The Union Carbide India Limited plant to produce pesticides was in poor condition: various pipes were corroded and several safety systems were not fully operational. A tank containing MIC lost the pressurized nitrogen gas needed to pump out the MIC. Water entered the tank. Kalelkar and Little (1988) argue at length that the probable cause of water entering the tank was sabotage on the part of a disgruntled operator and not an "inadvertent failure to place a slipblind during water-washing of lines near the process filters," a view seconded by Tomm Sprick, Director of the Union Carbide Information Center (Sprick 2022). Combined with iron from corroding pipes, the water set off an uncontrolled exothermic reaction. The five-fold increase in pressure in the tank was dismissed as a monitor malfunction. Pressure and temperature continued to increase. Ultimately, 40 tons of MIC leaked into the air, causing between 2000 and 16,000 ultimate deaths and over 500,000 injuries. (See Brittanica, Editors of Encyclopedia 2008; Diamond 1985; Hanley 2022.) In 1989, Union Carbide paid a settlement of $470 million to the Indian government, which then assumed all further responsibility for clean-up. According to the Council on Scientific and Industrial Research, established by the government of India in 1942:

It appears that the factors that led to the toxic gas leakage and its heavy toll existed in the unique combination of properties of MIC and from the features of design of the plant. Storage of large quantities of such a material for unnecessarily long periods in single large tanks was made possible by the facilities installed. Insufficient caution in design, choice of materials of construction and in instruments, together with lack of facilities for safe quick disposal of materials showing instability contributed to the event and to the adoption of guidelines and practices in operation and maintenance. The combination of conditions for the accident were inherent and extant. Some inputs of integrated scientific analysis of the chemistry, design and controls relevant to the manufacture would have avoided or lessened considerably the extent of damage.

CSIR 1985

Case 75: 1986 Chernobyl Nuclear Facility Meltdown, Chernobyl, Ukraine

Considered the worst nuclear accident in history, the Chernobyl disaster, it is generally agreed, resulted from a combination of design flaws and poorly trained personnel. The accident occurred during a test of the facility's cooling system after a power outage. After emergency shutdowns, the reactor continues to run at 7% power and hence needs to be cooled. It was theorized that during a power outage, rotational energy from the steam turbine would cover the one-minute gap before backup diesel generators were sufficiently powered up to run the main pump. Testing the system to see if the reactor could be cooled during a power outage, workers shut down reactor 4's emergency core cooling system and removed control rods from the core. The reactor's power level was supposed to be above 700 MW and the steam turbine generator at full speed before the experiment began. Steam for the turbine would then be shut off to see if the turbine could provide enough power to cover the gap until the diesel generators were up to speed, at which point the steam turbines would freewheel down. The test was delayed until 1:23 am, meaning it was the night shift, not the day shift, that carried out the test. The required initial conditions were not met, but the test proceeded anyway. Due to reduced reactor power, neutron absorbing xenon-135 was produced (rather than xenon-136). For reasons that are not entirely clear, power unexpectedly dropped precipitously. Operators made several mistakes in dealing with the situation, resulting in two explosions, the first of which appears to have been nuclear, which propelled radioactive material into the air and eventually across much of Europe. Controversy remains about the extent of the casualties and the details of what went wrong.

Brittanica, Editors of the Encyclopedia 2022; USNRC 2022

Case 76: 1995 Sampoong Superstore Collapse, Seoul, South Korea

Originally designed as an apartment complex, Sampoong decided to switch the design to a superstore. "Although using a building of this size as a department store went against zoning regulations, Lee circumvented this by ordering the addition of a skating rink on an originally unplanned-for fifth floor" (Trosper 2021). The size and number of support beams were reduced in order to accommodate escalators. Rather than a steel frame or steel columns, the structure was supported by concrete columns whose steel reinforcing rods were thinner than legal regulations demanded. During construction, the skating rink was abandoned in favor of restaurants heated by water pipes beneath the floor, adding additional weight. A forty-five-ton air conditioning system was placed on the roof. Responding to neighbors' noise complaints, Sampoong moved the air conditioners to a different part of the roof using pullies and rollers rather than cranes, resulting in cracks in the roof. Structural engineers warned of an immediate danger of collapse, but the Board chose not to evacuate the structure, which would result in loss of sales. The AC units fell through the roof, collapsing the column beneath them. Each floor then collapsed onto the floor below. 502 were killed, 937 were injured, and 1500 were trapped. Rescuing the survivors took as long as sixteen days (Trosper 2021 and Marshall 2015). The Chairman of Sampoong was subsequently convicted of negligent homicide.

Case 77: 1997 Maccabiah Pedestrian
Bridge Collapse, Tel Aviv, Israel

A pedestrian bridge built for the Maccabiah Games collapsed as the Australian team was crossing to enter the arena, killing four of the athletes and injuring others. Subsequent investigations concluded that, instead of paying the Israeli Army to build the bridge, as past Organizing Committees had done, the Maccabiah Games Organizing Committee saved costs by contracting with a theatrical production company that had never designed or built a bridge. That company, in turn, subcontracted design and construction to two individuals not licensed to build bridges and whose firm had never constructed a bridge. Investigators concluded that the design was inadequate and that substandard materials (e.g., wire-bound rusty pipes) were employed. According to the Jewish Telegraph Agency, the Dotan public commission investigating the collapse concluded that "the engineer, Micha Bar Ilan, had never submitted an engineering plan for the bridge, did not design a bridge to meet the intended needs and did not properly oversee the work. The contracting company, Karagula-Ben Ezra, was faulted for doing shoddy work, using substandard materials and being unauthorized to build such a structure" (Jewish Telegraph Agency 1997). Tel Aviv's Magistrate Court convicted five individuals, including the engineer who designed the bridge.

Lavie 2000; Hanaor 1999

Case 78: 1998 Injaka Bridge Collapse,
Mpumalanga, South Africa

When the bridge collapsed during incremental launch, 14 were killed and 19 injured. "The primary cause of the collapse was...the positioning of temporary bearings on which the deck structure slid out during construction. These were located inside the permanent bearing positions which were to be under the box section webs. As a result[,] the temporary bearings punched through the structure and caused the collapse" (Editor 2002). Several other problems with the project were revealed. Cracks in the bridge decks, which should have been a huge red flag, were dismissed. The steel launch nose was insufficiently stiff due to material defects. Automatic pier deflection monitoring was not employed. "Reinforcement bunching was seen in the deck structure, caused by inadequate construction control. This meant that there was incorrect separation between top and bottom reinforcement in the bottom slab. Although there was evidence of poor workmanship it is not clear that it contributed to the collapse" (Oliver 2002). The designer of the bridge, Marelize Gouws (who died in the collapse), had just received her Bachelor degree and was not yet a registered professional engineer. Rolf Heese, who oversaw temporary works design, was also found to lack adequate experience. No design review took place.

Case 79: 1998 Pepcon Fire and Explosion, Nevada, USA

The Henderson, Nevada facility was used by Pepcon and American Pacific to manufacture sodium and ammonium perchlorate, a rocket fuel oxidizer. After the Challenger disaster, over 4000 tons of ammonium perchlorate were warehoused on site. Lambert and Alvarez (2003) claim that Pepcon "pioneered the use of Poly drums for convenience and for corrosion control knowing that the combination of organic materials with oxidizers was unwise." A member of the Henderson Fire Dept. claimed to have seen ammonium perchlorate on the ground at American Pacific. A natural gas pipeline also ran through the site. Also on site were bulk quantities of anhydrous ammonia, hydrochloric acid, and nitric acid. A series of three explosions on site killed two and injured 327. The cause of the explosion has been the subject of some dispute. U.S. Homeland Security and FEMA list the cause as a welding igniting a structure, whose fiberglass walls spread the fire. Pepcon officials cited leakage from the gas pipe as the cause. The United Steelworkers Union blamed the explosion on unsafe working conditions, including an inadequate fire suppression system and the failure to repair faulty electrical systems. Mniszewski (1995) suggests the fire began in a contaminated barrel of ammonium perchlorate that then ignited the fiberglass wall.

Routley 1988; Reed 1988

Case 80: 2004 Charles de Gaulle Airport Terminal Collapse, Roissy-en-France, France

Terminal 2E was a 450 feet long elliptical tunnel tube of reinforced concrete and steel tension struts. On May 23, a 110 foot section of the terminal collapsed, killing 4 people. Several design flaws might have been caught with adequate detailed analysis and design checking. 1) Terminal 2E "was a circular tube with air flowing outside of it. The outer casing of Terminal 2E was subjected to great temperature changes causing the outer steel to expand and contract" (Craven 2019). 2) The external tensile reinforcement was improperly placed: it should have been under the compressive side (Kaljas 2017). 3) The struts lacked strength and sheer stiffness and were inadequately anchored (Kaljas 2017). As a result of these factors, the tensile reinforcement was ineffective. 4) Redundant supports were lacking. "The collapse of Terminal 2E was a wake-up call for many firms to use file-sharing software such as BIM" (Craven 2019). Aéroports de Paris and three subcontractors were found guilty and fined by a Bobigny court.

Case 81: 2009 Lotus Riverside Complex Collapse, Shanghai, China

Heavy rainfall saturated the site's soil. Dirt excavated from under the building for a garage and dumped on a landfill caused the collapse of the adjacent river bank on the north side. Water flowed from the north toward the garage excavation in the south, destabilizing the foundation. The precast concrete pilings snapped and the building toppled over sideways.

Wang et al. 2017

Case 82: 2010 Deepwater Horizon Oil Spill, Gulf of Mexico

BP's Deep Horizon oil rig was located in the Gulf of Mexico, 66 km from the Louisiana coast. After a subcontractor, Halliburton, installed a concrete core to seal temporarily an exploratory well, a pulse of natural gas shot up, causing the drill pipe to buckle. The blowout protector intended to cap the well failed (Borunda 2020). Gas surged up the drill rig and exploded near the platform, causing the rig to sink two days later. 11 were killed. During the next three months, as much as 60,000 barrels of oil a day spilled into the Gulf, totaling an estimated 134 million gallons. According to Richard Pallardy, "a similar incident had occurred on a BP-owned rig in the Caspian Sea in September 2008. Both cores were likely too weak to withstand the pressure because they were composed of a concrete mixture that used nitrogen gas to accelerate curing" (Pallardy 2021). Various attempts to stop the leak and control damage met with varying degrees of success until a capping stack was installed 87 days later.

Case 83: 2012 Indian Power Grid failure (670 million people without power), Northern India

Most of India's electricity in 2012 was generated from coal in the Eastern section, although about a tenth came from hydroelectric sources in the North. The North, West, and South regions constituted the largest energy drains. Thus, India's power grid was a complex network of heavy power flows, managed by power utilities in each state. Power demand in India had been fast outgrowing supply. While China added 84 gigawatts to its power grid between 2006 and 2012, India added only 14 gigawatts. The complicated network of interacting agencies governing the use, production, and maintenance of electricity was severely taxed. However, "Grid operators and the [State Electricity Boards] believed that the system could be managed on the brink and would never fail" (Sen 2012). On July 30, a massive outage left 300 million without power. The next day, the system failed again, affecting 670 million individuals. Surgical operations were cancelled, passengers were stranded, and hundreds of miners were trapped in West Bengal. Riots resulted, with rioters blocking roads and assaulting or holding hostage energy company officials.

Case 84: 2018 Ponte Morandi collapse, Genoa, Italy

Riccardo Morandi designed the bridge over the Polcevera river near Genoa, which opened in 1967. The innovative design, by the engineer known for "bending concrete," differed in key respects from any previous bridge. The deck was made completely of reinforced concrete. Only 4 cables covered in pre-stressed concrete were used per tower. Cable-stays were made of reinforced stressed concrete rather than steel (Invernizzi et al. 2022). Because of its minimal use of steel and the protective vest covering the cables, it was touted as needed no or minimal maintenance. However, the traffic over the bridge increased markedly since the bridge opened as Genoa grew. Moreover, concrete corrodes over time with exposure to environmental pollution and salty air. The concrete vest made the cables difficult to inspect. In 1979 Morandi suggested interventions, which were not implemented. In 1992, instead of replacing corroded cables, new cables were placed around the old ones. In August 2018, cables snapped and a 650 foot section of the bridge collapsed, killing 43 and leaving 600 homeless. 59 individuals, including government officials and executives of Autostrade per l'Italia, which managed the bridge, are on trial. (The trial opened July 7, 2022 and is expected to continue into 2023.) Prosecutors claim that no maintenance work to reinforce the stays of pillar 9 was undertaken in the 51 years between the bridge's opening and the collapse (Landini and Parodi 2021).

Villani 2019; Pianigiani 2020

Case 85: 2021 Surfside Condominium building collapse, Florida, USA

Champlain Towers South, a twelve-story oceanfront condominium in Surfside, Florida, underwent a partial collapse, killing 98 persons. An engineering report in 2018 warned of "'major structural damage' to a concrete structural slab below its pool deck that needed extensive repairs," including "exposed, deteriorating rebar" and "abundant cracking and spalling" in columns, beams and walls (Burke and LaGrone 2021). The report attributed the damage to "waterproofing that was 'beyond its useful life'" and "warned that 'failure to replace the waterproofing in the near future will cause the extent of the concrete deterioration to expand exponentially'" (Schwartz and Mann 2021). Singhvi et al. (2021) note the following additional areas of concern: A 2020 video revealed water damage on the ceiling close to a beam. Several planters, adding tens of thousands of pounds additional weight, were not in the original design. Waterproofing for the pool deck was installed flat, not on a slope (which would have facilitated drainage). Extra beams supporting the ground level deck were removed from the original drawings. Three beams under the planters were shallow and lacked stirrups (reinforcing steel). In March 2022, victims and families of the collapse reached a tentative settlement with the condo association and two engineering firms.

REFERENCES

Blakemore, Erin (2020) "How America's Most Powerful Men Caused America's Deadliest Flood," *History.Com* https://www.history.com/news/how-americas-most-powerful-men-caused-americas-deadliest-flood

Borunda, Alejandra (2020) "We Still Don't Know the Full Impacts of the BP Oil Spill, 10 Years Later," *National Geographic* https://www.nationalgeographic.com/science/article/bp-oil-spill-still-dont-know-effects-decade-later

Brittanica, Editors of Encyclopedia (2008) "Bhopal Disaster," *Encyclopedia Britannica* https://www.britannica.com/event/Bhopal-disaster

Brittanica, Editors of Encyclopedia (2016) "Three Mile Island Accident," *Encyclopedia Britannica* https://www.britannica.com/event/Three-Mile-Island-accident

Brittanica, Editors of the Encyclopedia (2022) "Chernobyl Disaster," *Encyclopedia Britannica* https://www.britannica.com/event/Chernobyl-disaster

Burke, Peter and LaGrone, Katie (2021) "Engineering Report Showed 'Major Structural Damage' before Surfside Condo Collapse," *Associated Press (WPTV)* https://www.wptv.com/news/state/miami-dade/engineering-report-showed-major-structural-damage-before-surfside-condo-collapse

Craven, Jackie (2019) "The 2004 Collapse at Airport Charles de Gaulle: Scrutinizing the Architectural Process of Paul Andreu," *ThoughtCo* https://www.thoughtco.com/charles-de-gaulle-airport-terminal-collapse-3972251

CSIR (Council of Scientific and Industrial Research) (1985) "Report on Scientific Studies on the Factors Related to Bhopal Toxic Gas Leakage," https://bhopalgasdisaster.files.wordpress.com/2014/12/csir-report-on-scientific-studies-december-1985.pdf

Diamond, Stuart (1985) "The Bhopal Disaster: How It Happened," *NY Times* Jan 28, 1985 https://www.nytimes.com/1985/01/28/world/the-bhopal-disaster-how-it-happened.html

Editor (2002) "Inexperience and Error Led to Injaka Bridge Collapse," *New Civil Engineer* https://www.newcivilengineer.com/archive/inexperience-and-errors-led-to-fatal-injaka-bridge-collapse-2-08-08-2002/

Hanaor, Ariel (1999) "Joint Instability in Lattice Structures — Lessons from a Recent Collapse," *International Journal of Space Structures* 14:4, 257–267.

Hanley, Brian J. (2022) "What Caused the Bhopal Gas Tragedy? The Philosophical Importance of Causal and Pragmatic Details," *Philosophy of Science* 89:1 https://www.cambridge.org/core/journals/philosophy-of-science/article/what-caused-the-bhopal-gas-tragedy-the-philosophical-importance-of-causal-and-pragmatic-details/B0D7836EA17B49296DD5E92DD9FFD698

Invernizzi, Stefano, Montagnoli, Francesco and Carpenteri, Alfano (2022) "The Very High Cycle Corrosion Fatigue Study of the Collapsed Polcevera Bridge, Italy," *Journal of Bridge Engineering* 27:1 https://ascelibrary.org/doi/full/10.1061/%28ASCE%29BE.1943-5592.0001807

Jewish Telegraph Agency (1997) "Investigation Faults Many for Maccabiah Bridge Collapse," https://www.jta.org/1997/07/24/default/investigation-faults-many-for-maccabiah-bridge-collapse

Kalelkar, Ashok and Little, Arthur (1988) "Investigation of Large-Magnitude Incidents: Bhopal as a Case Study," *The Institution of Chemical Engineers Conference on Preventing Major Chemical Accidents*, London, May 1988 http://www.bhopal.com/document/case-study.pdf

Kaljas, Toomas (2017) "Reasons for the Charles de Gaulle Airport Collapse" *Journal of Civil Engineering and Architecture* 11, 411–419 https://www.davidpublisher.com/Public/uploads/Contribute/598bd316487ab.pdf

Lambert, H.E. and Alvarez, N.J. (2003) "The PEPCON Disaster – Causative Factors and Potential Preventive and Mitigative Measure," *U.S. Dept. of Energy/Lawrence Livermore National Laboratory* Proceedings of the Fourth International Seminar: Fire and Explosion Hazards, Londonderry, Northern Ireland, UK, September 8–12, 2003 https://www.osti.gov/servlets/purl/15005088

Landini and Parodi (2021) "Italian Prosecutors Prepare Possible Charges Based on Genoa Bridge Probe," *Reuters* https://www.reuters.com/world/europe/italian-prosecutors-prepare-possible-charges-after-concluding-genoa-bridge-probe-2021-04-22/

Lavie, Mark (2000) "Five Convicted in Israeli Bridge Fall," *AP News* https://apnews.com/article/d3d84e08cbcf886eb25e87434acfa706

Marshall, Colin (2015) "Learning from Seoul's Sampoong Dept. Store Disaster," *The Guardian* https://www.theguardian.com/cities/2015/may/27/seoul-sampoong-department-store-disaster-history-cities-50-buildings

Mniszewski, K. R. (1995) "The Pepcon Plant Fire/Explosion: A Rare Opportunity in Fire/Explosion Investigation," Triodyne Inc. Safety Brief 10:3.

Morin, Charles and Fischer, Chad R. (2006) "Catastrophe: Kansas City Hyatt Hotel Skyway Collapse," *Journal of Failure Analysis and Prevention* 6:2, 5–11.

NBS (National Bureau of Standards) (1982) *US Dept. of Commerce* "Investigation of the Kansas City Hyatt Regency Walkways Collapse," https://www.govinfo.gov/content/pkg/GOVPUB-C13-14d3d8005fa5ca3b6bfbff927e5f644c/pdf/GOVPUB-C13-14d3d8005fa5ca3b6bfbff927e5f644c.pdf

Newson, Timothy and Delatte, Norbert J. (2011) "Case Methods in Civil Engineering Teaching," *Canadian Journal of Civil Engineering* 38:9, 1016–1030.

NPS (2022) "The South Fork Dam," *National Park Service* https://www.nps.gov/jofl/learn/historyculture/the-south-fork-dam.htm

Oliver, Antony (Editor) (2002) "Inexperience and Errors Led to Fatal Injaka Bridge Collapse," *New Civil Engineer* https://www.newcivilengineer.com/archive/inexperience-and-errors-led-to-fatal-injaka-bridge-collapse-2-08-08-2002

Pallardy, Richard (2021) "Deepwater Horizon Oil Spill," *Encyclopedia Britannica* https://www.britannica.com/event/Deepwater-Horizon-oil-spill

Pianigiani, Gaia (2020) "Poor Maintenance and Construction Flaws Are Cited in Italy Bridge Collapse," *NY Times* Dec. 22, 2020 https://www.nytimes.com/2020/12/22/world/europe/genoa-bridge-collapse.html

Reed, Jack (1988) "Analysis of the Accidental Explosion at Pepcon, Henderson, Nevada, May 4, 1988," *U.S. Dept. of Energy* https://www.osti.gov/servlets/purl/6610302

Routley, J. Gordon (1988) "Fire and Explosions at Rocket Fuel Plant Henderson, Nevada," *Department of Homeland Security* https://www.usfa.fema.gov/downloads/pdf/publications/tr-021.pdf

Royal Commission (1971) *Report of the Royal Commission into the Failure of West Gate Bridge* (Melbourne: C.H. Rixon) https://www.westgatebridge.org/sites/default/files/downloads/report-of-royal-commission.pdfe

Schwartz, Matthew and Mann, Brian (2021) "Engineers Found Florida Condo Had 'Major Structural Damage' Before It Collapsed," *NPR* https://www.npr.org/sections/live-updates-miami-area-condo-collapse/2021/06/26/1010542570/florida-condo-structural-damage-engineers-report-surfside-miami

Sen, Sandip (2012) "How and Why the Indian Power Grid Collapsed," *The Economic Times* https://economictimes.indiatimes.com/blogs/Whathappensif/how-and-why-the-indian-power-grid-collapsed/

Singhvi, Anjali, Weiyi Cai, Mike, Gröndahl, Mika and Patanjali, Karthik (2021) "The Surfside Condo Was Failing. Here's a Look Inside," *New York Times* https://www.nytimes.com/interactive/2021/09/01/us/miami-building-collapse.html

Sprick, Tomm, Union Carbide Information Center (2022) Private Correspondence 4/21/2022.

Staff (2006) "Hyatt Regency Walkway Collapse," *Engineering.Com* https://www.engineering.com/story/hyatt-regency-walkway-collapse

Texas A & M University (2009) "The Kansas City Hyatt Regency Walkways Collapse," Department of Philosophy and Department of Mechanical Engineering http://ethics.tamu.edu/wp-content/uploads/sites/7/2017/04/HyattRegency.pdf

Trosper, Jaime (2021) "Death and Calamity: The Sampoong Department Store Collapse Explained," *Interesting Engineering* https://interestingengineering.com/death-and-calamity-sampoong-department-store-collapse-explained

USNRC (United States Nuclear Regulatory Commission) (2018) "Backgrounder on the Three Mile Island Incident," https://www.nrc.gov/reading-rm/doc-collections/fact-sheets/3mile-isle.html

USNRC (United States Nuclear Regulatory Commission) (2022) "Backgrounder on Chernobyl Nuclear Power Plant Accident," https://www.nrc.gov/reading-rm/doc-collections/fact-sheets/chernobyl-bg.html

Villani, P. (2019) "Dynamics and Causes of the Collapse of the Morandi Viaduct in Genoa," *XXVI World Road Congress*, Abu Dhabi, UAE, October 6–10, 2019 https://re.public.polimi.it/retrieve/handle/11311/1102900/414271/IP0294-Villani-E-Full.pdf

Wang, W.D., Li, Q., Hu, Y., Shi, J.W. and Ng, C. W. W. (2017) "Field Investigation of Collapse of a 13-Story High-Rise Residential Building in Shanghai," *Journal of Performance of Constructed Facilities* 31:4, 04017012.

WSDOT (2004) "Tacoma Narrows Bridge History-Bridge-Lessons from Failure," *Washington State Dept. of Transportation* https://wsdot.wa.gov/TNBhistory/bridges-failure.htm

Appendix IV: Two Sample Suggestions

Organizations, whether governmental, non-profit, or commercial, are typically structured and focused on particular goals, such as building bridges or making money. Special structures within an organization can thus be useful in ensuring that ethical concerns are given proper attention. Below are two suggestions organizations might implement to this end: incorporating a community advisory board and creating an ethical ombudsperson.

FORMATION OF AN ENVIRONMENTAL AND COMMUNITY ISSUES ADVISORY BOARD (ECIAB)

The purpose of the ECIAB is to include citizens in the process of making decisions that affect community interests, such as the environment and safety, since community values are an important factor in making such decisions ethically. The ECIAB is advisory only, so its deliberations should be full, frank, informed, and free of company/organizational pressure. Its job is not to make technical assessments but to speak for community values in the decision-making process. The advice of the ECIAB, while not determinative, provides important input to the organization in making its decisions. To meet this goal, it is crucial that the ECIAB includes representatives of the entire spectrum of the community. Moreover, special efforts should be made to represent groups likely to object to the organization's decisions. However, frankness is feasible only if the members of the ECIAB understand the importance of keeping confidential the sensitive information they are given and can be trusted not to violate the demands of confidentiality.

A viable ECIAB might include a state representative, a member of the clergy, a representative of a parents' group, a company engineer, a representative of an environmental group, a representative of an animal rights' group, a representative of the business community, and representative citizens from the full range of socio-economic, racial, ethnic, disability status, gender and sexual orientation, and religious segments of the region.

This list is tentative and should be tailored to fit the particular circumstances of the region. (For example, if the region affected consists of a city surrounded by farms, both the urban and rural populations should be represented.)

ETHICAL OMBUDSPERSON

Organizations should employ an individual, with special training in ethics, who is independent of all other chains of command. The ombudsperson's role is to give advice to anyone in the company who seeks it and to speak up for ethical considerations in any matter affecting the company/ organization. The independence of the ombudsperson is crucial, for only so can any member of the company. feel free to consult the ombudsperson fully and frankly about any ethical problem that might arise, without fear of reprisal or negative consequences, and only so can the ombudsperson feel fully free to speak frankly and forcefully. Again, the ombudsperson's role is advisory: the ombudsperson has no decision-making power. However, the opinions and arguments of a well-trained and respected ombudsperson will have considerable weight.

Additional duties of the ombudsperson might include running ethics training sessions for the organization as well as publication and professional activity in areas of professional ethics.

List of Cases

Case 1: The Overheard Remark (Chapter 1)
Case 2: Hurricane Katrina and the New Orleans Levee System (Chapter 2)
Case 3: Delta Company Relocation (Chapter 3)
Case 4: Chemical N in a Lipstick (Chapter 4)
Case 5: Drug P and Pancreatic Cancer (Chapter 4)
Case 6: Amtrak/Conrail Collision (Chapter 4)
Case 7: The Dalkon Shield (Chapter 4)
Case 8: Narbitrol Emulsifier (Chapter 4)
Case 9: Fukushima Daiishi (Chapter 4)
Case 10: Quebec Bridge (Chapter 4)
Case 11: Sandbagging a Rival (Chapter 4)
Case 12: Economical Cooling System (Chapter 4)
Case 13: The Tardy Employee Chapter (Chapter 5)
Case 14: Informing Employees of Outside Opportunities (Chapter 5)
Case 15: Promoting "My Kind of Person" (Chapter 5)
Case 16: The Ford Pinto Case (Chapter 5)
Case 17: He-Man Cigarettes (Chapter 5)
Case 18: Giving Credit When Due (Chapter 5)
Case 19: Freedom of Speech (Chapter 5)
Case 20: Banqiao Dam Failure (Chapter 6)
Case 21: When to Take Bids (Bidding 1) (Chapter 6)
Case 22: "Adjusting" the Records (Chapter 6)
Case 23: Keeping a Promise (Chapter 6)
Case 24: Cadmium and Outflow Sampling (Chapter 7)
Case 25: Going Easy on Safety Assessments (Chapter 7)
Case 26: Installing a Rapid Transit System (B.A.R.T.) (Chapter 7)
Case 27: Challenger Disaster (Chapter 7)
Case 28: The DC-10 Cargo Door (Chapter 7)
Case 29: Cost Overruns (C-5A Galaxy) (Chapter 7)
Case 30: Surry Nuclear Power Plant Facility (Chapter 7)
Case 31: Competence (Chapter 7)
Case 32: Tianjin Warehouse Explosion (Chapter 7)
Case 33: GM Ignition Switch (Chapter 7)
Case 34: Volkswagen Emissions (Chapter 7)

Case 35: Sales Honesty (Chapter 7)

Case 36: Trade Secrets (Chapter 8)

Case 37: When to Cease Bidding (Bidding 2) (Chapter 8)

Case 38: The Expense-Paid Trip (Chapter 8)

Case 39: The Personal Discount (Chapter 8)

Case 40: Discussing One Vender with Another (Chapter 8)

Case 41: Using an Unsuccessful Bidder's Idea (Chapter 8)

Case 42: Owning Stock (Chapter 8)

Case 43: Personal Use of Company Facilities (Chapter 8)

Case 44: "Rescuing" from the Garbage Dump (Chapter 8)

Case 45: Copying Company Software (Chapter 8)

Case 46: Taking Home a Pencil (Chapter 8)

Case 47: Publicizing Your Work (Chapter 8)

Case 48: Sexual Innuendos in Advertising and Session Titles (Chapter 9)

Case 49: Moonlighting (Chapter 9)

Case 50: Dismissal for Non-Work Related Conduct (Chapter 9)

Case 51: Giving Reasons for Dismissal (Chapter 9)

Case 52: Hiring Away (Chapter 9)

Case 53: Complaints about One's Successor Part 1 (Chapter 9)

Case 54: Complaints about One's Successor Part II (Chapter 9)

Case 55: Silver Bay and Reserve Mining (Chapter 10)

Case 56: Environmentally Harmful Products (Toxinal) (Chapter 10)

Case 57: Environmentally Harmful Manufacturing (Chapter 10)

Case 58: Environmental Accidents (Chapter 10)

Case 59: Waste Materials (Chapter 10)

Case 60: Using Non-Renewable Resources (Chapter 10)

Case 61: Indonesia Textiles (Chapter 11)

Case 62: Sub-Saharan Plant Safety (Chapter 11)

Case 63: Paying a Bribe (Chapter 11)

Case 64: Genetic Engineering of Mosquitos (Chapter 12)

Case 65: Vaginal Device for Urinary Incontinence Exercises (Chapter 12)

Case 66: Inadequate Security System (Chapter 13)

Case 67: Expanding a Faceprint Database (Chapter 13)

Case 68: Reviewing by an Unregistered Subordinate (Chapter 14)

Case 69: South Fork Dam (Appendix III)

Case 70: Tacoma Narrows Bridge (Appendix III)

Case 71: West Gate Bridge (Appendix III)

Case 72: Three Mile Island (Appendix III)

Case 73: Hyatt Regency Walkway (Appendix III)

Case 74: Bhopal (Union Carbide) (Appendix III)
Case 75: Chernobyl (Appendix III)
Case 76: Sampoong Superstore (Appendix III)
Case 77: Maccabiah Bridge (Appendix III)
Case 78: Injaka Bridge (Appendix III)
Case 79: Pepcon (Appendix III)
Case 80: Charles de Gaulle Airport (Appendix III)
Case 81: Lotus Riverside Complex (Appendix III)
Case 82: Deepwater Horizon (Appendix III)
Case 83: Indian Power Grid (Appendix III)
Case 84: Ponte Morandi (Appendix III)
Case 85: Surfside Condominium (Appendix III)

Index

Access
  biotechnology, 305, 317, 320n10
  information technology, 333,
    348–350
Accountability, principles of, *see*
    Institutional responsibilities
Advertising, 247, 355–359
Ahimsa, 125
Alger, Christensen, and Olmsted, 205,
    213, 218, 220, 224n3, 250
Amadei, B., 290
Amoralism, 36, 373, 374n6
Appropriate technology, 290–292; *see also*
    Lesser developed regions
Aral Sea, 97
Aristotle, 78, 156, 369, 372
Arguments, 30–31
Arkin, R., 348
Association, duty of, 178–179
Aswan High Damn, 281
Attributionism, 159
Autonomous technology, 346–348
Autonomy, 17, 145–146, 164–165, 292,
    337–339, 342, 345; *see also*
    Respect for persons

Baggini, J., 5
Bah Aba, M., 291
Barem, M., 214–215
Barriero, S., 240
Bayles, M., 61
Bazelon, D. 105n7
Bentham, J. 370
Berenbach and Broy, 329, 351n1
Bias, 116, 200, 209, 219, 221, 236, 237, 238,
    244, 246, 249, 325, 334–335;
    see also Ideal of integrity
Biasetti, P. and de Mori, B., 21
Bidding, 174, 175–176, 209, 216–220, 360
Bielefelt, A., 5
Bille, E., 241
Bird, S.J., 158

Blackburn, S., 374
Blair, M. 72n13
Blotnick, S., 231–232
Bodnar, C. et al., 56
Bohannon, J., 236
Boisjolais, R., 40, 193–195
Bonnefon, J., et al., 347–348
Bottom-line thinking, problems with, 7,
    229–230
Bowen, W.R., 317–318
Brey, P., 335, 342, 344
Bribes, 27, 178, 201, 217, 219, 289–290,
    294–295, 360
Brooks, J., 237–238
Brzezinski, M., 237
Bt-maize, 311–312
Budinger, T. and M., 5

Cambridge Analytica, 343
Cantor, G., 320n3
Caplan, A., 299
Capurro, R., 344
Chesterton, G., 53
Climate change, *see* Environment,
    climate change
Codes of ethics, 5, 7, 28, 150n3, 177–180,
    365–368
Coercive bargaining position, 62, 70n5
Cole, B.M., 236
Community, 15, 25, 51, 53, 64, 69,
    70n4, 78, 79, 80, 83–84, 91,
    92–95, 96, 109, 113, 121,
    126–127, 130, 143, 145–146, 169,
    179, 199, 223, 249, 288, 290, 315,
    336–337, 389
  biotic, 100
  workplace, 6–7, 8, 9, 10–11, 109, 113,
    114, 115, 129, 186–187, 215, 222,
    228, 229, 231, 242, 243, 244,
    245, 246, 249, 250, 251
Competence, 178, 197–198
Compleat engineer, 54–59

Confidentiality, 137, 144, 173, 174, 185, 187, 191, 211, 212–213, 223, 389; see also Bidding; Trade secrets
Conflict of interest, 26, 209–212, 217, 220–222, 224n1–2, 360
Confucius, 124, 369
Conly, S., 145
Consequentialism, see Promoting good consequences
Consumer life, 12–14
Contingency fees, 360
Cort, R., 230
Counterexamples, 31–33
CRISPR-Cas9, 297, 306–309, 318
Critical theory, 373
Cropley, D., 91
Cultural appropriation, 53–54, 69n3
Cummins, E., 291
Cut-throat workplace, 10–11

Dancy, J. 373
Daniels, N., 302, 303
Dao, the, 126
Davis, M., xix, 61, 105n3, 179
DDT, 284–285, 292–293
De George, R., 148n3
DeCew, J., 337
Deciding in the face of uncertainty, 85–87, 89–90, 156–157, 272–274
Deontology, 371
Derogatory remarks about other engineers, 360–361
Design ethics, 52–53
Design objectives, 50–51
Designer babies, 306–309
Desjardins, J. and McCall, J., 68–69
Dessler, G., 245
Digital divide, see Access, information technology
Disability and technology, 314–315, 320n11
Dobos, N., 288
Doorn N. and van de Poel, I., 158, 159
Dual-investor theory, 67–68, 72n13, 117
Dual-use, see Responsibility
Duska, R., 186

Edwards, R.G., 313
Egoism, 370, 374n1–2

Emerson, C., et al., 237
Engineering
    definition of, 50–52
    models of 63–64, 179
    process 59–60
Enhancement/treatment distinction, 301–305, 319–320, 320n4–6
Environment, 96–107, 255–278, 299
    anthropocentric view, 100
    caretaker view, 100
    climate change, 255–261
    conservation and preservation, 264, 286
    deep ecology, 100
    ecocentric, 100
    ecocentric views, 100–101
    ecofeminism, 98–99
    ethic, environmentalism as, 97–98
    Gaia theory, 98, 106n11
    geocentric, 100
    last man argument 97–98, 99
    nature, what is, 101–102, 106n15, 107n17
    Nestle baby formula, 287
    overlord view, 100
    partnership with nature, 96, 97, 99, 101–104, 106–107n16
    pollution, 261–262, 272–275, 286–287
    prudence, environmentalism as, 97
    recycling, reusing, reducing, 267–271, 278n1
    resource and factor income costs, 265–266
    responses to environmental problems, 271–272, 274–277
    sentiocentric, 100
    sustainability, 262–264
    Transcendentalists, 106n13
    virtue theory, 99, 106n12
Epstein, R. and Hundert, E., 70n4
Equality, diversity, and inclusion, 234–238, 293, 335
Ethical decision-making
    five step process, 26–27, 37–40
    flow chart for, 40, 42–44
    like buying a car, 22–23
    like legal reasoning, 23

preparation for
  becoming aware of sensitive
    situation, 27–28
  ethics toolkit, 28–29
  pre-planning responses, 28
Ethical matrix, 21
Ethics
  codes of, 177–180
  is good business, 7
  makes for more productive engineers, 8
Etzioni, A., 72n13
Ewing, D., 234

Facial recognition technology (FRT),
    341–342, 351
Facilitators, *see* Supervisors, types of
Favoritism, 241–244
Feminism, 98–99, 337, 369, 373
Firmage, D.A., 233
Five-step process, *see* Ethical
    decision-making
Fleming, J., 6
Flexner, A., 70n4
Florman, S. 69n1, 69n2, 105n9, 107n17, 190
fMRI, 318
Ford Pinto, 7, 19n3
Freeman, R.E., 67, 68
Friedman, B., 327–328, 335
Friedman, M., 67–68, 71n9, 72n11,
    205–6n4
Future generations, 161–162, 292

Gallagher, M., 53–54
Gandhi, M. 139
Garrett, R., and Lewis, J., 332, 334
Genetic testing, 315–317
Global warming, *see* Environment,
    climate change
Glover, J., 319–320n2
GMOs, 297, 298, 299, 309–313, 320n8
Goland, M., 70n4
Golden rule, 112, 125, 130–133, 149n8–9
Good life, 153–154, 180n1–2
Gregory, D., and Travers, E., 217
Gyekye, K., 127

Haigh, M. 5–6
Hanson, K., 5
Hargrove, E., 100

Harris, C., 61, 294
Hashi, A., 128
Hill, T., 99
Hiring and promotion, 9, 244–246
  hiring away, 214, 246–247, 250
Hobbes, Thomas, 13
Hoke, T., 41–42
Hursthouse, R. and Pettigrove, G.,
    372

Ideal of integrity, *see* Favoritism
Indigeneering, 54
Institutional duties, 61–62, 168–170, 186
Institutional responsibilities, 167–168
Intervening wills, doctrine of, 165–166
Intuitionism, 372
Invisible hand argument, *see* Shareholder
    model of business

Jack, H., 51
Jones, D., 6
Joseph, L., 98
Junzi, 126

Kammapatha, 126
Kant, I., *see* Deontology; Respect for
    persons; Universality
Kapuska, S., 3
Kass, L., 298, 313
Katrina, hurricane, 40–42
Keleher, L., 281
Kemper, J., and Sanders, B., 50
Kizza, J., 348–349
Koen, B.V., 56, 59

Law, obeying, xvii, 28, 199–201
Layton, E., 70n4
Leave the world no worse, duty to, 64,
    68, 116–122, 148n2, 162–163,
    186, 292
Leevers, H., 235, 236
Leopold, A., 100
Lesser developed regions, 281–295, 313
Loui, M. and Miller, K., 5, 61
Loven, J., et al., 236
Lowry, S., and McPherson, G., 334
Luegenbiehl, H., 145
Lurie Y., and Mark, S., 325, 328
Lynch, W., and Kline, R., 190

Mack, E., 374
Mackie, J.L., 374
Maghari, B. M. and Ardekani,
        A. M., 309
Margulis, L., 106n11
Martin, M. and Schinzinger, R., 64, 69n2,
        81, 105n3–4, 213
McGinn, R., 61
Means, evaluating, 104n2
Mentors, see Proteges
Metz, T., 128
Mill, J. S.,370
Mino Pimatisiwin, 126–127
Mohammed, J.A., 125
Moonlighting, 247–248
Moral considerability, 99–101
Moral beauty,154–156
Moral edifice, 25, 33–35
Moral precedent, duty to set, see
        Proclamative principle
More, V., 234

Nader, R., 205
Nanji, A., 128
Newton, L., 70n4, 71n7
Nissenbaum, H., 327, 335, 340–341
Nissim, K., and Wood, A., 337
Non-fraternization rules, see Favoritism
Nonlocality of internet, 326, 345
Nordli, B., 233
Norlock, K., 373
Nozick, R., 147, 374n3

Ombudsperson, ethical, 389
Order-givers, see supervisors, types of
Ottoson, G. 7

Particularism, 373
Passmore, J., 100
Patents and copyrights, 215–216, 223,
        224n3, 333–334
Pavlovic, K.R., 45n1
Peirce, B., 240
Perlman, B. and Varma, R., 11
Personal use, 222–223
Peters, T. and Austin, N., 230, 231
Peterson, M., 5, 189, 343
Picture building, 33–35
Plato, 370

Poulsen, E., 235
Png, I.P.L., and Samila, S., 214
Precautionary principle, 156–158
Privacy, 113, 143, 212, 234, 298, 316–317,
        319, 335–343, 344, 346, 351,
        351n2
Prisoner's dilemma, 94–95
Proclamative principle, 137–138, 139, 179
Profession, models of, 61–63, 70n4, 71n7
Promoting good consequences, 128–130,
        149n6–7, 160–161, 179, 181n8,
        186, 370–371, 374n3
Proteges, 232, 238
Public outreach, duty of, 64, 178, 179
Pumphrey, P.V., 120
Putnam, H., 313

Qur'an, 125, 128

Rachels, J., 45, 347, 369
Rawls, J., 181n8, 372
Raynal, M., 201–202
Recognition, arguments of, 104–105
Records, keeping accurate, 198–200,
        200–202
Relativism, 35–37, 45n2, 372
Reproductive technologies, new, 313–314,
        320n9
Respect for persons, 106n12, 109, 140–141,
        146, 229, 371
Responsibility, 158–160, 292, 333
        hosts' for users, 343–346
Revelle, P. and Revelle, C., 103, 276
Reviewing the work of other engineers,
        361–362
Rights, 141–144
        moral, 143, 149n14
        of natural objects, 100–101, 106n14
Rincon R., and Yates, N., 235
Risk, 57, 292; see also Safety
        balancing against benefit, 80–82
        extent of, 79–80
        nature of, 82
        publicizing, 82–84
Roadstrum, W.H., 233
Robots, 317–318
Ross, A., and Athanassoulis, N., 50
Routley, R. see Environment, last man
        argument

Rules, 24–25
  limits of, 5
  when to break, 170–176, 181n7
Ryder, M., 348

Safety, 22, 24–25, 57–58, 79–90, 105n3–4,
    161–162, 178, 181n9, 294, 298,
    299, 362
Sales, consulting and adversarial, 203–205;
    see also Advertising
Sandel, M., 298, 299, 307–309, 320n7
Savulescu, J., 307
Schooley, S.,237
Self-regulation, need for, 63
Sexual harassment, 238–241, 247
Shareholder model of business, 66–67, 71n9,
    72n12, 120–122, 205–206n4
Shaw, W., and Barry, V., 210–211
Shermer, M., 320n4
Shew, A., 315
Shutte, A., 126
Silver, L., 299
Singularity, 325
Sinnott-Armstrong, W., 371
Slippery slope, 300
Snoeyenbos N., and Almeder, R., 245
Social convergence, 153
Social contract theory, 371–372
Social justice, *see* Access; Equality,
    diversity, and inclusion;
    Sustainability
Software
  content (data), 334–335
  failure, 330–333
  security, 340–341
Sparrow, R., 348
Speech rights, 141, 143, 144, 179, 248, 344
Socialism, 71n8
Stakeholder as shareholder, *see*
    Dual-Investor theory
Stakeholder model of business, 67
Stroh's Brewery, 68–69
Sullivan, W., 70n4
Supervisors
  dealing with subordinates, 112–116,
    129, 144, 177, 233–238, 248–250,
    250–252
  types of, 229–232
Synergism, 6, 181n4

Tavani, H., 337, 340–341
Technology, 50, 52, 91, 158, 163, 300, 308
  appropriate, 290–292
  as practical wisdom, 75–79, 104n2
  autonomous, 346–348
Telehealthcare, 317, 319
Television, 75–78
Thermosets, 266
Thomas, R., 237–238
Thomson, J.J., 337
Tiefel, H., 313
Timmons, M., 369
Total quality management (TQM), 229
Townsend, F., 41
Trade secrets, 84, 144, 187, 212, 213–215,
    220, 223–224, 224n3
Treating others fairly and well, 17,
    112–116, 187
Tredgold, T. 69n2
Trinity test, 161, 181n3
Trolley problem, *see* Technology,
    autonomous

Ubuntu, 126
Unger, S., 197
Universality 133–137, 149n10–13, 202, 371
Universalizability, *see* Universality
Utilitarianism, *see* Promoting good
    consequences

Valla, L.G., et al., 236
Vallor, S. and Narayanan, A., 328–329
Values, 22, 160, 202, 223, 243, 264
  engineering profession, of, 75–107
  life of, 14–15
  personal, 153–154
  value sensitive design (VSD), 325–326
Van Gorp, A. and van de Poel, I., 52
Van Zyl, L., 372
Veach, C., 5
Virtues, 156, 369, 372, 374n5; *see also*
    Environment, virtue theory
Volkov, M., 241

Warren, K., 99
Watson, Julia, 54
Watson, James, 300
Weak Samaritan Principle, 163–164, 292
Webley, S. and More, E., 7

Westrum, R., 87, 190
WhatsApp, 344
When to fight a battle, 109–111, 186, 189
Whistleblowing, 185–197, 205n1–3
Whitbeck, C., 38
Wilkinson, 230
Wilson, B., et al., 237
Wittgenstein, L., 154–155

Wong, D., 126, 372
Work relations, types of. 227–228

Yip, C. et al., 236
Young, James O., 53, 69n3
Young, John A., 95–96

Zagzebski, L., 374n5